Anaerobic Wastewater Treatment

Other Titles by the Author

Biology of Wastewater Treatment
Second Edition
ISBN: 978-1-86094-328-7
ISBN: 978-1-86094-332-4 (pbk)

Fixed-Film Reactors in Wastewater Treatment
ISBN: 978-1-78634-924-8

Natural Wastewater Treatment Systems and Sustainability
ISBN: 978-1-80061-083-5

Activated Sludge: Developments and Sustainable Solutions
ISBN: 978-1-80061-387-4

Anaerobic Wastewater Treatment

NICHOLAS F GRAY
Trinity College, University of Dublin, Ireland

World Scientific

NEW JERSEY • LONDON • SINGAPORE • BEIJING • SHANGHAI • TAIPEI • CHENNAI

Published by

World Scientific Publishing Europe Ltd.
57 Shelton Street, Covent Garden, London WC2H 9HE
Head office: 5 Toh Tuck Link, Singapore 596224
USA office: 27 Warren Street, Suite 401-402, Hackensack, NJ 07601

British Library Cataloguing-in-Publication Data
A catalogue record for this book is available from the British Library.

ANAEROBIC WASTEWATER TREATMENT

ISBN 978-1-80061-721-6 (hardcover)
ISBN 978-1-80061-722-3 (ebook for institutions)
ISBN 978-1-80061-723-0 (ebook for individuals)

For any available supplementary material, please visit
https://www.worldscientific.com/worldscibooks/10.1142/Q0507#t=suppl

Desk Editors: Gregory Lee/Amanda Yun

Typeset by Stallion Press
Email: enquiries@stallionpress.com

Dedicated to Lucy, Cat and Becky

'To you it's just crap, to me it's bread and butter.'
Spike Milligan
Recollections of the Latrine Orderly

Preface

Achieving sustainable wastewater treatment is one of the most important engineering challenges facing society, as the need to mitigate the dual problems of population growth and global warming becomes critical. The traditional roles of protecting surface and ground waters, preserving biodiversity, and preventing pathogen transfer have all become more challenging as wastewaters become more complex with new chemicals that are increasingly inhibitory or toxic to treatment. Today, in fulfilling the needs of sustainability, circularity has become a defining layer on top of our existing environmental and health priorities with carbon harvesting, recovery of metals, and other useful products, and, of course, reuse of water — all achievable expectations. Traditionally, wastewater treatment has relied on separation and high-energy aerobic processes, with a large portion of the carbon in the wastewater lost as carbon dioxide (CO_2) through aerobic metabolism, with anaerobic treatment occasionally used for sludge stabilization. To achieve sustainability, the emphasis of treatment has to change to primarily anaerobic treatment, with aerobic treatment restricted to effluent polishing where membrane filtration technologies are unable to achieve the final effluent quality, although many natural treatment options are available with low carbon footprints.

Traditionally, anaerobic digestion was a slow process, operating at low microbial concentrations requiring large reactors and subject to many operational problems. But today's modern designs, coupled with a greater understanding of the microbial basis of anaerobic treatment, especially of the methanogenic archaea, have seen a rapid development in compact, rapid, and highly efficient processes — many utilizing granular sludges.

Modern wastewater treatment has embraced the circular economy ideology, becoming increasingly sustainable. Among the huge range of valuable byproducts that can be generated by fermentation are biomethane, bioplastics, biodiesel, and volatile fatty acids (VFAs). VFAs are a valuable raw material used in the food, chemical, and pharmaceutical sectors, as well as a carbon source for a wide range of bio-based compounds. All these byproducts are harvesting the carbon in the waste for reuse, preventing direct CO_2 release into the atmosphere. The disadvantage of using methane as a fuel is that CO_2 is produced when it is utilized — this can be overcome by producing hydrogen instead, which, when used, produces only water. Biohydrogen (Bio-H_2) is a carbon-free renewable fuel that is mainly generated using hydrocarbons or gas. Today, there are increasing opportunities to use anaerobic fermentation to produce Bio-H_2, making it a truly renewable fuel. Sulfate reducing bacteria (SRB) can be used to recover metals from wastewaters, safeguarding our future supplies of important metals as well as protecting the environment from heavy metal pollution. The key advantage to using anaerobic treatment in the recovery of metals is that even at very low concentrations, they can be effectively recovered, producing highly insoluble salts at low capital and operating costs. Metal sulfide nanoparticles are an exciting byproduct that has a wide range of uses, from the manufacturing of batteries, LEDs, and solar cells to electroplating and use in environmental remediation, biomedicine, and agriculture.

We tend to think of nutrient removal as requiring an aerobic step, but today, both nitrogen and phosphorous can be effectively treated using anaerobic technologies, including anaerobic ammonium oxidation (anammox) and sulfur-based autotrophic denitrification (SAD). Membrane technology can improve the efficiency of anaerobic wastewater treatment by retaining solids and biomass within the reactor while also ensuring a high-quality final effluent, which is particle-free with extremely low pathogen and antibiotic gene concentrations.

The major advance in anaerobic treatment is due to our rapidly growing knowledge of the key bacteria and archaea involved in this syntrophic process. Research is also exposing the critical role of quorum sensing and signal molecules, which is allowing researchers to develop new and more efficient fermenter designs.

It is estimated that between 3 and 4% of global electricity supplies are consumed by wastewater treatment plants (WWTPs) being primarily required for aeration systems, pumping, and sludge dewatering. The objective of wastewater sustainability is to obtain net zero in terms of energy consumption and, if possible, even become net positive. However, as effluent standards increase and loadings to WWTPs increase, more energy is required. Carbon neutrality is only possible using natural treatment systems or anaerobic processes, with the latter being the only option for centralized wastewater treatment systems. The primary objective of sustainable wastewater treatment is to ensure that greenhouse gases are not released directly from unit processes. When anaerobic systems are used instead of traditional aerobic-based treatment processes, the advantages can be directly related to significant reductions in carbon footprint and take us ever closer to real carbon neutrality.

Concerning the quote in the frontispiece, which has been used in all the editions and reprints of *Biology of Wastewater Treatment*, which this series replaces. I was working late one evening, hand-drawing the illustrations for the first edition as there were no computer graphics available at that time. While listening to the radio, I heard Spike Milligan utter that phrase, which seemed to apply so aptly not only to myself but to all wastewater engineers and scientists. When I wrote to him to ask for permission to use it, he replied that he was delighted that he belonged to "a greater brotherhood" and so allowed the phrase to be freely reproduced. This series of books is dedicated to all those very special men and women throughout the world who are part of that very special family of wastewater operators, managers, designers, manufacturers, and researchers.

Nick Gray
November 2024

Acknowledgments

I am very grateful to the many people who have so very kindly provided me with information, especially manufacturers. I would especially like to thank all the publishers, organizations, and companies who have given me permission to reproduce or modify copyright material in the text, in particular, the Chartered Institution of Water and Environmental Management. This book is all about those amazing researchers who continue to develop and refine the technologies and those companies who make those dreams a reality. I would like to particularly thank Dr. Amelia Rotaru of the University of Syddanski, Denmark for allowing me to reproduce the images of her anaerobic work bench, and Dr.-Ing Sören Weinrich for allowing me to reproduce his equations of hydrolysis. My special thanks go to Amanda Yun, Senior Editor at World Scientific Publishing, for all her help and patience and especially to Gregory Lee, who has been such an enormous help and support in getting this book, and indeed this series, into print.

Contents

Section A

Basis of Anaerobic Treatment

1

Introduction to Anaerobic Treatment

1.1 Scope of Anaerobic Treatment

Anaerobic processes are used to treat strong organic wastewaters (BOD >500 mg L^{-1}) and for the treatment of organic wastes, including primary and secondary sludges from domestic wastewater treatment, industry, and agriculture. Strong organic wastes generated by the agricultural and food industry sectors, often in large quantities, provide particularly difficult treatment challenges. When treating wastewaters containing large quantities of biodegradable organic matter, conventional aerobic treatment is beset with numerous operational difficulties. For example, the difficulty of maintaining aerobic conditions, especially if the wastewater has a high concentration of suspended solids, sludge bulking, inability to take high BOD (biochemical oxygen demand) or COD (chemical oxygen demand) loadings, high operational and energy costs, and a high production of biomass as wasted sludge that requires subsequent treatment and disposal. In contrast, anaerobic treatment, although slower, offers a number of attractive advantages in the treatment of strong organic wastes, such as a high degree of purification, the ability to treat high organic loads, production of a small quantity of excess sludge, which is normally very stable, and the production of a wide range of potential byproducts, such as methane, hydrogen, volatile fatty acids (VFAs), and much more. Our current need to reduce carbon emissions has resulted in a reappraisal of the carbon and nitrogen cycles in wastewater treatment. During aerobic treatment, high energy usage is required primarily from aeration, and where this comes from fossil fuels, then this results in a high carbon footprint. Also, as the majority of the carbon comprising the COD or BOD loading is oxidized to produce carbon dioxide (CO_2) and water, this increases the carbon footprint per m^3 of wastewater treated. In contrast, anaerobic treatment requires far less energy, as no aeration is required, and all the carbon in the wastewater and sludge is retained for recovery and reuse, thereby significantly reducing carbon emissions. It is only through anaerobic treatment that net zero for centralized wastewater treatment plants (WWTPs) can be achieved. Unlike aerobic systems, complete removal of organic matter is not always possible anaerobically, and subsequent aerobic treatment of anaerobic effluents is normally necessary, which indicates the potential of combined anaerobic and aerobic units in series, although the use of natural treatment systems or membrane filtration are low-carbon alternatives (Gray, 2022). The most widely used aerobic treatment process, activated sludge, has an energy requirement of 0.7–1.2 kWh kg^{-1} COD removed at a loading rate of between 0.5 and 2.0 kg COD m^{-3} d^{-1}, compared to loading rates for anaerobic treatment, which are as high as 25 kg m^{-3} d^{-1}, making anaerobic processes far more compact and easier to maintain (Table 1.1). Energy usage for the anaerobic process is generally ten times less than activated sludge, with most of that supplied by combined heat and power (CHP) units operated by the

Table 1.1. Typical ranges for operational parameters for the main anaerobic bioreactor designs (Zieliński *et al.*, 2023). Reproduced under the Creative Commons Attribution (CC BY) license (https://creativecommons.org/licenses/by/4.0/).

Bioreactor	Normal Load (kg COD/m^3/d)	Hydraulic Retention Time (day)	Solid Retention Time (day)
CSTR	0.25–4.0	10–60	10–60
Anaerobic filter	1–40	0.5–12	20
Fluidized bed reactor	1–100	0.2–5	30
UASB	10–30	0.5–7	20
Hybrid reactor	1–100	0.2–5	20
AnMBRs	0.5–12.5	3–30	20–40

Table 1.2. The advantages and disadvantages of anaerobic treatment compared to aerobic treatment.

Advantages	**Disadvantages**
Low operational costs	High capital costs Generally require heating
Low sludge production	Long retention times required (>24 h)
Reactors sealed giving no odor or aerosols	Corrosive and malodorous compounds produced during anaerobiosis
Sludge is highly stabilized	Not always as effective as aerobic stabilization for pathogen destruction
Methane gas produced as end product	Hydrogen sulfide also produced
Low nutrient requirement due to lower growth rate of anaerobes	Reactor may require additional alkalinity
Can be operated seasonally	Slow growth rate of anaerobes can result in long initial startup of reactors and recovery periods.
Rapid start-up possible after acclimation	Only used as pretreatment for liquid wastes

methane produced during the process so that most anaerobic processes achieve net zero or better in terms of carbon emissions. The advantages and disadvantages of anaerobic treatment are summarized in Table 1.2, while the typical carbon balances of both aerobic and anaerobic systems are compared in Fig. 1.1.

1.2 The Anaerobic Microbial Process

Anaerobic heterotrophic bacteria are either obligate (i.e., unable to grow in the presence of oxygen) or facultative (i.e., can adapt to environments, either with or without oxygen). The latter form a bridge between obligate aerobes and obligate anaerobic species. Growth and metabolism of obligate anaerobes are inhibited by oxygen, with oxygen toxicity depending on redox potential (E_h), partial pressure, composition of the substrate, growth rate, and cell density.

The traditional role of anaerobic heterotrophs is in sludge digestion, converting unstable sewage sludge (biosolids) into a more stabilized form. Digestion normally takes place in specially constructed

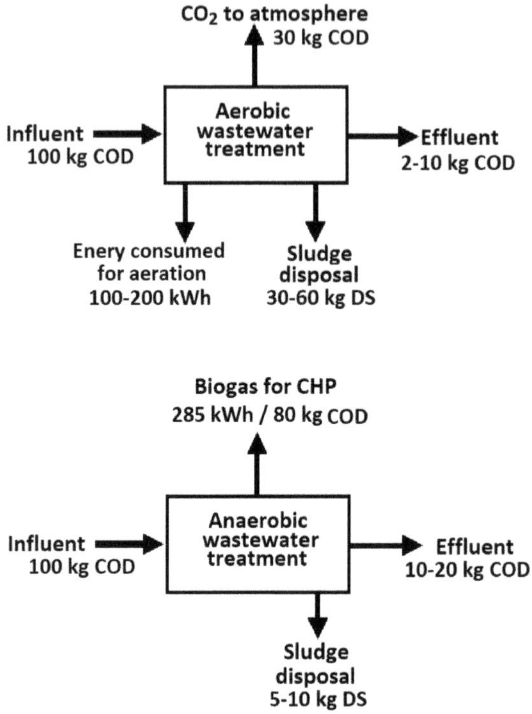

Figure 1.1. Comparison of the mass balance of carbon utilization in aerobic and anaerobic processes.

reactors, although anaerobic digestion also occurs in the sludge blanket of waste stabilization ponds. The presence of anaerobic activity in aerobic treatment processes is generally undesirable, with the commonest effects of anaerobiosis being the production of foul odors and interference of floc settling due to denitrification occurring in the sedimentation tank after the aeration basin in the activated sludge process.

There is a complex consortium of microorganisms involved in the anaerobic degradation of high molecular weight organic compounds to methane. Predominately bacterial and archaea, although protozoa and fungi have been reported (Finlay and Fenchel, 1991), anaerobic breakdown of organic matter is complex, involving synergistic reactions between different groups. There are four groups involved in this synergistic relationship: (1) hydrolytic bacteria, (2) fermentative

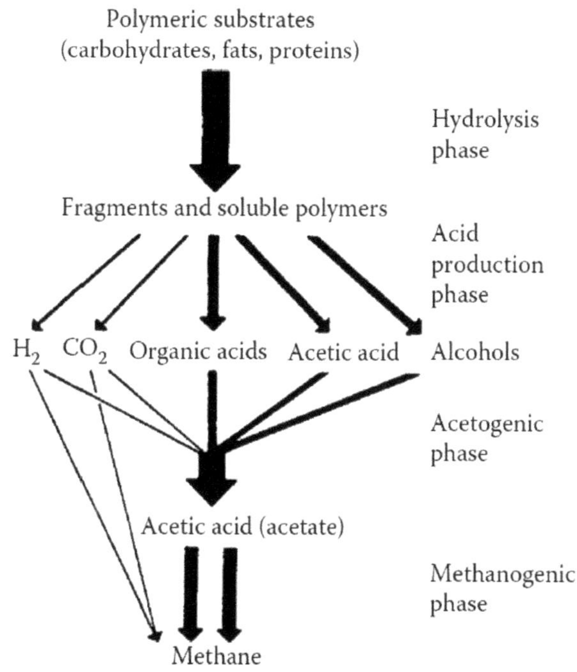

Figure 1.2. Major steps in anaerobic decomposition.

acidogenic bacteria, (3) acetogenic bacteria, and (4) methanogens (Fig. 1.2; Table 1.3). These are reviewed in Chap. 2.

Anaerobic degradation occurs in the absence of oxygen, with methanogens adversely affected by even trace amounts of oxygen. Anaerobic biomass can form a granular sludge structure under certain circumstances (e.g., up-flow anaerobic sludge blanket (UASB) reactor), and where this occurs, the methanogens are protected from the effect of inhibitory compounds within the granules (Kato *et al.*, 1993) (Chap. 5). The basic difference between aerobic and anaerobic oxidation is that in the aerobic system, oxygen is the ultimate hydrogen acceptor with a large release of energy, but in anaerobic systems, the ultimate hydrogen acceptor is nitrate, sulfate, or an organic compound with a much lower release of energy. Briefly, the process of anaerobic decomposition involves four discrete stages (Fig. 1.3). The first stage is the hydrolysis of high molecular weight carbohydrates, fats, and proteins that are often insoluble by enzymatic action, into

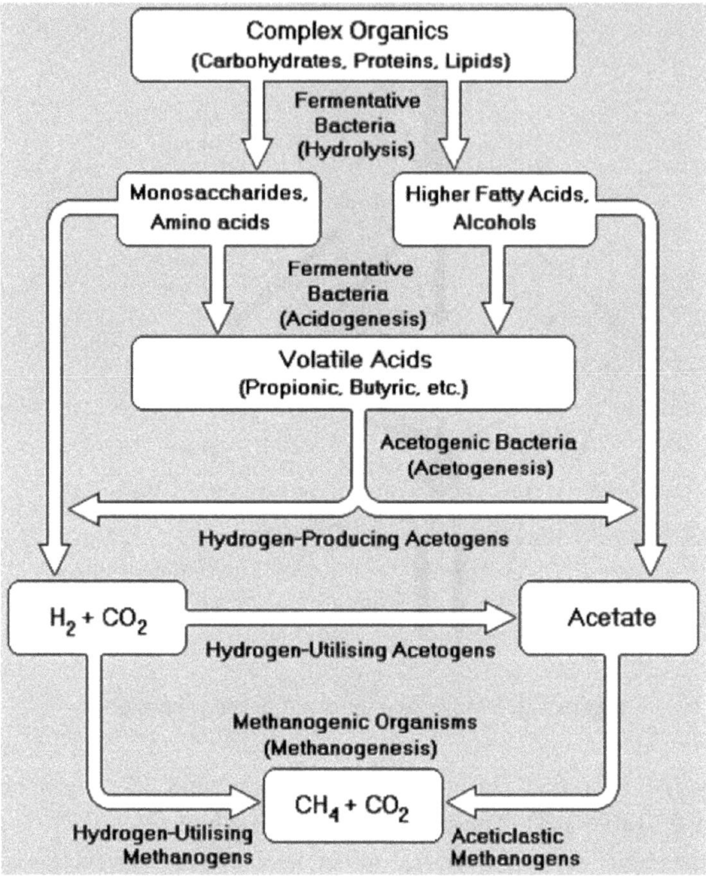

Figure 1.3. Key microbial groups involved in biochemical steps in anaerobic digestion.

Table 1.3. Main conversion steps and key microorganisms in anaerobic digestion (Zieliński *et al.*, 2023). Reproduced under the Creative Commons Attribution (CC BY) license (https://creativecommons.org/licenses/by/4.0/).

Step	Substrates	Products	Species
Hydrolysis	Protein, fats, carbohydrates	Amino acids, fatty acids, polyols, simple sugars	
Acidogenesis	Hydrolysis products	Fatty acids: acetic acid, propionic acid, butyric acid, isobutyric acid,	*Aerobacter, Alcaligenes, Clostridium, Flavobacterium,*

Table 1.3. (*Continued*)

Step	Substrates	Products	Species
		valeric acid, isovaleric acid, caproic acid, lactic acid Alcohols: ethanol, methanol, propanol; H_2, CO_2	*Lactobacillus, Micrococcus, Streptococcus, Pseudomonas, Lactobacterium*
Acetogenesis	Earlier products	Acetics, H_2, CO_2	*Syntrophobacter wolinii, Sytrophomonos wolfei, Clostridium* spp., *Peptococcus anaerobius, Lactobacillus,* and *Actinomyces*
Methanogenesis	Acetogenesis products	Biogas (CH_4, CO_2)	*Methanobacterium, Methanobacillus, Methanococcus, Methanosarcina. Methanosaeta*

soluble polymers. The second stage involves the acid-forming bacteria, which convert the soluble polymers into a range of organic acids (acetic, butyric, and propionic acids), alcohols, hydrogen, and carbon dioxide. Acetic acid, hydrogen, and carbon dioxide are the only end-products of the acid production that can be converted directly into methane by methanogenic archaea. A third stage is where organic acids and alcohols are converted to acetic acid by acetogenic bacteria. It is in the fourth and final stage, which is perhaps the most sensitive to inhibition, where methanogenic archaea convert the acetic acid to methane. Although methane is also produced from hydrogen and carbon dioxide, about 70% of the methane is produced solely from acetic acid. Obviously, the methanogenic stage is totally dependent on the production of acetic acid, and so it is the third stage, the acetogenic phase, which is normally the rate-limiting step in biogas production. The biochemistry of anaerobic decomposition is fully explored in Sec. 2.1.

1.3 Anaerobic Unit Processes

There is a large range of diverse anaerobic processes available for treating wastewater and solids, including anaerobic lagoons, digesters, and filters. Present anaerobic technology can be divided into two broad categories:

Flow-through systems for the digestion of concentrated wastes, such as animal manures or sewage sludges that have a solids concentration in the range of 2–10% — these include conventional completely mixed reactors (Fig. 1.4a), which are used primarily for sewage

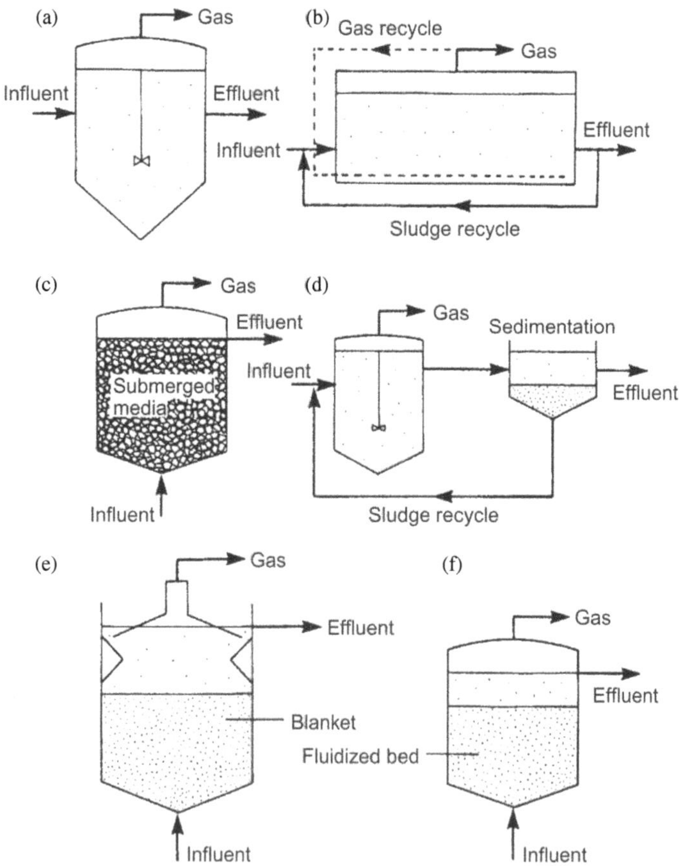

Figure 1.4. Schematic representation of digester types. Flow-through digesters ((a)–(b)) and contact systems ((c)–(f)) (adapted from (Casey, 1981)).

sludges, and plug flow reactors, which are used to a limited extent for the digestion of animal manures (Fig. 1.4b).

Wastewaters with a lower solids concentration are treated anaerobically by ***contact systems***, in which the wastewater is brought into contact with an active microbial biomass that is retained within the system. Contact systems are primarily used for treating industrial effluents and are widespread in France, the Netherlands, and Germany. Of the 400 European anaerobic systems treating industrial wastewaters, 278 are employed in treating food, drink, and fermentation wastes, and 48 in treating paper wastes. In flow-through systems, the residence time of the waste (i.e., the hydraulic retention time (HRT)) and microbial biomass (i.e., sludge or solids residence time (SRT)) in the unit are the same, whereas in a contact system, the SRT is far greater than the HRT of the wastewater. The biomass is retained within the reactor in a number of ways, e.g., by allowing the anaerobic microorganisms to develop as an attached film on a static medium or to develop as flocs maintained in suspension, either by mechanical mixing or by the upward flow of effluent through the reactor. By using a system similar to the activated sludge process (contact (continuously) stirred tank reactors (CSTRs)), the anaerobic biomass is present as suspended flocs, which are recovered in a separate settlement chamber and recycled back to the main reactor (Fig 1.4d). This process can also take place within a single reactor, where the depth of the sludge blanket is controlled by the up-flow rate of the influent wastewater (Fig. 1.4e). Attached anaerobic films can be on a static filter medium of natural stone or plastic, similar to a percolating filter (Fig. 1.4c), whereas fluidized or expanded beds incorporate a fine grained medium such as sand (Fig. 1.4f). Media reactors are completely flooded to ensure anaerobic conditions and, like the sludge blanket process, are operated in the up-flow mode. All these processes are able to capture carbon by methanogenic conversion to methane.

Anaerobic treatment is an effective method for the complete treatment of many organic wastes, especially animal wastes and organic effluents from the food processing industries. The organic substrate is degraded in the absence of oxygen to carbon dioxide and methane, with only a small amount of bacterial growth. Approximately 90% of the available chemical energy, in the form of organic material, is retained as methane production. Examples of methane yield from dairy wastewaters are given by Zieliński *et al.* (2023) (Table 1.4).

Apart from the economic value of the methane gas produced, anaerobic treatment has many advantages over aerobic treatment processes, such as less biomass produced per unit of substrate utilized (the lower biomass production means a lower requirement for nitrogen, phosphorus, and other nutrient and growth factors), and higher organic loadings are possible as anaerobic processes are not limited by oxygen transfer rates and the lower constructional and operational costs compared with aerobic processes. However, the major disadvantage in temperate and colder climates is the elevated temperatures required to maintain microbial activity at a reasonable level.

In 1994, there were over 2,570 anaerobic digesters in Europe, with 60% of the sewage sludge produced by member states before final disposal. There were a further 400 digesters treating industrial effluents, 400 treating farm slurry, and 200 treating landfill leachate (Nyns, 1994). So, anaerobic digestion is processing almost one billion tonnes of European waste each year. By 2020, this had risen to 19,000 digesters, producing 167 TWh of biogas. There are also a further 727 plants producing 26 TWh of biomethane operating over 18 European countries, with 232 operating in Germany, 131 in France, and 10 in the United Kingdom (UK). Both biogas and biomethane are produced by anaerobic digestion; the difference is that biogas comprises 60% methane and 40% carbon dioxide and is used in CHP systems while carbon dioxide, hydrogen sulfide, and water are removed from biomethane so that it has the same properties as natural gas and can be loaded directly into the grid. Biogas and biomethane production is expected to rise to 467 TWh by 2030 and double by 2050. Ireland, for example, aims to replace 20% of its natural gas with biomethane by 2030, primarily from the wastewater and agricultural sectors (European Gas Association, 2023; Energy Ireland, 2024).

Because so many different groups of microorganisms are involved, the digestion process is not easily modeled, especially as the methanogenic archaea have a much lower growth rate than the acid-producing bacteria. However, as the conversion of volatile acids to biogas is

Table 1.4. Examples of different anaerobic treatment processes treating dairy wastewaters showing the HRT, organic loading rate (OLR) and performance (Zieliński *et al.*, 2023). Reproduced under the Creative Commons Attribution (CC BY) license (https://creativecommons.org/licenses/by/4.0/).

Reactor Type	Wastewater	HRT (d)	OLR (kgCOD/m^{3}d)	Initial COD (mg O_2/dm^3)	COD Removal (%)	CH_4 Yield (m^3CH_4/kg COD)
Anaerobic filter	Whey	1–5	1–4	5–20	72–90.2	0.089–0.28
	Whey	10–15	2.2–3.3	33	90	NA
	Synthetic whey	1	1–10.2	1–10.2	85–93.8	0.158–0.35
Biofilm-supported CSTR	Synthetic dairy	10	NA	NA	46–59	NA
Fluidized bed reactor	Dairy	7.5–11.3	10	3.2	85–98	0.37
	Simulated milk	0.75–5	2.22–31	10–77.5	78	0.27
	Ice-cream wastewater	1	1–12	5	70–90	0.157–0.193
Hybrid UASB	Dairy	0.25	8–20	5	65–93	0.25–0.31
	Dairy	1.9	0.97–2.82	1.9–5.34	91–97	0.27–0.359
Hybrid reactor	Dairy	0.5–1	1–2	1–2	64–76	0.03–0.24
Anaerobic moving	Milk permeate	9.7–116.5	0.5–6.5	50–75	57.9–77	0.341
biofilm reactor	Whey	0.6–17	1–21	6–17	70–97	0.33
Anaerobic sequencing biofilm batch reactor (ASBBR)	Whey	0.33	2–12	1–6	76–95	NA

NA: not available.

generally considered to be the rate-limiting step of the overall reaction, the methanogenic phase is usually used for modeling purposes. The majority of models are applicable to the continuously stirred tank reactor (CSTR), as they are the most widely used and operate closest to the steady-state, which is easier to model. However, the development of dynamic models has demonstrated the need for more accurate ways of examining the operational problems of digesters. Steady-state models are usually derived from one or two differential equations that treat the digester contents as a single substrate and the bacteria as a single population. Among the best-known models of this type are those using Monod kinetics. Of particular interest are models based on Contois kinetics developed by Chen and Hashimoto (1978, 1980; Vavilin *et al.*, 2001). The dynamic models available all use Monod kinetics and normally consider the interactions between several substrates and bacterial populations. This results in a system of differential equations that can only be solved by computer. Andrews and Graef (1971) considered that relating substrate concentration to the specific growth rate (the Monod function) was invalid for anaerobic reactors, as the volatile acids not only acted as substrate for methanogenic bacteria but were also inhibitory at higher concentrations. They replaced the Monod function with an inhibition function (Haldane, 1930) so that:

$$\mu = \mu_m/[1 + K_s/s + s/K_i]$$

where K_i is the inhibition function. Whereas this early dynamic model only included a single substrate and organism, more recent models of this type consider more than one substrate or more than one bacterial population (Hill and Barth, 1977; Hill and Nordstedt, 1980; Batstone *et al.*, 2002a). Anaerobic digestion model No. 1 (ADM1) includes multiple steps that describe both biochemical processes (e.g., disintegration from homogeneous particles to carbohydrates, proteins, and lipids; extracellular hydrolysis of these particulate substrates to sugars, amino acids, and long chain fatty acids (LCFA), respectively; acidogenesis from sugars and amino acids to VFAs and hydrogen; acetogenesis of LCFA and VFAs to acetate; and separate methanogenesis steps from acetate and hydrogen/carbon dioxide), as

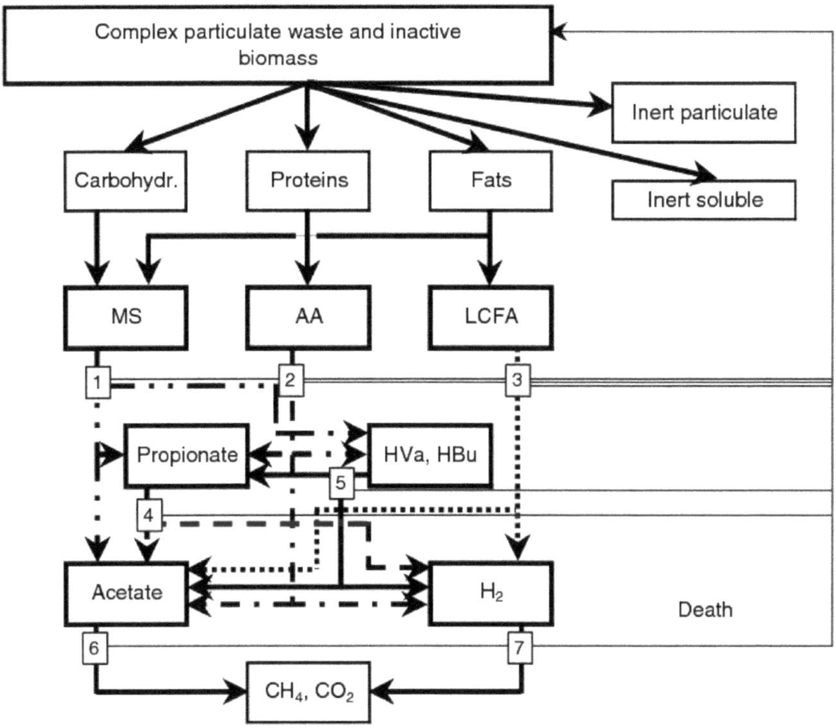

Figure 1.5. The IWA anaerobic digestion Model No 1 (ADM1) (Batstone *et al.*, 2002b). (1) Acidogenesis from monosaccharides (MSs), (2) acidogenesis from amino acids (AAs), (3) acetogenesis from long chain fatty acids (LCFAs), (4) acetogenesis from propionate, (5) acetogenesis from butyrate (HBu) and valerate (HVa), (6) acetoclastic methanogenesis, and (7) hydrogenotrophic methanogenesis. Reproduced with permission of IWA Publishing.

well as physicochemical processes (e.g., ion association and dissociation and gas-liquid transfer) (Fig. 1.5). The model comprises a set of differential and algebraic equations using 26 dynamic state concentration variables and 8 implicit algebraic variables per reactor vessel or element. It can also be implemented using differential equations only when 32 dynamic state concentration variables are used (Batstone *et al.*, 2002b). The ADM1 model has been extensively modified by researchers over the past decade to reflect changes in substrate complexity and process upgrading, especially in relation to fermentation and biogas production, which has been excellently reviewed by

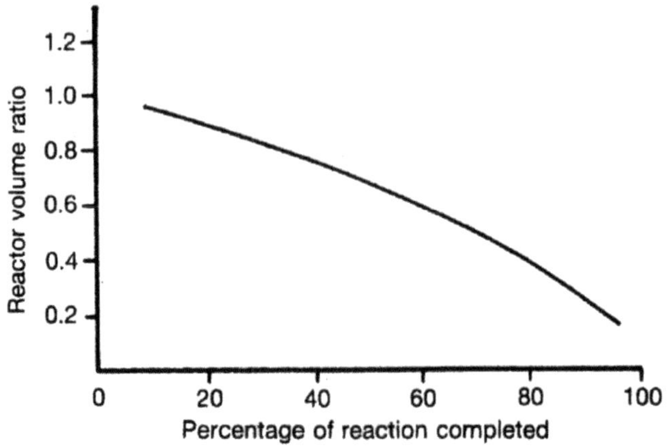

Figure 1.6. The ratio of plug flow reactor volume to completely mixed flow reactor volume for a first-order reactor process.

Mo *et al.* (2023). The use of models in anaerobic digestion has also been reviewed by Emelu *et al.* (2022), who explore the development of single-stage to multistage modeling of anaerobic digestion.

In plug-flow systems, there is a substrate concentration gradient between the inflow and outflow ends of the reactor. As the rate of reaction is dependent on the substrate, the average rate of reaction in a plug-flow system is theoretically higher than for a continuous stirred tank reactor (CSTR). Casey (1981) illustrates the significance of this difference in terms of the reactor volume required for a first-order reaction process (Fig. 1.6). This shows that the advantage of the plug-flow reactor over the CSTR increases with the required degree of reaction completion. Although a number of contact systems can be considered as plug-flow processes, those treating domestic sewage sludge are largely of the completely mixed type. Modeling of individual processes is further explored when looking at specific process designs.

1.4 The Role of Anaerobic Wastewater Treatment

When wastewater treatment was pioneered in the late 1890s, it was to bring to an end the misery of waterborne diseases such as cholera and typhoid that were killing millions of people globally, especially in urban areas, as well as changing how society could operate. The rapid expansion of sewerage schemes in Victorian times, followed by the rapid development of wastewater treatment between 1910 and 1960, coupled with secure water supplies and water treatment, especially chlorination, achieved not only the elimination of almost all waterborne diseases in the developed world but also a massive improvement in surface water quality in terms of organic and ammonia pollution. During the latter part of the 20th century, priorities in terms of wastewater treatment included the recovery of useful byproducts and the removal of nitrogen and later phosphorous to control eutrophication. The EU directive of wastewater treatment set new treatment objectives in the 1990s, including the phasing out of sewage sludge dumping at sea and the increasing controls of its use on agricultural land, requiring sludges to receive advanced treatment, which included thermal hydrolysis, which made anaerobic digestion more effective and stable as a process. The directive also expanded the use of anoxic and anaerobic zones in what had been strictly aerobic-activated sludge processes for biological nutrient removal bringing anaerobic treatment into the mainstream of wastewater treatment. With the final acceptance of global warming and the need to reduce carbon emissions, combined with the problem of failing sewer infrastructure and aging treatment plants, and exacerbated by ever-increasing loading due to population expansion, the need to improve wastewater treatment capacity as well as efficiency in terms of sustainability became urgent. So, for the first time in 150 years, wastewater engineers are beginning to look at new concepts in treatment, driven by the need to improve efficiency and reliability, reduce carbon dioxide emissions and energy consumption, expand the scope of material recycling, and the production of renewable energy and other low carbon byproducts. The circular economy concept has been an important tool in achieving increasingly sustainable wastewater treatment, and anaerobic processes are at the heart of this new revolution (Zieliński *et al.*, 2023).

The following chapters give the reader a broad overview of modern anaerobic wastewater treatment, starting with the key metabolic pathways (Chap. 2). Gaseous products of digestion include methane, hydrogen sulfide, ammonia, volatile organic compounds (VOCs), and siloxanes, all of which are explored in terms of their formation, impacts on health, and environmental effects (Chap. 3). Treatment processes are broken down into flow-through (*conventional*) systems and contact (*high-rate*) systems. Flow-through systems include more traditional designs such as septic and Imhoff tanks used primarily for individual houses or very small community systems and anaerobic lagoons, which are widely used by the food processing industries as well as wastewaters in warmer countries (Chap. 4). Sludge digestion is now the norm for sewage sludge treatment and stabilization, with many WWTPs using regional sludge digestion plants that cater for a number of small to medium-sized WWTPs producing small quantities of waste sludge. Considered a difficult and often unreliable process with the gas often flared off up to the 1970s, sludge digestion has now evolved into a reliable process, from which many other anaerobic processes have developed with many other sectors, especially agriculture, taking advantage of waste solids, including food wastes, to generate biogas (Chap. 5). Modern anaerobic wastewater treatment has been made possible through the development of contact systems, from biofilm processes using carriers to dispersed growth systems, such as the well-established UASB and the evolution of granular-based reactors (Chaps. 6 and 7). As society makes the slow transition from fossil fuels, anaerobic wastewater and sludge treatment technologies are both contributing to the production of renewable fuels, primarily through methane production for use at WWTPs or by cleaning to remove hydrogen sulfide, carbon dioxide, excess water, and other impurities to produce biomethane that can then be fed directly into the national gas networks (Chap. 8). So, although anaerobic processes are already making a huge contribution toward a cleaner and more sustainable environment, future prospects are extremely exciting as our understanding of how quorum sensing can be used to enhance gas yields, the production of biohydrogen as a new highly efficient fuel, or the production of new materials from volatile acid production, and the recovery of other materials such as metals (Chap. 9). But perhaps it is our improved understanding of the microorganisms that carry out this complex synergistic set of reactions that comprise anaerobic digestion and our ability to now rapidly identify them and the community structures that provide us with the unique communities that are able to deliver such a wide range of services transforming WWTPs from pollution control or waste disposal centers to resource recovery hubs. The future of anaerobic wastewater treatment is very exciting and will change wastewater treatment for the better.

References

Andrews, J.F. and Graef, S.P. (1971) Dynamic modelling and simulation of the anaerobic digestion process. In *Anaerobic Biological Treatment* (ed. R.F. Gould). American Chemical Society, pp. 126–162.

Batstone, D.J., Keller, J., Angelidaki, I., Kalyuzhnyi, S.V., Pavlostathis, S.G., Rozzi, A., Sanders, W.T.M., Siegrist, H. and Vavillin, V.A. (2002a) *The IWA Anaerobic Digestion Model No. 1 (ADM1)*. IWA Publishing, London, UK.

Batstone, D.J., Keller, J., Angelidaki, I., Kalyuzhnyi, S.V., Pavlostathis, S.G., Rozzi, A., Sanders, W.T.M., Siegrist, H. and Vavillin, V.A. (2002b) The IWA anaerobic digestion model No. 1 (ADM1). *Water Science and Technology*, **45**, (10), 65–73.

Casey, T.J. (1981) Developments in anaerobic digestion. *Transactions of the Institute of Engineers in Ireland*, **105**, 25–32.

Chen, Y.R. and Hashimoto, A.G. (1978) Kinetics of methane fermentation. *Biotechnology and Bioengineering Symposium*, **8**, 269–282.

Chen, Y.R. and Hashimoto, A.G. (1980) Substrate utilization kinetic model for biological treatment processes. *Biotechology and Bioengineering*, **22**, 2081–2095.

Emebu, S., Pecha, J. and Janáćová, D. (2022) Review on anaerobic digestion models: Model classification and elaboration of process phenomena. *Renewable and Sustainable Energy Reviews*, **160**, 112288. https://doi.org/10.1016/j.rser.2022.112288

Energy Ireland (2024) https://www.energyireland.ie/the-future-of-anaerobic-digestion-in-ireland/

European Gas Association (2023) The 'European Biomethane Map 2020' shows a 51% increase of biomethane plants in Europe in two years. https://www.europeanbiogas.eu/the-european-biomethane-map-2020-shows-a-51-increase-of-biomethane-plants-in-europe-in-two-years/#:~:text=European%20Biomethane%20Map'.-,The%20analysis%20of%20the%20data%20collected%20shows%20that%20the%20number,countries%20producing%20biomethane%20in%20Europe.

Finlay, B.J. and Fenchel, T. (1991) An anaerobic protozoan, with symbiotic methanogens, living in municipal landfill material. *FEMS Microbial Ecology*, **85**, 169–180.

Gray, N.F. (2022) *Natural Wastewater Treatment Systems and Sustainability*. World Scientific Publishing, London.

Haldane, J.B.S. (1930) *Enzymes*. Longman, London.

Hill, D.T. and Barth, C.L. (1977) A dynamic model for simulation of animal waste digestion. *Journal of the Water Pollution Control Federation*, **49**, 2129–2143.

Hill, D.T. and Nordstedt, R.A. (1980) Modelling techniques and computer simulation of agricultural waste treatment processes. *Agricultural Wastes*, **2**, 135–156.

Kato, M., Field, J.A. and Lettinga, G. (1993) The high tolerance of methanogens in granular sludge to oxygen. *Biotechnology and Bioengineering*, **42**, 1360–1366.

Mo, R., Guo, W., Batstone, D., Makinia, J. and Li, Y. (2023) Modifications to the anaerobic model no. 1 (AMD1) for enhanced understanding and application of the anaerobic treatment process — A comprehensive review. *Water Research*, **244**, 120504. https://doi.org/10.1016/j.watres.2023.120504

Nyns, E.J. (1994) *A Guide to Successful Industrial Implementation of Biomethanisation Technologies*. The Institute Wallon, Namur, Belguim.

Vavilin, V.A., Rytov, S.V., Lokshina, L.Y., Rintala, J.A. and Lyberatos, G. (2001) Simplified hydrolysis models for the optimal design of two-stage anaerobic digestion. *Water Research*, **35**, (4), 4247–4251.

Zieliński, M., Kazimierowicz, J. and Dębowski, M. (2023) Advantages and limitations of anaerobic wastewater treatment — Technological basics, development directions, and technological innovations. *Energies*, **16**, 83. https://doi.org/10.3390/en16010083

2

Anaerobic Metabolism and Microorganisms

2.1 Introduction

A wide range of anaerobic microorganisms are found in the environment, with obligate species associated with anaerobic conditions, including deep soil sediments, waterlogged soils, deep lakes, and marine environments. Wastewater also contains anaerobes from the human gut, especially *Bacteroidetes* spp.

Anaerobic bacteria are classified as facultative, obligate, and aerotolerant anaerobes.

Facultative anaerobes can grow in both aerobic and anaerobic environments, being able to metabolize energy in the form of adenosine triphosphate (ATP), either by respiration (aerobic) or by fermentation (anaerobic). ATP production is much lower by fermentation than by aerobic respiration, but sufficient is produced by fermentation to provide anaerobes with sufficient cellular energy. They form a bridge between obligate aerobes and obligate anaerobic species. Growth and metabolism of obligate anaerobes are inhibited by oxygen, with oxygen toxicity depending on the redox potential (E_h), partial pressure, composition of the substrate, growth rate, and cell density.

Obligate anaerobes only grow in the absence of oxygen and are generally inhibited or even killed when oxygen is present. ATP is generated by fermentation or anaerobic respiration. Many obligate anaerobic bacteria produce hydrogen sulfide.

Aerotolerant anaerobes are not inhibited by the presence of oxygen but can only generate ATP by fermentation (Pavlostathis and Giraldo-Gomez, 1991). Aerobic respiration constantly produces reactive oxygen species (ROS) as a byproduct, which are highly damaging to cellular structures but which are broken down cellularly by the production of enzymes, such as superoxide dismutase, peroxidase, and catalase. Obligate anaerobes are generally unable to produce these enzymes, which is the cause of their destruction when exposed to aerobic environments where ROS are present. In contrast, aerotolerant anaerobes are able to produce superoxide dismutase and NADH oxidase/NADH peroxidase but not catalase, which give them the ability to tolerate oxygen (Hansson and Häggström, 1984; Brioukhanov and Netrusov, 2004; 2007). The most common ROS are superoxide (O_2^-), peroxides (H_2O_2), and hydroxyl radical (OH•). ROS formation starts with molecular oxygen being reduced to superoxide:

$$O_2 + e^- \rightarrow O_2^- \tag{2.1}$$

The superoxide is disproportionated (i.e., dismutation) to hydrogen peroxide:

$$2H^+ + O_2^- + O_2^- \rightarrow H_2O_2 + O_2 \tag{2.2}$$

Partial or full reduction of hydrogen peroxide then follows:

$$H_2O_2 + e^- \rightarrow HO^- + OH\bullet \tag{2.3}$$

$$2H^+ + 2e^- + H_2O_2 \rightarrow 2H_2O \tag{2.4}$$

Aerotolerant anaerobes produce superoxide dismutase (SOD) that breaks down superoxide anions to hydrogen peroxide:

$$2O_2^- + 2H^+ \rightarrow H_2O_2 + O_2 \tag{2.5}$$

Singlet oxygen ($^1O_2$) and hypochlorite ion (OCl^-) are also types of ROS.

The American Type Culture Collection (ATCC) not only holds one of the largest reference collections of microorganisms (i.e., viruses, bacteriophages, bacterial, archaea, fungi, yeasts, cyanobacteria, and protists), it also provides a wide variety of aerobic and anaerobic microbial species for bioremediation, biocontrol, and biofuel production. They also produce a wide range of standardized protocols and methods for handling and growing anaerobes. Similar collections are found worldwide; for example, the Leibniz Institute (DSMZ) in Germany has over 3,500 holdings of anaerobes, including over 2,000 type strains.

Anaerobes require very specific growth conditions, which are devoid of all oxygen in the case of obligate anaerobes. Anaerobic jars or chambers that contain chemical packs to remove any oxygen and release carbon dioxide are most widely used. Anaerobic chambers usually have sealed gloves that allow technicians to work within the chamber from the outside, preventing any ingress of air (Fig. 2.1). The handling, growth, and identification of anaerobes is explained in a report prepared by the Leibniz Institute (DSMZ, 2023), with growth media specific to particular genera or species.

A recent development in taxonomic identification for anaerobic bacteria and archaea is the *Microbial Database for Activated Sludge* (MIDAS) developed by the Center for Microbial Communities at Aalborg University in Denmark (https://www.midasfieldguide.org/guide) (Dueholm *et al.*, 2022). It is a remarkable resource that includes details of both the physiology and ecology of the key species and taxonomic details. It allows close examination of how environmental and biogeographical factors affect anaerobic community structure in digesters. The database includes the results of full-length 16S rRNA gene sequencing of bacteria and archaea collected from 285 full-scale anaerobic digesters from around the world. It also allows sequences to be classified directly online. The most recent version of the database, MIDAS-5, can identify 692 genera and 1013 species (Dueholm *et al.*, 2024).

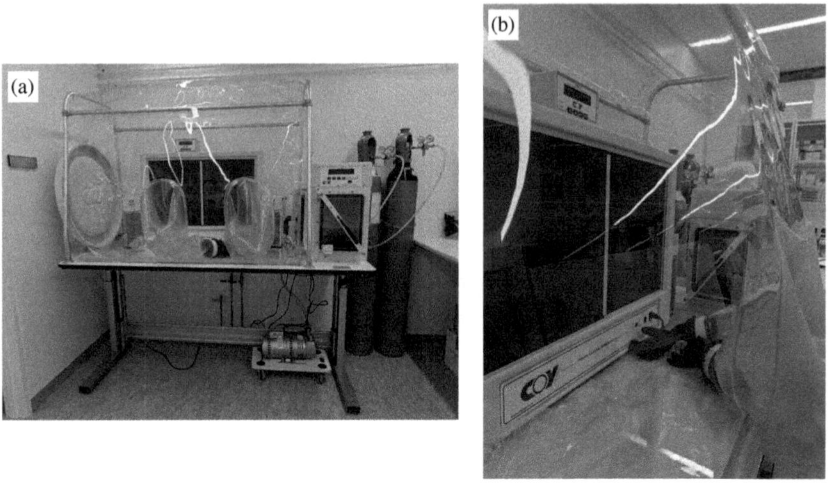

Figure 2.1. Anaerobic chambers are used to ensure oxygen-free environments when culturing and transferring anaerobic microorganisms. (a) Front view and (b) side view showing the inside chamber. Reproduced with the kind permission of Professor Amelia-Elena Rotaru, University of Sydney.

2.2 Anaerobic Degradation of Organic Matter

Anaerobic digestion can be carried out in a wide range of reactor types, treating both wastewaters and sludges. However, it is perhaps easiest to understand the process of anaerobic degradation by considering its use in domestic wastewater sludge treatment using digestion. The settled sludge, from both primary and secondary settlement tanks, is converted from its highly degradable odorous state to a stable and disposable product that neither smells nor undergoes further decomposition on storage. The main constituents of raw sludge are protein, fats, and polysaccharides, both soluble and insoluble, often in particulate form. Typical chemical analysis of sewage sludge is protein (25%), cellulose and lignin (25%), and fats (20%) (Gray, 2017). If anaerobic breakdown is complete, then the end-products will be methane, carbon dioxide, hydrogen sulfide, ammonia, water, and new bacterial cells, although volatile organic compounds and siloxanes may also be produced (Chap. 3). However, incomplete breakdown due to environmental conditions or inhibitors can result in highly polluting intermediate products being formed, such as volatile fatty acids (VFAs) and alcohols, as well as ammonia.

The process of anaerobic breakdown of organic matter comprises four discrete stages: hydrolysis, acid formation, and acetogenesis (i.e., the non-methanogenic phases), followed by methane formation (methanogenic phase) (Fig. 1.3). This requires a complex consortium of microorganisms, which can be classified as fermentative bacteria, hydrogen-producing acetogenic bacteria, hydrogen-consuming acetogenic bacteria, carbon-dioxide-reducing methanogens, and aceticlastic methanogens.

It is convenient to think of these stages as different trophic levels, and although all four stages normally occur simultaneously within a digester, the microorganisms involved at each stage are metabolically dependent on each other. For example, the methanogenic archaea require the catabolized end-products of the acid-forming bacteria. However, the latter species would eventually become inhibited by their own end-products if these were not degraded in turn by the methanogens. Although bacteria are the major group of microorganisms involved in anaerobic breakdown, archaea, fermentative ciliate and flagellate protozoa, and some anaerobic fungi also occur. The process does not readily occur in the presence of electron acceptors such as oxygen, sulfate, or nitrate. Energy transformation is by the ATP system, with energy being stored by the reaction of adenosine diphosphate (ADP) and inorganic phosphate (P_i) to form ATP. The energy conserved in the pyrophosphate bond is used by splitting ATP into either ADP + P_i or adenosine-5-monophosphate (AMP) and pyrophosphate (PP_i).

2.2.1 *Hydrolysis*

The particulate solids present in the waste stream cannot be assimilated by anaerobic microorganisms directly but need to have the large organic polymers, which comprise these solids, broken down into simpler soluble compounds that can be readily absorbed by the fermentative bacteria. Nearly all organic wastes are amenable to being anaerobically digested, although solid and semi-solid organic wastes, as well as those comprising large particle sizes, are pretreated using physicochemical methods to ensure maximum hydrolysis (Sec. 5.3). Domestic wastewaters and sludges contain high percentages of biodegradable solids comprising proteins, carbohydrates (including celluloses and lignins), and lipids. So, in the first stage of anaerobic digestion, the major substrates in the sludge or wastewater are hydrolyzed, with the proteins converted to peptides and amino acids, fats to glycerol and long-chain fatty acids (LCFAs), and polysaccharides to mono- or disaccharides (Fig. 1.3), which are subsequently used by the acidogenic bacteria. Wastewaters and sludges, including excess biomass from bioreactors, are rich in bacteria, including pathogens. Hydrolysis results in the death of a large proportion of any aerobic cells present, including pathogens, followed by cell lysis that releases the cellular contents back into solution.

As the process of anaerobic digestion involves four discrete steps, each mediated by a consortium of specialist bacteria, archaea, and other microorganisms, it is not surprising that the rate-limiting step for the overall process is hydrolysis, on which the reactor design is largely based. The rate of hydrolysis depends primarily on temperature and pH, with the hydrolytic fermentative bacteria extremely temperature sensitive (Veeken and Hamelers, 1999). Noike *et al.* (1985) compared the rates of substrate utilization for a range of substrates and found that the rate of hydrolysis of cellulose was the rate-limiting step overall in digestion.

Reactor design has to consider the minimum reactor temperature and pH but also substrate composition, substrate residence time, the size of particulate matter resent, and the concentration of end-products (e.g., VFAs and ammonia), which can inhibit hydrolysis if allowed to build up. The rate of hydrolysis is largely dependent on the nature of the organic matter being used as the main substrate. In turn, this will determine the relative composition of the breakdown products. Hydrolysis is modeled as a first-order process with coefficient rates of between 0.1 and 0.3 d^{-1} (Batstone *et al.*, 2002).

The extracellular enzymes responsible for hydrolysis (i.e., hydrolases) are produced by hydrolytic fermentative bacteria, with the products of hydrolysis used as their substrate.

Proteins are complex macromolecules comprising 20 different amino acids, the sequence of which determines both the structure of the protein and its properties. Proteins are more difficult to hydrolyze than either carbohydrates or lipids. They are broken down into smaller building blocks, such as polypeptides, oligopeptides, or even individual amino acids, by proteolytic extracellular enzymes called proteases, which are produced by only a small proportion of the bacteria present during anaerobic degradation. The majority of bacteria are able to utilize these smaller soluble peptides or the amino acids, which are absorbed directly across the cell wall and are broken down intracellularly. Protease production is far in excess of that required, even though protease-producing bacteria represent such a small percentage of the total bacteria present overall. Estimates indicate that this over-production could be in the order of 50 times more than is required (Hattingh *et al.* 1967), with the most active proteolytic bacteria, the spore-forming *Clostridium* spp. In anaerobic sludge, proteolytic bacteria can reach concentrations of 6.5×10^7 mL^{-1}, of which 65% of the isolates examined by Toerien (1970) were spore formers, 21% cocci, and the remainder non-sporing rods and bifid-like bacteria. The rate of breakdown is dependent primarily on the solubility and overall structure of the proteins, although temperature and pH are also important. Duong *et al.* (2022) observed that increasing VFA concentration reduced protein hydrolysis rates while carbohydrate concentration did not. An example of a simple oligopeptide comprising just two amino acids (i.e., a dipeptide) being hydrolyzed by protease into its component amino acids is shown below:

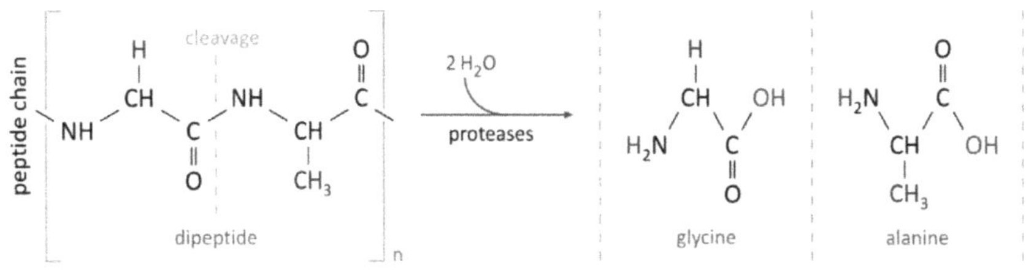

Fats and oils (i.e., lipids) are common in wastewaters and are esters of fatty acids, the vast majority of which are triacylglycerol (commonly known as triglycerides), which are triple esters of glycerol. Hydrolysis of lipids is mediated by enzymes known as lipases or esterases, produced by lipolytic bacteria, splitting triacylglycerol ($C_6H_8O_6$) into glycerol and three long chain fatty acids (LCHAs), as shown below:

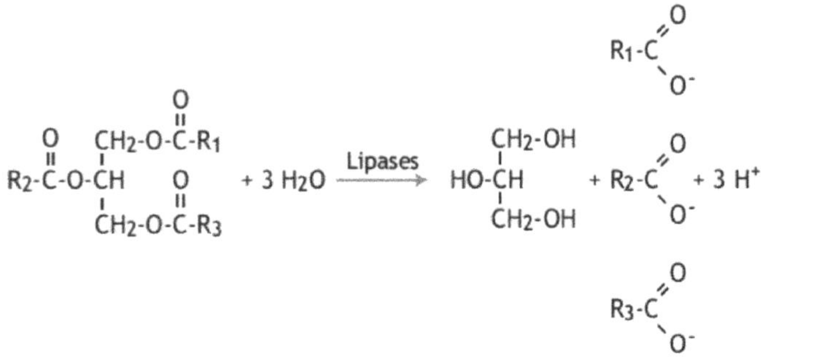

Triacylglycerol Glycerol Long Chain Fatty Acids

Hydrolysis of fats is usually completely achieved during digestion, although more slowly compared to other substrates. Relatively little is known about the lipolytic bacteria, even though they have been shown to be highly effective in anaerobic digesters. They are present in densities of up to 7×10^4 mL^{-1}, and the addition of used vegetable oil to digesters to enhance gas production is practiced in some countries.

Carbohydrates are found both as simple sugars or monosaccharides that can be directly absorbed by bacteria or common disaccharides (e.g., sucrose and maltose), and complex oligosaccharides and polysaccharides, which need to be broken down into their component monosaccharides by hydrolysis by glucosidases such as cellulase produced by cellulolytic bacteria before being absorbed by bacteria. Wastewaters are particularly rich in long chain polysaccharides, including pectin, starch, and celluloses. The hydrolysis of the disaccharide cellobiose ($(C_6H_7(OH)_4O)_2O$) is shown below:

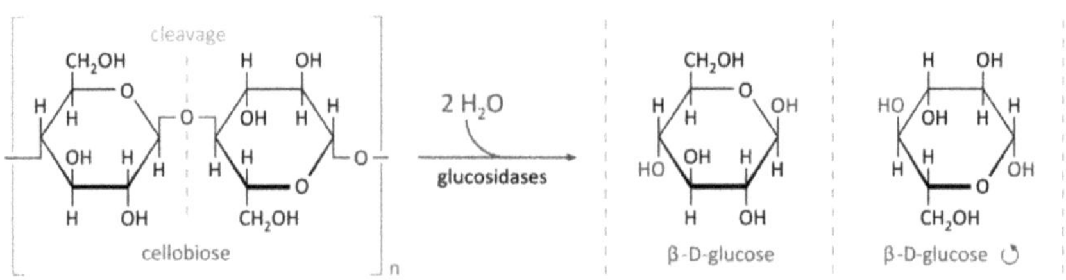

Disaccharides are hydrolyzed very rapidly; however, the hydrolysis of polysaccharides takes much longer. Agricultural wastewaters and wastes are rich in lignocellulosic compounds, which can only be partially hydrolyzed, as lignin is not broken down under anaerobic conditions (Brune and Sahm, 1981).

Anaerobic cellulolytic bacteria are present in anaerobic sludge at concentrations of between 10^4 and 10^5 mL^{-1} and are predominantly Gram-negative coccobacilli, with *Bacteroides ruminicola* being a particularly common species. Other species, including a Gram-positive species forming curved rods, which formed short chains, were originally isolated by Hobson and Shaw (1971). In sewage sludge, the ability to hydrolyze starch is the most common activity of these bacteria (Tables 1.3 and 3.2).

2.2.2 *Acidogenesis or Fermentation*

The heterogeneous groups of facultative and anaerobic bacteria, which are responsible for hydrolysis, are also responsible for acid formation. In this second stage, the hydrolyzed substrate is converted to organic acids and alcohols, with new cells also being produced. Various biochemical pathways are utilized, including fermentation and β-oxidation. There is very little stabilization of the substrate in terms of BOD or COD removal, with the products of acid fermentation being large organic molecules. Obligate anaerobic bacteria occur in much larger numbers than facultative bacteria, with ratios of 1:100 and numbers in the order of 10^7–10^8 mL^{-1} not uncommon (Kirsch, 1968). Major acid-forming organisms include *Bacillus* sp., *Micrococcus* sp., and *Pseudomonas* sp. (Table 1.3). Mono- and disaccharides, LCFA, glycerol, amino acids, and short-chain peptides provide the main carbon source for growth, with saturated fatty acids, carbon dioxide, and ammonia being the main end-products. In practice, it is the monosaccharides and amino acids that are the most abundant substrates. Alcohols, aldehydes, and ketones are also produced but only in minute quantities (Fig. 1.3). The concept of pyruvate as the pivotal compound in metabolism was discussed earlier. When no external electron acceptor is present, as is the case in acid fermentation, pyruvate can undergo several alternative reactions that regenerate NAD from

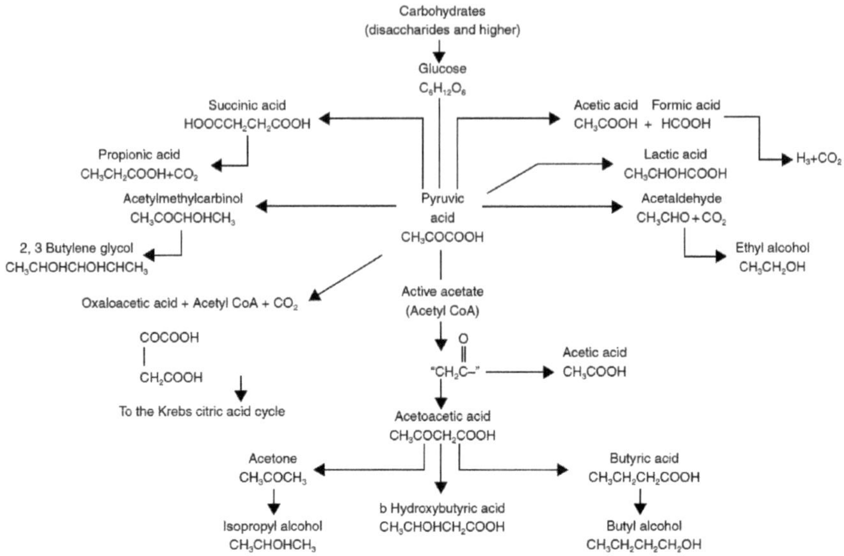

Figure 2.2. Pyruvic acid is a pivotal compound in metabolism from which many other compounds can be metabolized, depending on the organism and environmental conditions.

NADH (Fig. 2.2). Acetic, propionic, butyric, and lactic acids are the most frequently produced end-products during fermentation. Propionic and LCFAs are degraded by an intermediate microbial group called the obligate hydrogen-producing acetogenic bacteria, whereas other acid-producing bacteria, referred to as the homacetogenic bacteria, produce acetic and sometimes, other acids (McInerney *et al.*, 1980).

Fermentation of monosaccharides

Monosaccharide fermentation is by two pathways, the Meyerhof-Parnas (EMP) or Entner Doudoroff (ED) pathways. The final products via acetyl-CoA are C3 (lactate or propionate), C2 (acetate), C4 (butyrate), or, occasionally, C6 (caproate) products. The most common products formed during fermentation are acetate, butyrate, and ethanol. The main degradation pathways of glucose during acidogenesis are:

$$\underset{Glucose}{C_6H_{12}O_6} + 2H_2O \rightarrow \underset{Acetic\ acid}{2CH_3COOH} + 2CO_2 + 4H_2 \quad (2.6)$$

$$\underset{Glucose}{C_6H_{12}O_6} \rightarrow \underset{Butyric\ acid}{CH_3[CH_2]_2COOH} + 2CO_2 + 2H_2 \quad (2.7)$$

$$\underset{Glucose}{C_6H_{12}O_6} \rightarrow \underset{Lactic\ acid}{2CH_3CH(OH)COOH} \quad (2.8)$$

$$\underset{Glucose}{C_6H_{12}O_6} + 2H_2 \rightarrow \underset{Propionic\ acid}{2CH_3CH_2COOH} + 2H_2O \quad (2.9)$$

$$\underset{Glucose}{C_6H_{12}O_6} \rightarrow \underset{Propionic\ acid}{CH_3CH_2COOH} + \underset{Acetic\ acid}{CH_3COOH} + 2CO_2 + 2H_2 \quad (2.10)$$

$$\underset{Glucose}{C_6H_{12}O_6} \rightarrow \underset{Ethanol}{2CH_3CH_2OH} + 2CO_2 \quad (2.11)$$

$$\underset{Glucose}{C_6H_{12}O_6} + H_2O \rightarrow \underset{Ethanol}{CH_3CH_2OH} + \underset{Acetic\ acid}{CH_3COOH} + 2CO_2 + 2H_2 \quad (2.12)$$

$$\underset{Glucose}{C_6H_{12}O_6} + H_2O \rightarrow \underset{Ethanol}{CH_3CH_2OH} + \underset{Latic\ acid}{CH_3CH(OH)COOH} + 2CO_2 \quad (2.13)$$

In these reactions, glucose is used as an electron acceptor to produce oxidized (e.g., acetate) or reduced products (e.g., propionate, ethanol) (Angelidaki *et al.*, 2011). The EMP and ED pathways can coexist, as shown by Jojima *et al.* (2021), who studied the ethanol-producing bacteria *Corynebacterium glutamicum*.

Fermentation of amino acids

Two different amino acids generally degrade together in a coupled oxidation-reduction reaction known as the Stickland fermentation, with one amino acid oxidized while the other is reduced. Uncoupled amino acid degradation is also possible, e.g., glutamate fermentation. Winter *et al.* (1987) give the example of coupled fermentation of alanine with glycine by the Stickland reaction, with alanine oxidized and glycine reduced:

$$CH_3CHNH_2COO^- + 2H_2O \rightarrow CH_3COO^- + CO_2 + NH_3 + 4H^+ + 4e^-$$

Alanine *Acetate* (2.14)

$$2CH_2NH_2COO^- + 4H^+ + 4e^- \rightarrow 2CH_3COO^- + 2NH_3^- \quad (2.15)$$

Glycine *Acetate*

2.2.3 *Acetogenesis*

Acetate is synthesized during acetogenesis either by the reduction of carbon dioxide or from organic acids. Hydrogen-producing acetogens oxidize ethanol and organic acids to acetate, hydrogen, and carbon dioxide.

$$CH_3CH_2OH + H_2O \rightarrow CH_3COO^- + H^+ + 2H_2 \quad (2.16)$$

Ethanol *Acetate*

$$CH_3[CH_2]_2COO^- + 2H_2O \rightarrow CH_3COO^- + H^+ + 2H_2 \quad (2.17)$$

Butyrate *Acetate*

$$CH_3CH[OH]COO^- + 2H_2O \rightarrow CH_3COO^- + HCO_3^- + H^+ + 2H_2$$

Lactate *Acetate* (2.18)

$$CH_3CH_2COO^- + 3H_2O \rightarrow CH_3COO^- + HCO_3^- + H^+ + 3H_2 \quad (2.19)$$

Propionate *Acetate*

$$CH_3[CH_2]_3COO^- + 2H_2O \rightarrow CH_3COO^- + \rightarrow CH_3CH_2COO^- + H^+ + 2H_2 \quad (2.20)$$

Valeriate *Acetate*

These reactions are generally endergonic under normal operating conditions, and to make them exergonic, the resulting hydrogen must be continuously removed. This is done by hydrogen-utilizing archaeal methanogens (i.e., hydrogenotrophic methanogenesis), which require a close relationship between the bacteria and archaea-creating interspecies hydrogen transfer requiring a close physical proximity between the two (i.e., syntropy). The closer the inter-microbial distance, the greater the specific growth rates.

Hydrogenotrophic methanogenesis $4H_2 + 2CO_2 \rightarrow CH_4 + 2H_2O$ (2.21)

This relationship also allows the degradation of fatty acids that cannot be broken down by acetogens or methanogens alone (Angelidaki *et al.*, 2011). An example of syntrophic acetogenesis and hydrogenotrophic methanogenesis is shown below:

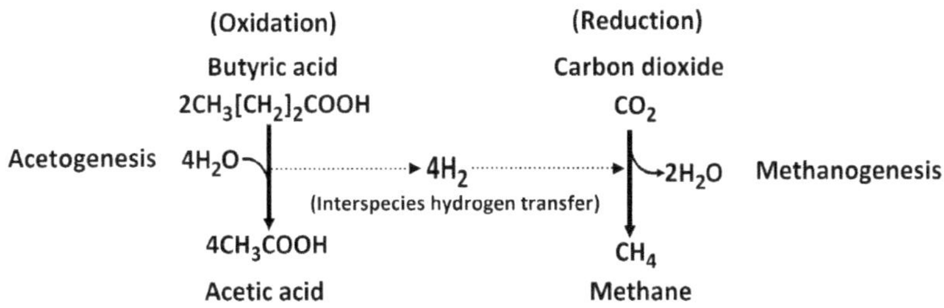

At low temperatures or acidic conditions, hydrogen-utilizing acetogens (homoacetogens) can compete with the hydrogen-utilizing archaea for the available hydrogen where they reduce CO_2 to acetate:

Homoacetogenesis $4H_2 + 2CO_2 \rightarrow CH_3COO^- + H^+ + 2H_2O$ (2.22)

Hydrogen can also be removed by sulfate-reducing bacteria, although this can impact methane production by reducing the available substrate for hydrogenotrophic methanogenesis.

Sulfate reduction $4H_2 + SO_4^{2-} + H^+ \rightarrow H^+ + HS^- + 4H_2O$ (2.23)

2.2.4 *Methanogenesis*

There are three principle anaerobic pathways in methane formation: (i) Aceticlastic methanogenesis uses acetate to form methane, (ii) hydrogenotrophic methanogenesis uses hydrogen and carbon dioxide, while (iii) methylotrophic methanogenesis utilizes methylated C1 compounds, such as methanol (CH_3OH) or methylamines (CH_3NH_2). These pathways are represented below:

$$\text{(i) } CH_3COO^- + H^+ \rightarrow CH_4 + CO_2 \quad (2.24)$$

$$\text{(ii) } 4H_2 + CO_2 \rightarrow CH_4 + 2H_2O \quad (2.25)$$

$$\text{(iii) } 4CH_3OH \rightarrow 3CH_4 + CO_2 + 2H_2O \quad (2.26)$$

In practice, it is aceticlastic and hydrogenotrophic methanogenesis that are the principle methane formation pathways in the digestion of wastewater sludges, representing 70 and 30% of the overall methane generation respectively. Many hydrogenotrophic methanogens can also use formate. Further details of methane formation and microbial kinetics of methanogens are given in Sec. 3.2.1.

During methanogenesis, complete stabilization of the substrate occurs, which is more efficient than complete aerobic stabilization, with only gaseous end-products formed. Methane is an ideal end-product as it is non-toxic, easily separated from the site of production, is not very soluble, inert under anaerobic conditions, and can be readily collected for use as an energy source. In the overall anaerobic fermentation of carbohydrates to carbon dioxide and methane, equal volumes of each gas (**a**) are produced.

$$(C_6H_{12}O_6)_x \times \mathbf{a}H_2O \rightarrow \mathbf{a}C_6H_{12}O_6 \rightarrow 3\mathbf{a}CH_4 + 3\mathbf{a}CO_2 \qquad (2.27)$$

The carbon dioxide produced only partially escapes as gas because, unlike methane, it is relatively soluble in water. It also reacts with any hydroxide ions (OH^-) in the system to produce bicarbonate ions (HCO_3^-). The evolution of carbon dioxide gas is, therefore, a function of factors such as pH, bicarbonate concentration, temperature, and substrate composition. The biodegradable protein is deaminated to produce ammonia, which reacts with water:

$$NH_3 + HOH \leftrightarrow NH_4^+ + OH^- \qquad (2.28)$$

This is the major source of hydroxide ions, which react with the carbon dioxide evolved during methanogenesis to form bicarbonate ions:

$$CO_2 + HOH \leftrightarrow H_2CO_3 \leftrightarrow H^+ + HCO_3^- \qquad (2.29)$$

$$H_2CO_3 + OH^- \leftrightarrow HCO_3^- + HOH \qquad (2.30)$$

Therefore, the protein content of the wastewater substrate will significantly affect the quantity of carbon dioxide actually released from solution as well as the buffering capacity of the system in terms of bicarbonate. The portion of carbon dioxide incorporated in the bicarbonate ion is eventually removed from the reactor in the liquid rather than in the gas phase.

The current taxonomy classifies methanogens, which are from the domain Archaea and not bacteria, into six well-established orders: Methanobacteriales, Methanocellales, Methanococcales, Methanomicrobiales, Methanosarcinales, and Methanopyrales, excluding the order Methanomassiliicoccales, which is methylotrophic (Buan, 2018) (Sec. 3.2.1). The order Methanopyrales is hyperthermophilic and closely related to both *Methanobacteriales* and *Methanococcales* (Tables 2.1 and 3.1). There is currently only one known species, *Methanopyrus kandleri* (Liu, 2010). Methanogens are unusual in that they are composed of many species with very different cell morphology. For example, long filaments (Methanosarcinales), short rods (Methanobacteriales and Methanopyrales), and coccoidal cells (Methanococcales and Methanomicrobiales).

Acetoclastic methanogenesis that uses acetate to form methane is the most important pathway for methane production (Sec. 3.2.1) and is carried out predominantly by the Methanosarcinales. Methylotrophic methanogenesis is mainly carried out by archaea from the Methanosarcinales, although *Methanosphaera* spp. (Methanobacteriales) can also use methylated C1 compounds. Hydrogenotrophic methanogenesis uses hydrogen and carbon dioxide and is carried out by members of the Methanobacteriales, Methanocellales, Methanococcales, Methanomicrobiales, and Methanopyrales (Tables 2.1 and 3.1) (Angelidaki *et al.*, 2011). Digesters support a wide range of methanogens, with the

Table 2.1. Anaerobic pathway, temperature, and pH ranges for the main six orders of methanogens. Adapted from (Angeliidki *et al.*, 2011).

Order	Predominant pathway	Temperature range (°C)	pH range
Methanobacteriales	Hydrogenotrophic	15–60	6.1–8.0
Methanocellales	Hydrogenotrophic	25–40	6.5–7.8
Methanococcales	Hydrogenotrophic	20–88	4.5–9.8
Methanomicrobiales	Hydrogenotrophic	15–60	6.1–8.0
Methanosarcinales	Aceticlastic, hydrogenotrophic	10–70	4.0–10.0
Methanopyrales	Hydrogenotrophic	84–110	5.5–7.0

community structure influenced by a wide range of operating parameters and designs. For example, acetoclastic methanogen *Methanosarcina thermophila* and *Methanosaeta* are particularly common at higher (thermophilic) temperatures. A decrease in pH, caused by increasing the loading rate, which raises the VFA concentration, favors the acid tolerant methanogens, such as *Methanobrevibacter acididurans.*

2.2.5 *Importance of Redox Potential*

Microaerobic or anaerobic conditions are required for fermentation and methanogenic reactions to occur, not only in the treatment of wastewaters and waste sludges but in an enormous array of other fermentation processes, such as ethanol production. This is measured and controlled by the use of redox potential, also known as oxidation-reduction potential (ORP). ORP reactions in biological reactions are normally defined in terms of the loss and gain of hydrogen or electrons. Each oxidation is accompanied by a reduction and can be summarized as:

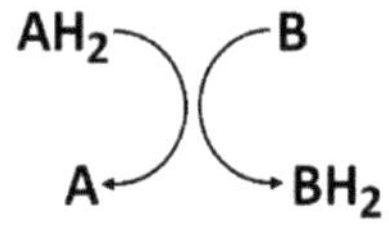

where AH_2 is the hydrogen donor, and B is the hydrogen acceptor. Each couple (AH_2/A or B/BH_2) has a tendency to either donate reducing equivalents and be oxidized ($AH_2 \rightarrow A$) or accept them and be reduced ($B \rightarrow BH_2$). When both couples combine in a complete redox reaction, the net flow of the reaction can be determined by the relative tendency of each couple to donate or accept reducing equivalents, which is the redox potential. This can be measured using a galvanic cell consisting of two electrodes connected by a conducting solution: oxidation occurs at the negative electrode (anode), and electrons are produced, whereas electrons are consumed and reduction takes place at the positive electrode (cathode). The redox potential is quantified by comparison with a standard redox couple. By convention, the standard redox couple is present at a hydrogen electrode consisting of a platinum electrode, with hydrogen ions in solution. In the presence of platinum as the catalyst, the reaction is:

$$H_2 \leftrightarrow 2H^+ + 2e^- \tag{2.31}$$

and the tendency to donate reducing equivalents, as electrons in this case, is measured as the voltage (potential) of the electrical current generated when the electrode is coupled in series with another redox couple electrode. Under standard conditions at 25°C, 1 atm of hydrogen, and at pH 0, the redox potential of the hydrogen electrode is zero. At pH 7, the potential of the redox couple $H_2/2H^+ + 2e^-$ is –420 mV. The symbol E_h is used for the redox potential under standard conditions. So fascinatingly, the ORP provides information on the activity of electrons within the liquid medium as well as at the cellular level.

A couple of lower redox potential will always donate reducing equivalents to a couple of higher potential and, during the oxidation of a substrate, reducing equivalents are transferred in the direction of increasing potential. This transfer is accompanied by the release of free energy, the magnitude of which is given by the standard free energy change,

$$\Delta G^{o\prime} = -nF\Delta E_h \tag{2.32}$$

where n is the number of electrons transferred in the reaction, F is the charge on one mole of electrons, which is Faraday's constant (96.649 kJV^{-1} mol^{-1}), and ΔE_h is the standard electrode potential. Biological systems have evolved to conserve this energy and convert it into biologically useful forms (i.e., oxidative phosphorylation).

In practical terms, ORP can be used to indicate which oxidative-reduction reactions will occur within a wastewater system and is particularly useful in the management of anaerobic systems (Fig. 2.3). The ORP gives a measure of the general condition of the liquid. Anaerobic processes similar to those found in sludge lagoons and digesters will have low values of E_h (< –200 mV), whereas aerobic processes will have higher values (> +50 mV). More precisely, values of E_h –150 mV to –420 mV are found in anaerobic environments, whereas aerobic environments vary between –200 mV and +420 mV.

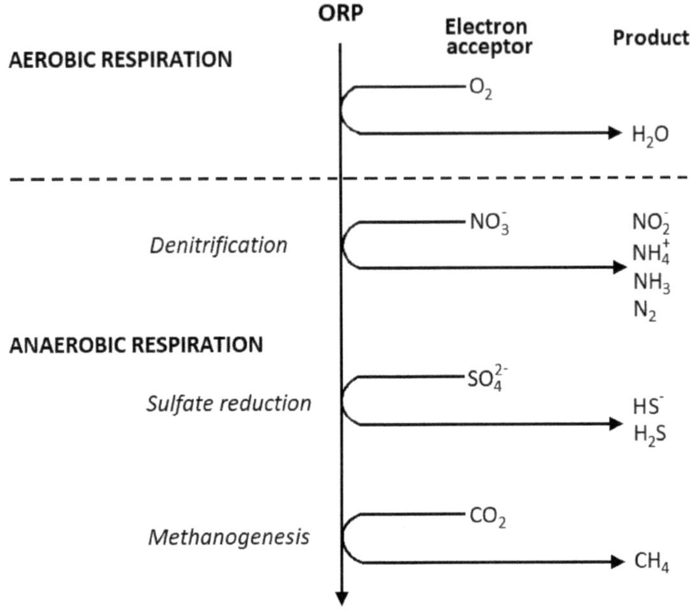

Figure 2.3. The redox potential indicates which inorganic electron acceptor is being used by the microbial community.

Facultative environments change from aerobic to anaerobic systems at about +100 mV. ORP is a more reliable measure of aerobic conditions than a dissolved oxygen concentration measurement, as it is often used in the control of environmental conditions in activated sludge tanks. Because odors are produced from anaerobic rather than aerobic processes, ORP can be used as a convenient measure to determine the onset of odor production during the treatment of high-strength wastes, such as animal wastes. The VFA concentration in the liquid phase can be controlled by using ORP and pH, as can the ratio of particular VFAs produced. This is widely used for the industrial production of acetic, propionic, and butyric acids by anaerobic fermentation using hydrolytic, acidogenic, and acetogenic bacteria (Chen *et al.*, 2022). So, the control of ORP is critical in all fermentation reactions, including wastewater anaerobic treatment and sludge digestion. At the intracellular level, when electrons are transferred, oxygen will be the prior acceptor, with the next acceptor utilized in

Table 2.2 Examples of the ORP of redox pairs in intracellular metabolism (Liu *et al.*, 2013). Reproduced with permission of Elsevier Publishing.

Redox pairs	**ORP E^h (mV)**
H+/H	-420
NAD+/NADH	-320
NADP+/NADPH	-315
Lipoate; ox/red	-290
Acetoacetate/3-hydroxybutyrate	-270
GSSG/2GSH	-240
Pyruvate/lactate	-190
Oxaloacetate/malate	-170
Fumarate/succinate	+30
Cytochrome b; Fe3+/Fe2+	+80
Ubiquinone; ox/red	+100
Cytochrome c1; Fe3+/Fe2+	+220
Cytochrome a; Fe3+/Fe2+	+290
Oxygen/water	+820

its absence, according to Table 2.2. In anaerobic fermentation, the key cofactors NAD+ (oxidized nicotinamide adenine dinucleotide and NADP+ (nicotinamide adenine dinucleotide phosphate) are able to accept electrons to support the metabolism of a large range of intracellular reactions. They are reduced to NADH and NADPH, respectively, so that the ratio of NADH/NAD+, in particular, determines the intracellular ORP (Liu *et al.*, 2013).

References

Angelidaki, I., Karakashev, D., Batstone, D.J., Plugge, C.M. and Stams, A.J.M. (2011) Biomethanation and its potential. *Methods in Enzymology*, **494**, 327–351.

Batstone, D.J., Keller, J., Angelidaki, I., Kalyuzhnyi, S.V., Pavlostathis, S.G., Rozzi, A., Sanders, W.T., Siegrist, H. and Vavilin, V.A. (2002) Anaerobic digestion model No. 1 (ADM1). Scientific and technical report: **13**, IWA Publishing, London.

Brioukhanov, A.L. and Netrusov, A.I. (2004) Catalase and superoxide dismutase: Distribution, properties, and physiological role in cells of strict anaerobes. *Biochemistry (Moscow)* **69**, 949–962. https://doi.org/10.1023/B:BIRY.0000043537.04115.d9

Brioukhanov, A.L. and Netrusov, A.I. (2007) Aerotolerance of strictly anaerobic microorganisms and factors of defense against oxidative stress: a review. *Prikladnaya Biokhimiya i Mikrobiologiya*, **43**, (6), 635–652. In Russian. https://pubmed.ncbi.nlm.nih.gov/18173105/

Brune, G. and Sahm, H. (1981) Anaerobe umsetzung von organisch hochbelasteten abwässern aus der zellstoffindustrie. *Wissenschaft und Umwelt*, **3**, S.124–S.126.

Buan, N.R. (2018) Methanogens: Pushing the boundaries of biology. *Emerging Topics in Life Science*, **2**, (4), 629–646. https://doi.org/10.1042/ETLS20180031

Chen, B., Rupani, P.F., Azman, S., Dewil, R. and Appels, L. (2022) A redox strategy to enhance propionic and butyric acid production during anaerobic fermentation. *Bioresource Technology*, **361**, 127672. https://doi.org/10.1016/j.biortech.2022.127672

DSMZ (2023) *Cultivation of Anaerobes.* Leibniz Institute DSMZ-German Collection of Microorganisms and Cell Cultures, Braunschweig-Süd, Germany. Available online: https://www.dsmz.de/fileadmin/Bereiche/Microbiology/Dateien/Kultivierungshinweise/englAnaerob.pdf

Dueholm, M.K.D., Nierychlo, M., Andersen, K.S., Rudkjøbing, V., Knutsson, S., Albertsen, M., Nielsen, P.H. (2022) MiDAS 4: A global catalogue of full-length 16S rRNA gene sequences and taxonomy for studies of bacterial communities in wastewater treatment plants. *Nature Communications,* **13**, 1908. https://doi.org/10.1038/s41467-022-29438-7

Dueholm, M.K.D., Andersen, K.S., Korntved, AK.C., *et al.* (2024) MiDAS 5: Global diversity of bacteria and archaea in anaerobic digesters. *Nature Communications,* **15**, 5361. https://doi.org/10.1038/s41467-024-49641-y

Duong, T.H., van Eekert, M., Grolle, K., Tran, T.V.N., Zeeman, G., and Temmink, H. (2022) Effect of carbohydrates on protein hydrolysis in anaerobic digestion. *Water Science and Technology,* **86**, (1), 66–79. https://doi.org/10.2166/wst.2022.200

Gray, N.F. (2017) *Water Science and Technology: An Introduction.* 4th edition. CRC Press, Boca Raton, FL.

Hansson, L. and Häggström, M.H. (1984) Effects of growth conditions on the activities of superoxide dismutase and NADH-oxidase/NADH-peroxidase in *Streptococcus lactis. Current Microbiology* **10**, 345–351. https://doi.org/10.1007/BF01626563

Hattingh, W.H.J., Kotze, J.P., Thiel, P.G., Toerien, D.F. and Siebert, M.L. (1967) Biological changes during the adaptation of an anaerobic digester to a synthetic substrate. *Water Research*, **1**, 255–277.

Hobson, P.N. and Shaw, B.G. (1971) In *Microbial Aspects of Pollution* (eds. G. Sykes and F.A. Skinner). Academic Press, London, pp. 103–121.

Jojima, T., Igari, T., Noburyu, R., Watanabe, A., Suda, A. and Inui, M. (2021) Coexistence of the Entner–Doudoroff and Embden–Meyerhof–Parnas pathways enhances glucose consumption of ethanol-producing *Corynebacterium glutamicum. Biotechnology for Biofuels and Bioproducts,* **14**, 45. https://doi.org/10.1186/s13068-021-01876-3

Kirsch, E.J. (1968) Studies on the enumeration and isolation of obligate anaerobic bacteria from digesting sludge. *Developments in Industrial Microbiology,* **10**, 170–176.

Liu, C.-G., Xue, C., Lin, Y.-H. and Bai, F.-W. (2013) Redox potential control and applications in microaerobic and anaerobic fermentations. *Biotechnoloy Advances*, **31**, 257–265. http://dx.doi.org/10.1016/j.biotechadv.2012.11.005

Liu, Y. (2010) Methanopyrales. In *Handbook of Hydrocarbon and Lipid Microbiology* (ed. K.N. Timmis). Springer, Berlin, Heidelberg. https://doi.org/10.1007/978-3-540-77587-4_47

McInerney, M.L., Bryant, M.P. and Stafford, D.A. (1980). Metabolic stages and energetics of microbial anaerobic digestion. In *Anaerobic Digestion* (eds. D.A. Stafford, B.I. Wheatley and D.E. Hughes). Applied Science Publishers, London, pp. 91–98.

Noike, T., Endo, G., Chang, J.-E., Yaguchi, J.-I. and Matsumoto, J.-I. (1985) Characteristics of carbohydrate degradation and the rate-limiting step in anaerobic digestion. *Biotechnology and Bioengineering*, **27**, 1482–1489. https://doi.org/10.1002/bit.260271013

Pavlostathis, S.G. and Giraldo-Gomez, E. (1991) Kinetics of anaerobic treatment: A critical review. *Critical Reviews in Environmental Control*, **21**, (5/6), 441–490. https://doi.org/10.1080/10643389109388424

Toerien, D.F. (1970). Population description of the non-methanogenic phase of anaerobic digestion. 1. Isolation, characterization and identification of numerically important bacteria. *Water Research*, **4**, 129–148.

Veeken, A. and Hamelers, B. (1999) Effect of temperature on hydrolysis rates of selected bio-waste components. *Bioresource Technology*, **69**, (3), 249–254. https://doi.org/10.1016/S0960-8524(98)00188

Winter, J., Schindler, F. and Wildenauer, F.X. (1987) Fermentation of alanine and glycine by pure and syntrophic cultures of *Clostridium sporogenes*. *FEMS Microbial Ecology*, **45**, 153–161. https://doi.org/10.1111/j.1574-6968.1987.tb02351.x

3

Methane and Other Volatile Products of Digestion

3.1 Introduction

Biogas produced from organic waste streams by anaerobic digestion is primarily a mixture of methane (CH_4) and carbon dioxide (CO_2) with hydrogen (H_2), nitrogen (N_2), and water vapor (H_2O) also present. However, a wide range of other compounds can also be present in biogas, several of which are potentially damaging to the environment and human health or cause damage over time to equipment, either by corrosion or abrasion. These include hydrogen sulfide (H_2S), ammonia (NH_2), siloxanes, and a wide range of volatile organic carbon compounds (VOCs), including terpenes and halogenated hydrocarbons (AOXs). Sludge digestion often produces high hydrogen sulfide concentrations, while wastewater anaerobic treatment is more associated with AOXs and siloxanes (Rasi *et al.*, 2007; Calbry-Muzyka *et al.*, 2022).

Traditionally, biogas has been used at wastewater treatment plants (WWTPs) to operate combined heat and power (CHP) engines, which provide electricity to supplement plant operation as well as heat for digestion and space heating. However, to be able to convert biogas into a more valuable and versatile energy source, it must be converted into either compressed natural gas (Bio-CNG) or liquefied natural gas (Bio-LNG). In order to be able to inject biogas directly into the gas grid, the purity of the methane must be 99%, with only trace contaminants left. This purified biogas is known as biomethane. The purification process comprises a number of steps, which may include activated carbon, bio-scrubbers, and membrane filtration and are discussed in detail in Chap. 8. Biomethane can also be used to produce methanol and hydrogen.

3.2 Methane

3.2.1 *Formation*

Methane is a naturally formed alkane discovered by Alessandro Volta in 1776. It is colorless, odorless, lighter than air, slightly soluble, and highly flammable. It is produced in anoxic and anaerobic habitats such as the sediments of lakes, rivers, and estuaries, as well as swamps, marshes, and peat bogs from where it diffuses into the atmosphere, although 60–70% is currently emitted from human activities that include livestock rearing, landfill, waste, oil and gas extraction, wastewater treatment, and leakage from gas networks. It is the third-largest energy resource after oil and coal. Methane is produced by anaerobic archaea by fermentation (digestion). As described in Chap. 2, anaerobic digestion comprises four discrete stages: hydrolysis, acid formation, and acetogenesis (the non-methanogenic phases), followed by methane formation (methanogenic phase) (Fig. 1.2), with the process of methanogenesis already explained in Sec. 2.2.4.

Methanogenic archaea conserve energy for adenosine triphosphate (ATP) synthesis through the production of methane gas via the Wolfe Cycle (Thauer, 2012). Until the late 1970s, methanogens were thought to be bacteria until Dr. Carl Woese, using genetic analysis, discovered they were in a completely new domain, which he named the *Archaea*. This means that the domain prokaryotes, which was originally composed solely of the bacteria, has been replaced by two new domains, the Bacteria and Archaea, alongside eukaryotes. What is exciting about archaea is that they are extremophiles, with methanogens so far isolated between pH 3.0 and 10.2, with temperatures from –2 to 122°C, and are from a wide range of salinities. They are strict anaerobes and obligate methane producers.

Table 3.1. There are actually seven methanogen orders, which are listed below, along with common species and methanogenesis pathway (Buan, 2018). Reproduced under the Creative Commons Attribution license CC BY-NC-ND 4.0 DEED.

Order	Representative organism	Methanogenesis pathways
Methanopyrales	*Methanopyrus kandleri*	Hydrogenotrophic
Methanococcales	*Methanococcus maripaludis*	Hydrogenotrophic
Methanobacteriales	*Methanobacterium thermoautotrophicum*	Hydrogenotrophic
Methanosarcinales	*Methanosarcina mazei*	Hydrogenotrophic, methylotrophic, carboxydotrophic, acetoclastic
Methanomicrobiales	*Methanospirillum hungatei*	Hydrogenotrophic
Methanocellales	*Methanocella paludicola*	Hydrogenotrophic
Methanomassiliicoccales	*Methanomassiliicoccus luminyensis*	Methylotrophic

The archaea responsible for methanogenesis in sludge digesters and anaerobic wastewater treatment processes are similar to those found in other anaerobic environments, including the digestive tract of animals. Most methanogens belong to the genera *Methanobacterium*, *Methanosarcina*, *Methanospirillum*, and *Methanococcus* (Table 3.1). The methanogenic genera isolated so far are limited to the catabolism of either one carbon (e.g., H_2/CO_2, CH_3OH, CO, HCOOH, CH_3NH_2) or two carbon compounds (e.g., CH_3COOH) (Sec. 2.2.4). Methanogens are unusual in that they are composed of many species with very different cell morphologies. They require a strict anaerobic environment for growth with a redox potential (ORP) below –300 mV and have simple nutritional requirements, primarily carbon dioxide, ammonia, and sulfide. Ammonia is the essential nitrogen source for growth, as no methanogenic species are known to utilize amino acids or peptides. While sulfide is the most common sulfur source, although some species can use cysteine instead. Methanogens contain a number of unique co-enzymes, for example, co-enzyme 420, which is involved in electron transfer instead of the usual ferredoxin, co-enzyme M, which is used in methyl transfer reactions, and factor B, which is required for the enzyme formation of methane from methyl co-enzyme M. The synthesis of ATP appears to be via electron transport linked to phosphorylation. The group is defined by not having muramic acid in their cell walls, unique co-enzymes, and unique oligo-nucleotide sequences of the 16S ribosomal RNA molecule (Buan, 2018). Research on methanogens has been slow compared to other fields of microbiology due to the culturing requiring oxygen-free culture media as well as a reducing environment with an ORP <50 mV.

There is a close relationship between gas production and the number of methanogens present in a digester, generally being present in numbers between 10^6 and 10^8 mL^{-1} (Table 3.2). It is from acetic and, to a lesser extent, propionic acid that the greatest percentage of methane is derived during methane fermentation, whereas formic acid fermentation and methane fermentation associated with *b*-oxidation of long chain fatty acids (LCFAs) probably account for most of the small percentage of methane not derived from acetic and propionic acids (Fig. 3.1).

Methane formation from acetic acid is a single-step process carried out by one group of methanogenic archaea:

$$\underset{\text{(acetic acid)}}{CH_3COOH} \rightarrow \underset{\text{(methane)}}{CH_4} + CO_2 \tag{3.1}$$

Table 3.2. Enumeration of anaerobic microbial populations in sewage sludge digesters (adapted from (Zeikus, 1980)).

Group	Numbers per mL^{-1}
Hydrolytic species	
Total	10^8–10^9
Proteolytic	10^7
Cellulolytic	10^5
Acetogenic species	10^6
Homoacetogenic species	10^5–10^6
Methanogenic species	10^6–10^8
Sulfate reducers	10^4

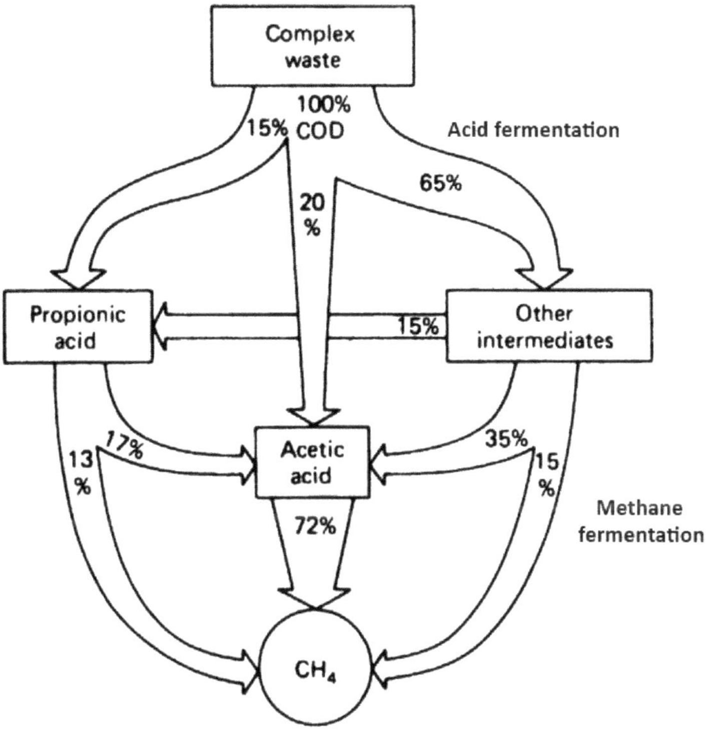

Figure 3.1. Pathways in methane formation of complex waste. Percentages represent the conversion of waste COD by various routes.

Methane fermentation of propionic acid is a two-step process involving two groups of methanogens, with acetic acid as the intermediate step:

Step 1:

$$\underset{\text{(propionic acid)}}{CH_3CH_2COOH} + 0.5H_2O \rightarrow \underset{\text{(acetic acid)}}{CH_3COOH} + 0.25CO_2 + \underset{\text{(methane)}}{0.75CH_4} \tag{3.2}$$

Step 2: $$\underset{\text{(acetic acid)}}{CH_3COOH} \rightarrow CO_2 + \underset{\text{(methane)}}{CH_4} \tag{3.3}$$

Overall: $$\underset{\text{(propionic acid)}}{CH_3CH_2COOH} + 0.5H_2O \rightarrow 1.25CO_2 + \underset{\text{(methane)}}{1.75CH_4} \tag{3.4}$$

The acid-fermenting archaea are tolerant to changes in pH and temperature and have a much higher rate of growth than the methane-fermenting archaea. This difference in growth rate, linked with a greater sensitivity to environmental factors, results in the methanogenic archaea becoming the major factor controlling the overall rate of anaerobic digestion. However, in the digestion of other wastes, the first stage of anaerobic digestion, the enzymic hydrolysis of polymers to monomers and cellulose hydrolysis, in particular, has been reported as the rate-limiting step in the conversion of cellulose to methane and in the digestion of household waste. The effect of environmental factors and the inhibition of toxic compounds in the anaerobic process are considered in Chap. 5.

The utilization of the organic fraction of wastewater is directly related to methane production and *vice versa*. Where the exact chemical nature of the sludge is known, the quantity of methane produced can be estimated by the following equation (Buswell and Mueller, 1952):

$$C_nH_aO_b + (n\text{-}a/4\text{-}b/2)H_2O \rightarrow (n/2\text{-}a/8\text{+}b/4)CO_2 + (n/2\text{+}a/8\text{-}b/4)CH_4 \qquad (3.5)$$

However, McCarty (1964) estimated the theoretical methane production from the complete stabilization of 1 kg of COD (chemical oxygen demand) as 0.348 m^3 at standard temperature and pressure.

It is useful to be able to calculate the proportion of substrate utilized either as energy or synthesis. Both the amount of electron acceptor required and the end-products produced can be calculated by knowing what proportion of the substrate is synthesized into new cellular material. As part of the substrate goes into energy formation and the remainder goes to cell synthesis, all the reacting material can be expressed as:

$$f_e + f_s = 1 \qquad (3.6)$$

where f_e is the fraction of electron donor used for energy and f_s is the fraction of electron donor used for cell synthesis. McCarty (1975) calculates the amount of substrate metabolized by bacteria to form energy and new cells by using balanced half equations. The overall reaction is constructed from three half-reactions (Table 3.3): one for

Table 3.3. Oxidation half-reactions. Reactants and products are given at unit activity except for $(H^+) = 10^{-7}$.

Equation	Half-reaction no.		$\Delta G°(W)$ * kcal per electron equivalent
Reactions for bacterial cell synthesis (R_c)			
	Ammonia as nitrogen source:		
1	$1/5CO_2 + I/20HCO_3^- + 1/20NH_4^+ + H^+ + e^-$	$= 1/20C_5H_7O_2N + 9/20H_2O$	
	Nitrate as nitrogen source:		
2	$1/28NO_3^- + 5/28CO_2 + 29/28H^+ + e^-$	$= 1/28C_5H_7O_2N + 11/28H_2O$	
Reactions for electron acceptors (R_a)			
	Oxygen		
3	$1/40_2 + H^+ + e^-$	$= 1/2H_2O$	–18.675
	Nitrate:		
4	$1/5NO_3^- + 6/5H^+ + e^-$	$= 1/10N_2 + 3/5H_2O$	–17.128
Reactions for bacterial cell synthesis (R_c)			
	Sulphate:		
5	$1/8SO_4^{2-} + 19/16H^+ + e^-$	$= 1/16H_2S + 1/16HS^- + 1/2H_2O$	5.085
	Carbon dioxide (methane fermentation):		
6	$1/8CO_2 + H^+ + e^-$	$= 1/8CH_4 + 1/4H_2O$	5.763

(*Continued*)

Table 3.3. (*Continued*)

Equation	Half-reaction no.		ΔG°(W) * kcal per electron equivalent
Reactions for electron donors (R_d)			
	Organic donors (heterotrophic reactions)		
	Domestic wastewater:		
7	$9/50CO_2 + 1/50NH_4^+ + 1/5OHCO_3^- + H^+ + c^-$	$= 1/5OC_{10}H_{19}O_3N + 9/25H_2O$	7.6
	Protein (amino acids, proteins, nitrogenous organics):		
8	$8/33CO_2 + 2/33NH_4^+ + 31/33H^+ + e^-$	$= 1/66C_{16}H_{24}C_5N_4 + 27/66H_2O$	7.7
	Carbohydrates (cellulose, starch, sugars):		
9	$1/4CO_2 + H^+ + e^-$	$= 1/4CH_2O + 1/4H_2O$	10.0
	Grease (fats and oils):		
10	$4/23CO_2 + H^+ + e^-$	$= 1/46C_8H_{16}O + 15/46H_2O$	6.6
	Acetate:		
11	$1/8CO_2 + I/8HCO_3^- + H^+ + e^-$	$= I/8CH_3COO^- + 3/8H_2O$	6.609
	Propionate:		
12	$1/7CO_2 + 1/14HCO_3^- + H^+ + e^-$	$= 1/14CH_3CH_2COO^- + 5/14H_2O$	6.664
	Benzoate:		
13	$1/5CO_2 + 1/30HCO_3^- + H^+ + e^-$	$= 1/30C_6H_5COO^- + 13/20H_2O$	6.892

	Ethanol:		
14	$1/6CO_2 + H^+ + e^-$	$= 1/2CH_3CH_2OH + 1/4H_2O$	7.592
	Lactate:		
15	$1/6CO_2 + 1/12HCO_3^- + H^+ + e^-$	$= 1/12CH_3CHOHCOO^- + 1/3H_2O$	7.873
	Pyruvate:		
16	$1/5CO_2 + 1/10HCO_3^- + H^+ + e^-$	$= 1/10CH_3COCCO^- + 2/5H_2O$	8.545
	Methanol:		
17	$1/6CO_2 + H^+ + e^-$		8.965
	Inorganic donors (autotrophic reactions)	$= 1/6CH_3OH + 1/6H_2O$	
	$Fe^{3+} + e^-$	$= Fe^{2+}$	–17.780
19	$1/2NO_3^- + H^+ + e^-$	$= 1/2NO_2^- + 1/2H_2O$	–9.430
20	$1/8NO_3^- + 5/4H^+ + e^-$	$= 1/8NH_4^+ + 3/8H_2O$	–8.245
21	$1/6NO_2^- + 4/3H^+ + e^-$	$= 1/6NH_4^+ + 1/3H_2O$	–7.852
22	$1/6SO_4^{2-} + 4/3H^+ + e^-$	$= 1/6S + 2/3H_2O$	4.657
23	$1/8SO_4^{2-} + 11/16H^+ + e^-$	$= 1/16H_2S + 1/16HS^- + 1/2H_2O$	5.085
24	$1/4SO_4^{2-} + 5/4H^+ + e^-$	$= 1/8S_2O_3^{2-} + 5/8H_2O$	5.091
25	$H^+ + e^-$	$= 1/2H_2$	9.670
26	$1/2SO_4^{2-} + H^+ + e^-$	$= 1/2SO_3^{2-} + 1/2H_2O$	10.595

the synthesis of new bacterial cells, which are assumed to be $C_5H_7O_2N$ (R_c), one for the electron acceptor (R_e), and one for the electron donor (R_d), which combine to give the relationship where R is the overall reaction:

$$R = f_s R_c + f_e R_e - R_d \tag{3.7}$$

The volume of bacteria biomass and methane produced in an anaerobic sewage sludge digester can be calculated using stoichiometry. Using the tables devised by McCarty (Tables 3.3 and 3.4), the

Table 3.4. Maximum cell yield $(f_s)_{max}$ (i.e., the maximum fraction of the electron donor used for synthesis) for various electron donors and acceptors.

Electron donor	Electron acceptor	(fs) max
Heterotrophic reactions		
Carbohydrate	O_2	0.72
Carbohydrate	NO_3	0.60
Carbohydrate	SO_4	0.30
Carbohydrate	CO_2	0.28
Protein	O_2	0.64
Protein	CO_2	0.08
Fatty acid	O_2	0.59
Fatty acid	SO_4	0.06
Fatty acid	CO_2	0.05
Glucose	O_2	0.79
Lactose	O_2	0.74
Sucrose	O_2	0.75
Glycine	O_2	0.52
Alanine	O_2	0.52
Propionate	O_2	0.58
Acetate	O_2	0.58
Methanol	NO_3	0.36
Methanol	CO_2	0.15

Table 3.4. (*Continued*)

Electron donor	Electron acceptor	(*fs*) max
Propionate	CO_2	0.07
Acetate	CO_2	0.06
Glucose	CO_2	0.27
Sewage sludge	CO_2	0.11
Autotrophic reactions		
S	O_2	0.22
S_2O_3	O_2	0.11
S_2O_3	NO_3	0.20
NH_4	O_2	0.10
H_2	O_2	0.24
H_2	CO_2	0.04
Fe	O_2	0.07

appropriate half-reactions can be selected with carbon dioxide as the electron acceptor:

$$R_c\text{: } 1/5CO_2 + 1/20HCO_3^- + 1/20NH_4^+ + H^+ + e^- = 1/20C_5H_7O_2N + 9/20H_2O \quad (3.8)$$

$$R_e\text{: } 1/8CH_4 + 1/4H_2O = 1/8CO_2 + H^+ + e^- \quad (3.9)$$

$$R_d\text{: } 1/50C_{10}H_{19}O_3N + 9/25H_2O = 9/50CO_2 + 1/50NH_4^+ + 1/50HCO_3^- + H^+ + e^- \quad (3.10)$$

Table 3.4 gives a $(f_s)_{max}$ value of 0.11 for sewage sludge, and as the digester has a mean cell residence time of 20 days, f_s is calculated as:

$$f_s = (f_s)_{max}\ 1 - [(0.8\ b\ \varphi_c)/(1+ b\ \varphi_c)] \quad (3.11)$$

where φ_c is the sludge age in days and the coefficient b represents the rate of cell death and decay, which is assumed to be 0.03.

$$f_s = 0.11\ (1 - [(0.8(0.03)(20)/1 + 0.03(20)]) = 0.077$$

As $f_s = 1 - f_e$,

$$f_e = 0.923$$

Therefore, the revised reactions using the relationship $R = f_s R_c + f_e R_e - R_d$ are

$R_c \times f_s$:

$$0.0154CO_2 + 0.0039HCO_3^- + 0.0039NH_4^+ + 0.007H^+ + 0.007e^- = 0.0039C_5H_7O_2N + 0.035H_2O \quad (3.12)$$

$R_e \times f_e$:

$$0.115CO_2 + 0.923H^+ + 0.923e^- = 0.115CH_4 + 0.231H_2O \quad (3.13)$$

R_d:

$$0.02C_{10}H_{19}O_3N + 0.36H_2O = 0.18CO_2 + 0.02NH_4^+ + 0.02HCO_3^- + H^+ + e^- \quad (3.14)$$

When added and normalized by dividing by 0.02, the overall reaction R is:

$$C_{10}H_{19}O_3N + 4.7H_2O = 0.195C_5H_7O_2N + 5.75CH_4 + 2.48CO_2 + 0.81NH_4^+ + 0.81HCO_3^- \quad (3.15)$$

Therefore, for each mole of sewage sludge, 0.195 mole of new cells are produced and 5.75 moles of methane is released.

A more precise value for $(f_s)_{max}$ can be calculated from basic constituents using Table 3.3. For example, if domestic wastewater contains 50% protein, 40% carbohydrate (CHO), and 10% fat, then $(f_s)_{max\ wastewater}$ of domestic wastewater for anaerobic digestion, where carbon dioxide is the electron acceptor, is:

$$(f_s)_{max\ wastewater} = (\%\ protein \times (f_s)_{max\ protein}) + (\%\ CHO \times (f_s)_{max\ CHO}) + (\%\ fat \times (f_s)_{max\ fat}) \quad (3.16)$$

$$(f_s)_{max\ wastewater} = (0.50 \times 0.08) + (0.40 \times 0.28) + (0.10 \times 0.05)$$

$$(f_s)_{max\ wastewater} = 0.157$$

3.2.2 *Health and Environmental Effects*

Methane (CAS: 74-82-8) is an extremely flammable gas that can displace oxygen, causing the risk of suffocation and explosion in low-lying areas and especially inside confined spaces. Explosions due to methane have occurred at atmospheric concentrations of between 5 and 15%. It is an odorless gas, so is not detected by smell, making it potentially very dangerous in the workplace, especially where anaerobic treatment processes or sludge digestion are employed at WWTPs.

People are frequently subjected to low levels of methane exposure due to the use of natural gas appliances or when walking in areas where it is naturally produced, such as wetlands. There are no associated health risks with low-level exposure. However, as the concentration in air increases, it reduces the amount of oxygen breathed in, causing a range of effects — initially rapid breathing, increased heart rate, facial flushing, balance problems, numbness leading to mood changes, slurred speech, vision problems, memory loss, and eventually nausea, vomiting, and then collapse and unconsciousness. Exposure has also been linked to pulmonary toxicity (Jo *et al.,* 2013). Excessive and prolonged exposure can be fatal (PHE, 2019). Being lighter than air, methane displaces the oxygen present, which is extremely serious in confirmed spaces as it will displace the oxygen from the ceiling downward. So, all confined spaces or low-lying areas, such as inside sedimentation or aeration tanks emptied for maintenance, especially when covered, should be checked before entry using a gas monitor to detect the presence of methane and sufficient oxygen (i.e., >18% oxygen), especially if there is residual wastewater or sludge present (CCOHS, 2023).

Global carbon dioxide emissions in 2022 from fuel combustion and industrial processes was 36.8 Gt (giga tonnes), while 589 Mt (million tonnes) of methane were released — 60% from manmade and 40% from natural sources (IEA, 2023). Although methane is far less abundant in the atmosphere than carbon dioxide at 1.91 compared to 417 ppm, respectively, in 2022, it has a significantly higher global warming potential (GWP). Current GWPs for methane vary over time at 84–87 over 20 years, falling to 27–36 over 100 years (UNEP, 2021). Due to the higher emission levels, nearly all climate action has focused on reducing carbon dioxide emissions, although there is growing concern over the role of methane in global warming and a general belief that more attention should be paid to reducing methane emissions overall. It is now thought that about half of the 1.1°C rise above pre-industrial temperatures is due to methane. Another factor is that methane has a relatively short half-life of 12 years compared to carbon dioxide at 300–1,000 years, so action on methane, with its high GWP and short half-life, could slow down global warming much faster in the short term than trying to focus solely on carbon dioxide reduction. Ocko *et al.* (2021) have predicted that removing 300 Mt of methane, which is equivalent to sequestering 25.8 billion tonnes of carbon dioxide over 20 years, could reduce global warming to 0.22°C by 2050, which would dampen the effects of global-warming-induced climate change.

Methane-removal technologies can be divided into those based on chemical oxidation or biological assimilation and have been reviewed by Wang and He (2023). However, in wastewater treatment and other waste processes, anaerobic digestion is used to capture carbon from the substrate in the form of methane, which is then thermally oxidized as a fuel releasing carbon dioxide. The amount of methane in the atmosphere has doubled since pre-industrial times, with the rate of release steadily increasing since 1980. So, while it is a key driver of global-warming-induced climate change, it is important to exploit methane's enormous potential as a valuable and sustainable resource (UNEP, 2021). Using methane as an energy source reduces overall greenhouse gas impact and is accepted as a carbon-capture method.

3.3 Hydrogen Sulfide

3.3.1 *Formation*

The production of hydrogen sulfide is the most common and well-known manifestation of anaerobiosis. Sulfide can be produced by anaerobic microorganisms in two ways: (1) Protein is broken down to amino acids, and those that contain sulfur, such as cysteine, cystine, and methionine, are degraded further, with sulfide being produced. Most anaerobic bacteria can produce sulfide from protein, including *Proteus*, *Bacteroides* spp., and some *Clostridium* spp. Although all can grow anaerobically, only *Bacteroides* spp., which can be present in feces at concentrations of up to 10^{10} g^{-1}, are obligate species. (2) However, in wastewater treatment systems, most sulfide is produced from sulfate reduction by the anaerobic sulfate-splitting bacterium *Desulfovibrio desulfuricans*, although species of the genus *Desulfotomaculum* are also routinely isolated from digesters. The former species is present in low numbers in sewage (60–600 mL^{-1}) but rapidly increases on storage up to concentrations in excess of 100,000 mL^{-1} after 14 days.

Sulfate reducing bacteria (SRB) only utilize a restricted range of carbon compounds, such as lactate and malate, and rely on the metabolic products of other anaerobic bacteria that can utilize more complex organic compounds (Fig. 3.2). SRB are found in a wide range of anaerobic environments, where there is a supply of sulfate, which they utilize instead of oxygen for respiration, and organic matter to produce compounds such as lactate. Sulfate is a major ion in seawater, and bacterial sulfate reduction is an important reaction in anoxic estuarine and marine environments. When heavy metals are present in the sediment and where sulfate-reducing bacteria are active, then the sulfide produced reacts to form insoluble metal salts; the black discoloration so often associated with anaerobic sediments in rivers is due largely to the formation of ferrous sulfides (Sec. 8.2.1). If no metals

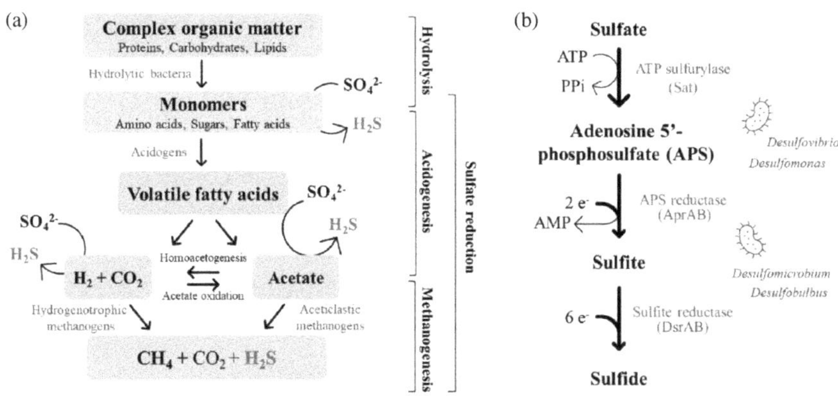

Figure 3.2. (a) Sulfide formation during anaerobic digestion and (b) the dissimilatory sulfate reduction pathway, including the main enzymes responsible (Jung *et al.*, 2022). Reproduced with permission of Elsevier Publishing.

are present, the sulfide escapes into the water column or the atmosphere as hydrogen sulfide gas (H_2S_g).

The sulfide can be used as a source of energy by SRB or as an electron donor by some of the photosynthetic bacteria. The nature of the sulfides produced during anaerobic digestion is dependent primarily on the temperature, pH, and sulfate loading, with dissolved sulfides S^{2-}, HS^-, and H_2S_{aq} produced along with gaseous hydrogen sulfide gas. The interaction between these forms of sulfide is shown below:

$$H_2S_{aq} \leftrightarrow HS^- + H^+ \tag{3.17}$$

$$HS^- + H^+ \leftrightarrow S^{2-} + 2H^+ \tag{3.18}$$

$$H_2S_{aq} + H_2S_g \tag{3.19}$$

The production of hydrogen sulfide is limited by the sulfate concentration in the wastewater or waste, which can range between 20 and 200 mg L^{-1} in domestic wastewater and 0.3 to 2.3% (wt%) in sewage sludge (Dewil *et al.*, 2008; Jung *et al.*, 2022). The higher the sulfur content of the waste stream, which comes mainly from proteins (Tian *et al.*, 2020), the greater the concentration of sulfide potentially produced. Expected hydrogen sulfide gas in biogas is normally 0.1 to 2% (v/v). Sulfides can be toxic to a number of anaerobes, especially methanogens, which can affect process operation, including the complete inhibition of the methanogenic step. Depending on actual operating conditions, sulfide inhibition of methanogenic activity has been reported at concentrations of between 100 and 800 mg L^{-1} as total dissolved sulfides ($S^{2-} + HS^- + H_2S_{aq}$) and between 50 and 400 mg L^{-1} of hydrogen sulfide gas. This is exacerbated by SRB competing directly with methanogens for key substrates such as acetate and hydrogen (Muyzer and Stams, 2008). The metabolism of the SRB has been extensively studied and reviewed (Li *et al.*, 2017; Zhang *et al.*, 2022).

3.3.2 *Health and Environmental Effects*

Hydrogen sulfide (CAS No. 7783-06-4) is a colorless gas with a smell of rotten eggs at low concentrations, which becomes less obvious as the concentration increases, as it paralyzes the olfactory senses.

The ability to detect the gas is lost at between 70 and 140 mg m^{-3} (50–100 ppm), so odor is not a safe indicator of overexposure. The mean odor threshold for the general population is just 0.0063 mg m^{-3} (0.0045 ppm) with safe concentrations <0.18 mg m^{-3} (<0.13 ppm), although lower values are recommended for continuous exposure. WWTPs and the sewerage system are potentially high-risk areas for hydrogen sulfide exposure, with concentrations in excess of 700 mg m^{-3} (≡502 ppm) reported (Locey, 2005). At low concentrations, it can cause eye irritation and tissue damage, dizziness, headaches, nausea, and respiratory problems. However, at high concentrations, it causes almost instant unconsciousness and rapid asphyxiation. The gas is heavier than air and tends to accumulate in underground tanks and manholes, where it displaces oxygen, leading to a highly toxic and potentially explosive environment, with concentrations in air above 4.3% being potentially explosive. Multiple deaths have occurred in underground pumping stations, sludge wells, sewers, and even septic tanks, so gas alert monitors should always be worn when working in confined spaces at the treatment plant, where the possibility of exposure to hydrogen sulfide, methane, carbon dioxide, carbon monoxide, or ammonia is possible. All potential risk areas should be well-vented, and warning signs posted with access restricted to those with gas alert monitors and, if necessary, a positive-pressure, self-contained breathing apparatus (Centre for Disease Control, 1977).

During the combustion of biogas, any hydrogen sulfide present is burnt to form sulfur dioxide:

$$H_2S + 1.5\ O_2 \rightarrow SO_2 + H_2O \qquad (3.20)$$

In the atmosphere, sulfur dioxide is oxidized to sulfur trioxide, which causes sulfuric acid to form when absorbed into water, causing acid rain.

3.3.3 *Corrosion*

Concrete corrosion can occur at dissolved sulfide concentrations as low as 0.1 mg S L^{-1}, becoming severe at 2.0 mg S L^{-1} (Hvitved-Jacobsen *et al.*, 2002). The release of hydrogen sulfide gas depends on the pressure, temperature, and pH. When produced in sewers by SRB, the gas can cause severe corrosion of the concrete pipework. The slightly acidic gas is absorbed into condensation water, which collects at the top or crown of the sewer or on its side walls. Here, the acidophilic sulfur oxidizing bacteria (SOB) in the surface biofilm, which are able to tolerate pH levels of <1.0, oxidize the hydrogen sulfide to sulfuric acid using atmospheric oxygen (Figs. 3.3 and 3.4):

$$H_2S + O_2 \xrightarrow{\textit{Acidithiobacillus thiooxidans}} H_2SO_4 + \text{energy} \qquad (3.21)$$

The sulfuric acid reacts with the alkalinity in the concrete to form a range of calcium salts, predominantly gypsum ($CaSO_4{\cdot}2H_2O$) (Eqs. (3.22), (3.23), and (3.24)) and ettringite, chemically known as

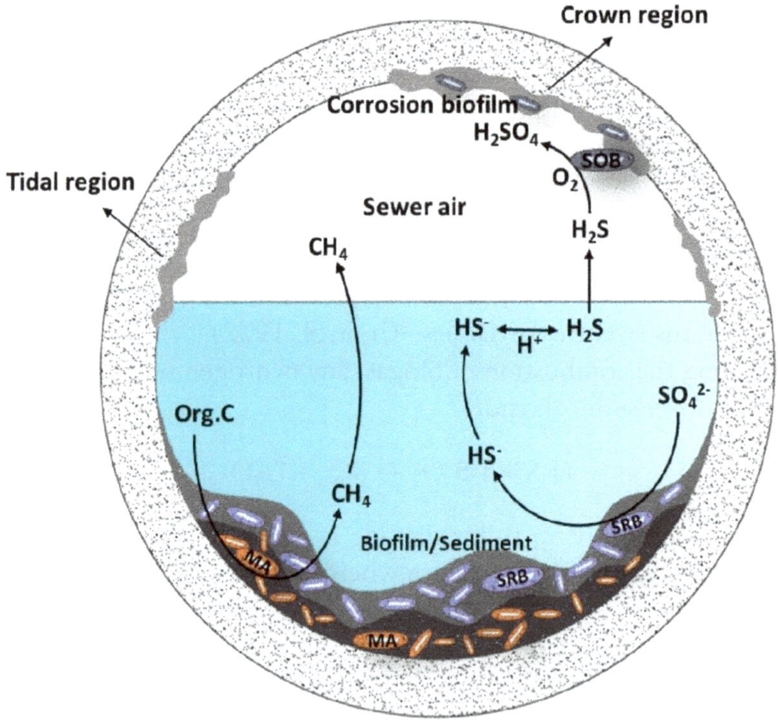

Figure 3.3. Crown erosion in concrete sewers due to microbial sulfide production, where SRB are sulfate reducing bacteria, SOB are sulfide oxidizing bacteria, and MA are methanogenic archaea (Zhang *et al.*, 2023). Reproduced with permission of Elsevier Publishing.

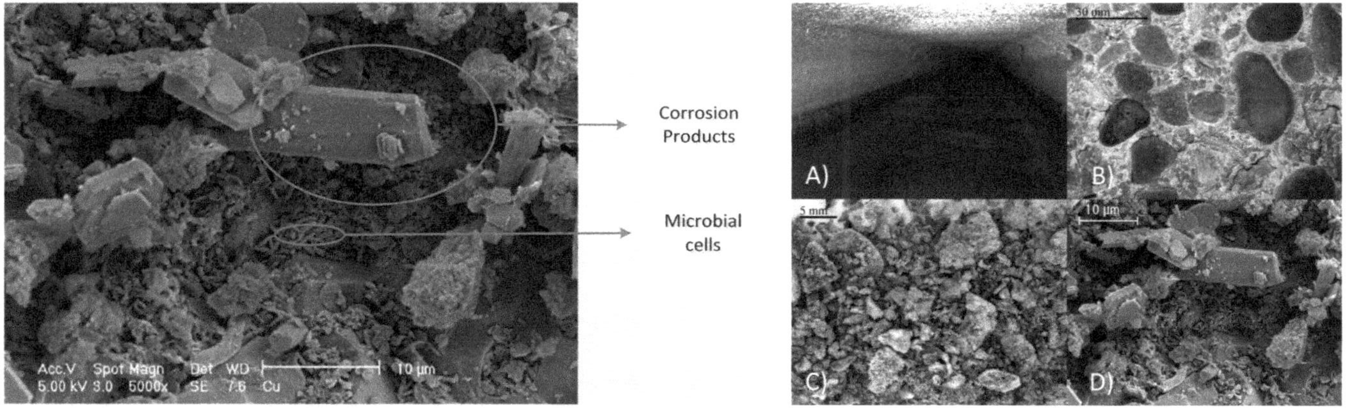

(a) (b)

Figure 3.4. An example of crown erosion in concrete sewers in Sydney, Australia. (A) Internal upstream view of a sewer showing erosion of the concrete ceiling. Sewer dimensions are 2 m (height) × 3 m (width) with a flow depth of 0.8–1 m. (B) Close-up of the corroded ceiling surface. (C) Scrapings of the corrosion layer. (D) Scanning electron microscopy image of the corrosion scrapings showing calcium salt crystals and acidophilic sulfur-oxidizing bacteria (Cayford *et al.*, 2012). Reproduced with permission of the Australian Centre for Water and Environmental Biotechnology. https://acweb.uq.edu.au/

calcium sulfoaluminate ($3CaO \cdot Al_2O_3 \cdot 3CaSO_4 \cdot 32H_2O$) (Eq. (3.25)). This leads to the breakdown of the concrete structure, making it weaker, leading to structural destabilization and cracking. Over time, the structural stability of the concrete pipe can become so weakened that it eventually collapses (Li *et al.*, 2017; Zhang *et al.*, 2023; Hvitved-Jacobsen *et al.*, 2013).

$$H_2SO_4 + CaCO_3 \rightarrow CaSO_4 + H_2CO_3 \qquad (3.22)$$

$$H_2SO_4 + Ca(OH)_2 \rightarrow CaSO_4 + 2H_2O \qquad (3.23)$$

$$H_2SO_4 + CaO \cdot SiO_2 \cdot 2H_2O \rightarrow CaSO_4 + Si(OH)_4 + H_2O \qquad (3.24)$$

$$3CaSO_4 + 3CaO \cdot Al_2O_3 \cdot 6H_2O + 26H_2O \rightarrow 3CaO \cdot Al_2O_3 \cdot 3CaSO_4 \cdot 32H_2O \qquad (3.25)$$

Crown erosion is particularly a problem in sewers receiving heated effluents, with wastewaters containing high sulfur content or suffering from saline intrusion and sewers that are inadequately vented. The problem is extremely critical for sustaining sewer infrastructure, with annual damage estimated at US$100 million in Australia and a staggering US$14 billion in the United States (Brongers *et al.*, 2002; Gutierrez *et al.*, 2016).

Corrosion-resistant pipe materials, such as vitrified clay or plastics, prevent corrosion in medium size sewers. In larger diameter sewers, where concrete is the only possible material, corrosion is reduced by ventilation, which expels the hydrogen sulfide and reduces condensation. Control methods include chemical addition to reduce sulfide formation or the removal of H_2S_{aq} after its formation. Inhibition of SRB is achieved using biocides, or the sulfides can be oxidized by oxygen injection or the addition of nitrate salts. Precipitation using iron salts converts the sulfides present into metal sulfides, which are insoluble, while increasing the pH by the addition of an alkali such as NaOH or $Mg(OH)_2$, reduces the H_2S_{aq} in the waste stream wastewater that can be converted to H_2S_g. The addition of chemicals can, however, result in problems with the biological oxidation of wastewater at the WWTP, especially where activated sludge is used. Chemical addition is a continuous process, making it extremely expensive in the medium to long term. Another alternative is coating concrete sewers with a protective layer, which is expensive and technically difficult, or replacing sewers with more corrosion-resistant concrete. Control strategies have been compared by Zhang *et al.* (2023), who also explore future options. Hydrogen sulfide is also highly corrosive to metal parts involved in handling and processing biogas.

There are numerous post-treatment processes for the removal of hydrogen sulfide gas in biogas, and these are explored in Sec. 8.2, with in-situ sulfide control particularly of interest as it is cheaper than adding a post-digestion process, although the *in-situ* technologies are at an early stage of development (Jung *et al.,* 2022). The easiest method of reducing sulfur in sewage sludge is to ensure the sludge is dewatered before it becomes anoxic or anaerobic, thus ensuring soluble sulfate is discharged with the effluent stream (Dewil *et al.*, 2008).

3.4 Ammonia

Nitrogen is normally absent from biogas and, if detected, is indicative of denitrification or an air leak. Ammonia can be produced during anaerobic digestion by the microbial reduction of proteins, and if in excess to nutritional requirements of the anaerobic microbiota, will be present in the biogas. Its presence can cause a number of operational problems, including process inhibition, reduced biogas production, reduced COD removal, malodor, and a reduction in the calorific value of biogas and so its energy density. It also results in corrosion to gas-processing equipment, including pipes and valves, due to reacting with H_2O to form a strong base, reducing efficiency and the life expectancy of equipment. During combustion, the ammonia in biogas is converted into polluting NO_x, which are also greenhouse gases, and released into the environment (Strik *et al.,* 2006).

Ammonia is essential to anaerobic microorganisms for the synthesis of amino acids, proteins, and nucleic acids for bacterial growth, and so should be in excess of 30–70 mg N L^{-1} but <200 mg N L^{-1} to maintain optimal activity, although, at higher concentrations, it becomes increasingly inhibitory. Ammonia also helps to neutralize organic acids produced during fermentation, helping to maintain a more neutral pH. Ammonia specifically inhibits the methanogenic phase, with reported critical threshold values between 1,500 and 7,000 mg L^{-1} (Rajagopal *et al.*, 2013). Inhibition is thought to be linked to the concentration of free ammonia nitrogen (FAN) rather than total ammonia nitrogen (TAN) (Jiang *et al.*, 2019). Where there is a high concentration of ammonia in the feedstock to the reactor, then it begins to appear in the biogas, with concentrations of up to 445 ppm recorded (Strik *et al.*, 2006).

In the liquid phase, ammonia is present either as ionized ammonia (NH_4^+) or free ammonia (NH_3) (Eq. (3.26)).

$$NH_3 + H^+ \leftrightarrow NH_4^+ \tag{3.26}$$

The relative fraction of each is dependent on both pH and temperature (Fig. 3.5):

$$FAN = TAN \times \left(1 + \frac{10^{-pH}}{10^{-\left(0.09018 + \frac{2729.92}{T(K)}\right)}}\right)^{-1} \tag{3.27}$$

Where FAN and TAN are the free and total ammonia nitrogen concentrations (mg L^{-1}) in the digestate, respectively, and T (K) is the temperature in °K. The more protein that is available, then the greater the pH leading to an increase in FAN, with temperature a key reason why ammonia inhibition is more often a problem in thermophilic than mesophilic digesters. The inhibitory process and adaptation of the microbial population to high ammonia concentrations are explored in depth by Jiang *et al.* (2019), including examples of available mitigation strategies, including adjusting the C:N ratio of the feed, air stripping ammonia gas from the digestate, and chemical precipitation of ammonia with magnesium and phosphorus to form struvite ($MgNH\,PO_4{\cdot}6H_2O$).

The role of ammonia in anaerobic digestion, including the effects on methanogens, is explored in detail in Sec. 5.4.1.

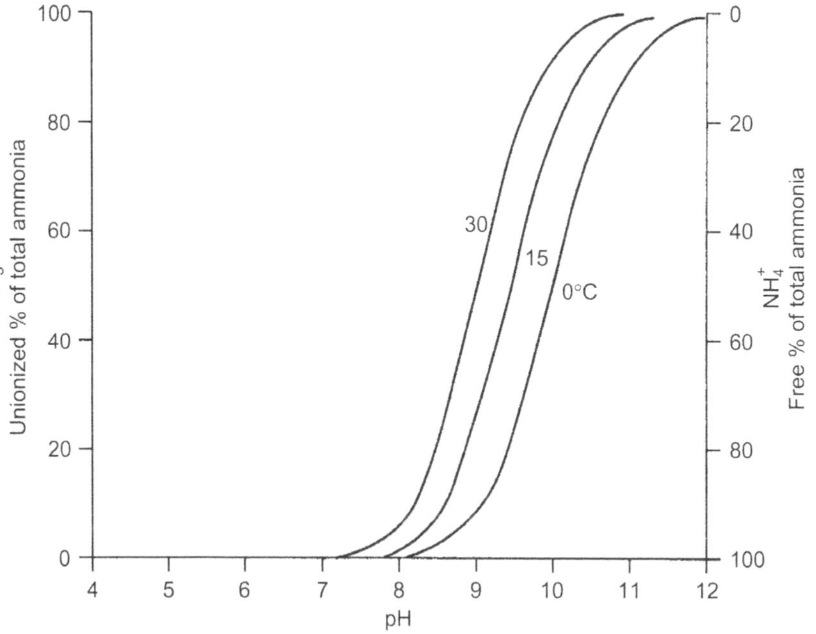

Figure 3.5. The general variation between the proportion of unionized to free ammonia at varying pH and temperatures.

3.5 Volatile Organic Compounds and Terpenes

3.5.1 *Volatile Organic Compounds*

Volatile organic compounds (VOCs) are a wide group of organic compounds with a high vapor pressure, meaning that at ambient temperature and pressures, they become gases. The main sources of VOCs are vehicle emissions, industrial processes using solvents, petroleum industries, chlorination, food extraction, paints, laundry printing, pharmaceutical industries, pesticide manufacture, and, increasingly, wildfires. The situation at WWTPs is more complex and depends on the nature and source of the wastewater streams being treated, while VOCs can be removed or generated during the treatment process, especially during anaerobic steps.

During anaerobic digestion, non-methane VOCs are produced in large quantities. These include terpenes, sulfur-containing compounds (e.g., dimethyl sulfide, carbon disulfide), oxygenated compounds (e.g., ethanol, acetone), halogenated hydrocarbons (e.g., ethyl chloride, chloroethylene, trichloroethylene), and aromatic hydrocarbons (e.g., benzene, toluene, xylene, and ethyl benzene). A breakdown of VOCs found at an anaerobic digestion plant is given in Fig. 3.6.

VOCs are a concern, as many of them are known carcinogens, with the remainder generally associated with other health-related problems. Environmental problems include odor generation during digestion, mainly due to sulfur-containing compounds and terpenes; ozone formation in the troposphere, resulting in photochemical smog; and the acceleration of ozone depletion in the stratosphere. VOCs can also absorb surface infrared radiation, enhancing global warming, although they have a short atmospheric life (Song *et al.*, 2019; David and Niculescu, 2021). The measurement of VOCs, especially in terms of air pollution monitoring, is based on four common and easily measured VOCs: benzene, toluene, ethylbenzene, and xylene, collectively known as the BTEX suite. BTEX monitors are now widely available and can be located around the WWTP or within buildings where digestion or sludge handling occurs.

While odors emitted during anaerobic digestion are primarily associated with volatile sulfur compounds, especially hydrogen sulfide, other non-sulfur VOCs also contribute to odor impact. These include amines, ketones, and a range of aromatic compounds, including indole, skatole, *ρ*-cresol ethylbenzene, and toluene (Chen *et al.*, 2004). Comparing VOC emissions from six different WWTPs in Australia, Fisher *et al.* (2017) found different sludge processing configurations resulted in distinct types and concentrations of VOCs, with the widest range and highest concentrations emitted from dewatered and stored biosolids. The most important odors detected were *ρ*-cresol and butanoic acid, with trimethylamine, indole, and phenol emitted from the dewatered and stored biosolids, and volatile fatty acids associated with the digester were also problematic. The authors concluded that effective odor management requires the monitoring of both volatile sulfur compounds and selected non-sulfur VOCs.

Zheng *et al.* (2020) examined the release of VOCs at an anaerobic digestion plant processing 300 tons of food waste per day over a period of 12 months and found that VOCs were released from all

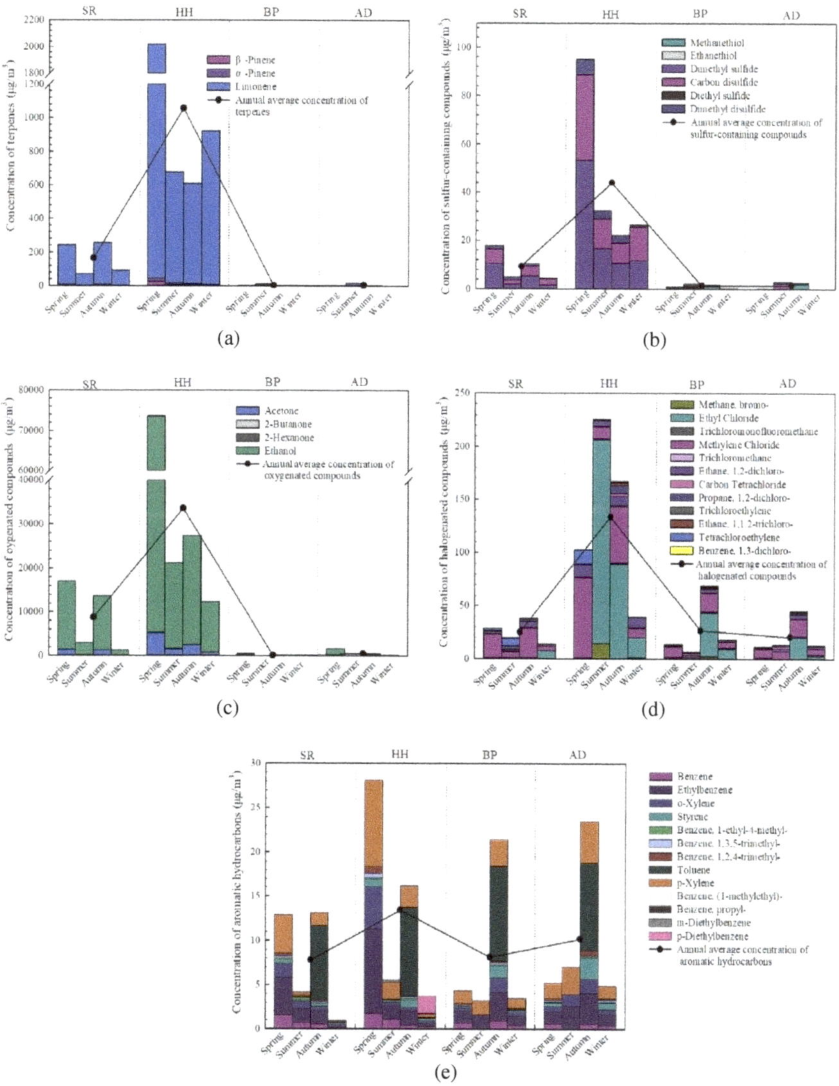

Figure 3.6. Seasonal concentration of VOCs from various process steps at a food waste anaerobic digestion system (i.e., SR (sorting crushing room), HH (hydrothermal hydrolysis unit), BP (biodiesel production unit), AD (anaerobic digestion unit)) showing (a) terpenes, (b) sulfur-containing compounds, (c) oxygenated compounds, (d) halogenated compounds, and (e) aromatic hydrocarbons (Zheng *et al.*, 2020). Reproduced with permission of Elsevier Publishing.

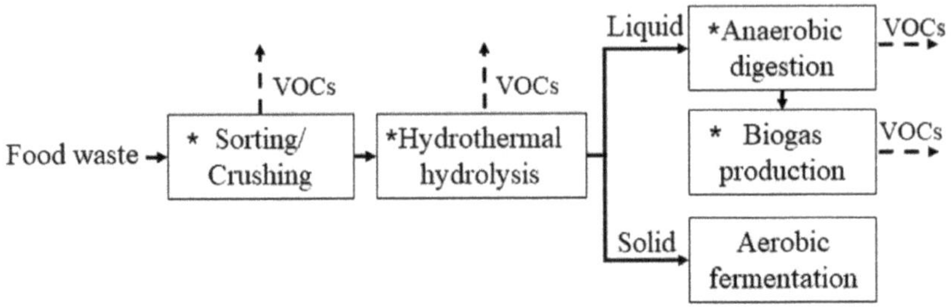

Figure 3.7. Process flow chart of a food waste anaerobic processing plant in Suzhou, China. Asterisks indicate VOC sampling points used in Fig. 3.4 (Zheng *et al.*, 2020). Reproduced with permission of Elsevier Publishing.

areas of the process (Fig. 3.7). Emissions varied seasonally, being highest in spring and lowest in winter, with concentrations of 3.49 to 104 mg m^{-3} recorded in the hydrothermal hydrolysis unit, 8.97 to 103 mg m^{-3} in the sorting/crushing room, 6.21 to 102 mg m^{-3} from the anaerobic digestion unit, and 2.01 to 102 mg m^{-3} in the biogas production unit (Fig. 3.6). Oxygenated compounds and terpenes accounted for >98% of total VOC emissions. They concluded that while emissions were below those associated with non-carcinogenic risk, workers should minimize the time spent in the hydrothermal hydrolysis unit. Some 380 different VOCs were recorded in a study comparing sewage sludge, thickened sludge, digested sludge, and dewatered sludge at a WWTP in France, of which 81 compounds were found in all sludge types (Haider *et al.*, 2022). Potential emissions of VOCs increased as the organic matter content rose, although dewatering digested sludge to a dry solids (DS) content of >60% did reduce emissions of sulfur compounds.

The use of VOCs produced by microorganisms (mVOCs) during anaerobic digestion as an early warning system of potential system failure has been proposed by Nie *et al.* (2023). Currently, physico-chemical parameters are used as indicators, such as pH, ORP, ammonia, alkalinity, and volatile fatty acids, but are not fully reliable. Using actual microbial metabolites as indicators could be far more precise, especially when used in conjunction with existing early warning parameters. They identified a number of potential pathways during digestion where mVOCs are produced and a number of potential candidate compounds (Fig. 3.8)

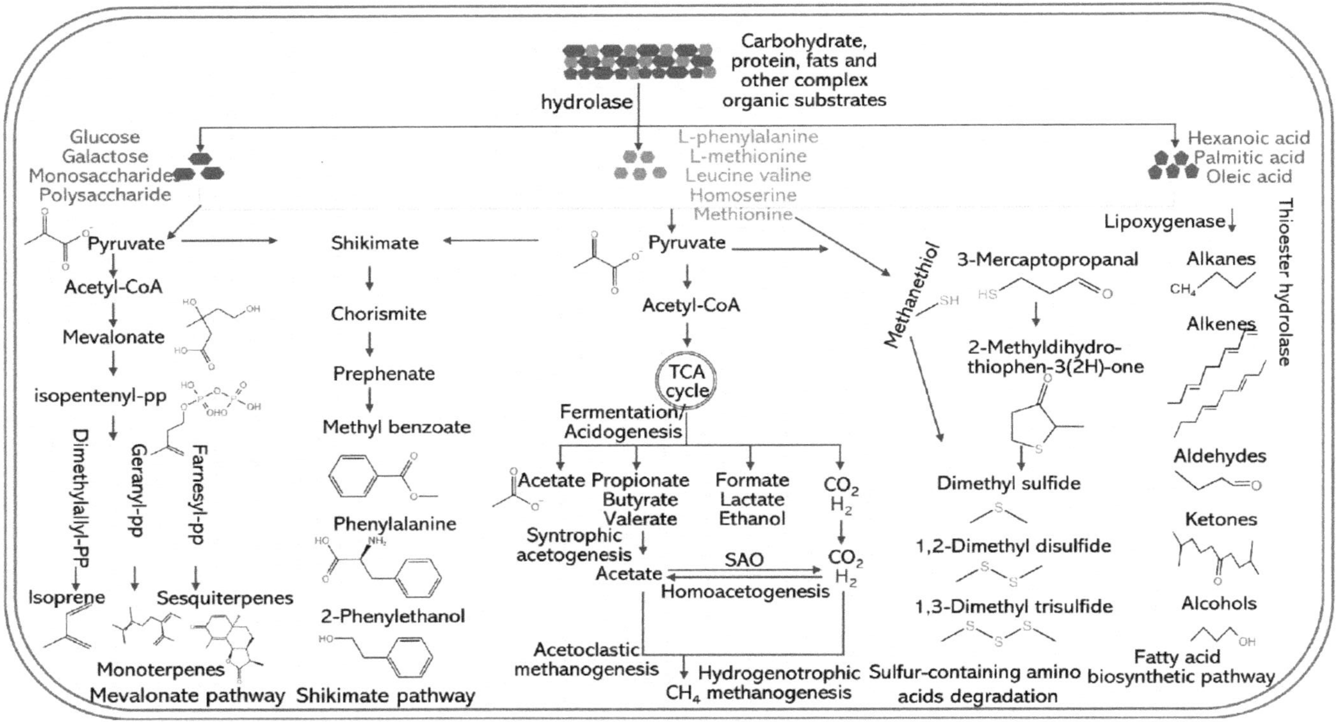

Figure 3.8. Potential pathways for the production of microbial VOCs during anaerobic digestion, which could be used as indicators of performance and operational stability (Nie *et al.*, 2023). SAO: synoptic acetate oxidation; TCA cycle: the tricarboxylic acid cycle. Reproduced with permission of Elsevier Publishing.

3.5.2 *Terpenes*

Terpenes are odorous compounds associated with specific plants such as citrus and coniferous species that find their way into the sludge feed.

In biogas, two terpenes are of importance, *ρ*-cymène and D-limonène, which are found in hundreds of different foods. For example, *ρ*-cymène ($C_{10}H_{14}$) is found in many widely consumed products, such as dairy produce, carrots, orange juice, raspberries, lemon oil, and many spices, while D-limonène ($C_{10}H_{16}$) is found in many plants, such as fennel, dill, celery, and lemons. It is also widely used as a food additive and in body-care products, and more recently, as a substitute for petroleum-based solvents in detergents and paints (Arrhenius *et al.*, 2017).

The increasing use of food waste co-digested with wastewater sludges to increase biogas yields has led to an increase in terpene levels in the gas. Terpenes are only a problem in biogas when it is converted to biomethane and must be removed before being pumped into the gas network. Terpenes can corrode the polyethylene pipework as well as rubber and plastic gaskets (Arrhenius *et al.*, 2017), result in condensation within the pipework, and also mask the sulfurous odor tetrahydrothiophene (THT), which is added to natural gas for leak detection. They can also cause workplace problems, both in terms of odors and potential health risks. Terpenes are also flammable, so if contaminated biogas is used in normal on-site CHP systems for heat recovery and electricity generation, they are destroyed as well as adding to the overall energy generated.

Five terpenes are regularly monitored in biomethane, *α*-pinène, *β*-pinène *ρ*-cymène, D-limonène, and 3-carene. They are analyzed using either a gas chromatography-thermal conductivity detector (GC-TCD) or, more recently, a gas chromatography-flame ionization detector (GC-FID) (Budiman *et al.*, 2015). Sampling and analysis of trace compounds in biogas and biomethane, including terpenes, are reviewed by Rasi *et al.* (2006; 2007) and Lecharlier *et al.* (2022). When food waste is the main source of digester feed, D-limonène is the main terpene present (>90%), with *ρ*-cymène found in small concentrations (1%). Both the concentration of specific substrates, especially citrus fruits, the consumption of which varies seasonally and

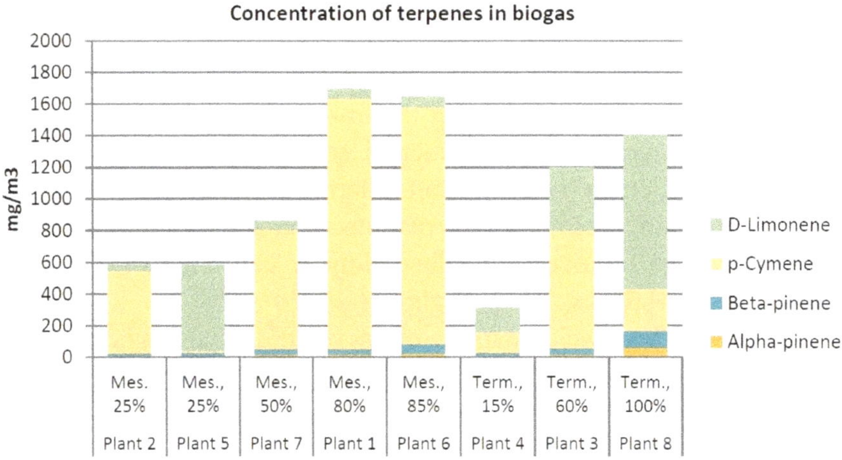

Figure 3.9. The concentration of main terpenes (mg m^{-3}) produced at eight different treatment plants using different proportions of food waste (%) in the feedstock using either mesophilic (Mes.) or thermophilic (Term.) digestion. There is a linear correlation between the proportion of food waste in the feedstock and the concentration of terpenes in the biogas of up to 70%, after which the concentration of terpenes levels out (Arrhenius *et al.*, 2017). Reproduced with permission of Energiforsk AB, Sweden. www.energiforsk.se

digester temperature determine the dominant terpene as well as the concentration. During digestion, D-limonène is largely converted to ρ-cymène, with the latter making up (>90%) of terpenes in the digestate. In thermophilic digestion, comparatively more D-limonène is normally found in biogas compared to mesophilic digestion, where the conversion of D-limonène to ρ-cymène should be more complete, resulting in higher concentrations of ρ-cymène evaporating into the biogas (Fig. 3.9).

Arrhenius *et al.* (2017) reported terpene concentrations in biogas generated from different percentages of food waste in the feedstock varied between 360 and 1,650 mg m^{-3}. After treatment, this was reduced to 20 to 240 mg m^{-3} after upgrading the biogas to biomethane using a variety of removal systems. Removal efficiency varied with pressure swing adsorption (PSA) (100%) and water scrubbers (83–96%), achieving higher removal efficiencies for terpenes than amine scrubbers (61–85%) (Chap. 8).

3.6 Siloxanes

Organosilicons (siloxanes) are a large group of water-soluble man-made synthetic polymers utilizing the linkage between silicon atoms and an organic group, usually a methyl group. They are widely used throughout society in cosmetics and toiletries, detergents, paper coatings, anti-foaming agents, textile manufacture, sealants, and cooking oils, and as food additives generally ending up concentrated in wastewater sludge (McBean, 2008; Paglini *et al.*, 2022).

Biogas can become contaminated by volatile organic silicon compounds (VOSiCs), and when these are oxidized during biogas combustion, microcrystalline silicon dioxide is formed, which has similar properties to glass. This can severely damage engines and turbines, increasing wear and blocking valves and pistons. Volatile methyl siloxanes (VMSs) are the most common VOSiCs found in digester gas, and these are shown in Table 3.5. The combustion of one of the commonest occurring VOSiCs in biogas, octamethylcyclotetrasiloxane ($[(CH_3)_2SiO]_4$), is shown in Eq. (3.28) with silicon dioxide formed.

$$[(CH_3)_2SiO]_4(g) + 16O_2 \rightarrow 4SiO_2(s) + 8CO_2(g) + 12H_2O(g) \quad (3.28)$$

Table 3.5. Chemical composition and key physicochemical properties of the most frequently occurring volatile methyl siloxanes in biogas (de Arespacochaga *et al.*, 2015). Reproduced with permission of Elsevier Publishing.

	Abbreviation	Chemical formula	Molecular weight (g/mol)	Boling point (°C)	Water solubility (mg/L at 25 °C)
Hexamethyldisiloxane	L2	$C_6H_{18}OSi_2$	162	107	0.93
Octamethyltrisiloxane	L3	$C_8H_{24}O_2Si_3$	237	153	0.034
Decamethyltetrasiloxane	L4	$C_{10}H_{30}O_3Si_4$	311	194	0.00674
Dodecamethylpentasiloxane	L5	$C_{12}H_{36}O_4Si_5$	385	232	0.000309
Hexamethylcyclotrisiloxane	D3	$C_6H_{18}O_3Si_3$	223	135	1.56
Octamethylcyclotetrasiloxane	D4	$C_8H_{24}O_4Si_4$	297	176	0.056
Decamethylcyclopentasiloxane	D5	$C_{10}H_{30}O_5Si_5$	371	211	0.017
Dodecamethylcyclohexasiloxane	D6	$C_{12}H_{36}O_6Si_6$	444	245	0.005
Trimethylsilanol	TMOH	C_3H_9SiOH	90	99	42,600

The highest concentrations of siloxanes are found in biogas derived from domestic and municipal wastewaters, with concentrations as high as 41 ppm. Lower values are recorded from landfill gas (4–9 ppm), with the least amount recorded in farm gas (Arnold and Kajolinna, 2010; Wasajja *et al.,* 2020). Removal of siloxane in biogas is discussed in Sec. 8.5.

References

Arnold, M. and Kajolinna, T. (2010) Development of on-line measurement techniques for siloxanes and other trace compounds in biogas. *Waste Management*, **30**, 1011–1017. https://doi.org/10.1016/j.wasman.2009.11.030

Arrhenius, K., Holmqvist, A., Carlsson, M., Engelbrektsson, J., Jansson, A., Rosell, L., Yaghooby, H. and Fischer, A. (2017) Terpenes in biogas plants digesting food wastes: Study to gain insight into the role of terpenes. Report: 2017:350. Energiforsk AB, Sweden. https://energiforskmedia.blob.core.windows.net/media/22208/terpenes-in-biogas-plants-digesting-food-wastes-energiforskrapport-2017-350.pdf

Brongers, M., Virmani, P. and Payer, J. (2002) *Drinking Water and Sewer Systems in Corrosion Costs and Preventative Strategies in the United States.* United States Department of Transportation Federal Highway Administration. http://impact.nace.org/documents/ccsupp.pdf

Buan, N. R. (2018) Methanogens: Pushing the boundaries of biology. *Emerging Topics in Life Sciences,* **2**, (4), 629–646. https://doi.org/10.1042/ETLS20180031

Budiman, H., Nuryatini and Zuas, O. (2015) Comparison between GC-TCD and GC_FID for the determination of propane in gas mixture. *Procedia Chemistry*, **16**, 465–472. https://doi.org/10.1016/j.proche.2015.12.080

Buswell, A.M. and Mueller, H.F. (1952) Mechanisms of methane fermentation. *Industrial Engineering Chemistry*, **44**, 550–552. https://doi.org/10.1021/ie50507a033

Calbry-Muzyka, A., Madi, H., Rüsch-Pfund, F., Gandiglio, M. and Biollaz, S. (2022) Biogas composition from agricultural sources and organic fraction of municipal solid waste. *Renewable Energy*, **181**, 1000–1007. https://doi.org/10.1016/j.renene.2021.09.100

Cayford, B.I., Dennis, P.G., Keller, J., Tyson, G.W., and Bond, P.L. (2012). High-throughput amplicon sequencing reveals distinct communities within a corroding concrete sewer system. *Applied and Environmental Microbiology*, **78**, 7160–7162. https://doi.org/10.1128/AEM.01582-12

CCOHS. (2023) *Methane*. Canadian Centre for Occupational Health and Safety, Hamilton, Ontario. https://www.ccohs.ca/oshanswers/chemicals/chem_profiles/methane.html

Centre for Disease Control. (1977) Criteria for a recommended standard: Occupational exposure to hydrogen sulfide. DHHS (NIOSH) Publication No. 77–158. National Institute for Occupational Safety and Health, Washington D.C. https://www.cdc.gov/niosh/docs/77-158/

Chen, Y.-C., Higgins, M., Murthy, S., Maas, N., Covert, K., Weaver, J., Toffey, W., Rupke, M. and Ross, D. (2004) Mechanisms for the production of odorous volatile aromatic compounds in wastewater biosolids. *Proceedings of the Water Environment Federation*, 2004, 540–553.

David, E. and Niculescu, V.-C. (2021) Volatile organic compounds (VOCs) as environmental pollutants: Occurrence and mitigation using nanomaterials. *International Journal of Environmental Research and Public Health*, **18**, 13147. https://doi.org/10.3390/ijerph182413147

de Arespacochaga, N., Valderrama, C., Raich-Montiu, J., Crest, M., Mehta, S. and Cortina, J.L. (2015) Understanding the effects of the origin, occurrence, monitoring, control, fate and removal of siloxanes on the energetic valorization of sewage biogas — a review. *Renewable and Sustainable Energy Review*, **52**, 366–381. https://doi.org/10.1016/j.rser.2015.07.106

Dewil, R., Baeyens, J., Roles, J. and van de Steene, B. (2008) Distribution of sulphur compounds in sewage sludge treatment. *Environmental Engineering Science*, **25**, (6). https://doi.org/10.1089/ees.2007.0143

Fisher, R.M., Le-Minh, N., Sivret, E.C., Alvarez-Gaitan, J.P., Moore, S.J. and Stuetz, R.M. (2017) Distribution and sensorial relevance of volatile organic compounds emitted throughout wastewater biosolids processing. *Science of the Total Environment*, **599–600**, 663–670. https://doi.org/10.1016/j.scitotenv.2017.04.129

Gutierrez, O., Jiang, G., Sharma, K. and Yuan, Z. (2016) Biofilm development in sewer networks. In *Aquatic Biofilm: Ecology, Water Quality and Wastewater Treatment* (eds. A.M. Romaní, H. Guasch and M. Dolors Balaguer). Caister Academic Press, Norfolk, UK, pp. 145–164.

Haider, K.M., Lafouge, F., Carpentier, Y., Houot, S., Petitprez, D., Loubet, B., Fosca, C. and Ciuraru, R. (2022) Chemical identification and quantification of volatile organic compounds emitted by sewage sludge. *Science of the Total Environment*, **838,** 155948, 663–670. https://doi.org/10.1016/j.scitotenv.2022.155948

Hvitved-Jacobsen, T., Vollertsen, J. and Matos, J.S. (2002) The sewer as a bioreactor — a dry weather approach. *Water Science and Technology*, **45,** (3), 11–24. https://doi.org/10.2166/wst.2002.0044

Hvitved-Jacobsen, T., Vollertsen, J. and Nielsen, A.H. (2013) *Sewer Processes: Microbial and Chemical Process Engineering of Sewer Networks*, 2nd edition. CRC Press, London.

IEA. (2023) IEA Global methane tracker. https://www.iea.org/reports/global-methane-tracker-2023

Jiang, Y., McAdam, E., Zhang, Y., Heaven, S., Banks, C. and Longhurst, P. (2019) Ammonia inhibition and toxicity in anaerobic digestion: A critical review. *Journal of Water Process Engineering,* **32**, 100899. https://doi.org/10.1016/j.jwpe.2019.100899

Jo, J.Y., Kwon, Y.S., Lee, J.W., Park, J.S., Rho, B.H. and Choi, W.-I. (2013) Acute respiratory distress due to methane inhalation. *Tuberculosis and Respiratory Diseases*, **74**, (3), 120–123. https://doi.org/10.4046/trd.2013.74.3.120

Jung, H., Kim, D., Choi, H. and Lee, C. (2022) A review of technologies for in-situ sulfide control in anaerobic digestion. *Renewable and Sustainable Energy Reviews*, **157**, 112068. https://doi.org/10.1016/j.rser.2021.112068

Lecharlier, A., Carrier, H. and Le Hécho, I. (2022) Characterization of biogas and biomethane trace compounds: A critical review of advances in in situ sampling and pre-concentration techniques. *Analytica Chimica Acta*, **1229**, 349174. https://doi.org/10.1016/j.aca.2022.340174

Li, X., Kappler, U., Jiang, G. and Bond, P.L. (2017) The ecology of acidophilic microorganisms in the corroding concrete sewer environment. *Frontiers of Microbiology*, **8**, 683. https://doi.org/10.3389/fmicb.2017.00683

Locey, B.J. (2005) Hydrogen sulphide. In *Encyclopaedia of Toxicology*, 2nd edition (ed. P. Walker), pp. 545–551. https://doi.org/10.1016/B0-12-369400-0/00503-2

McBean, E.A. (2008) Siloxanes in biogases from landfills and wastewater digesters. *Canadian Journal of Civil Engineering*, **35**, 431–436. https://doi.org/10.1139/L07-144

McCarty, P.L. (1964) Anaerobic waste treatment fundamentals. I. Chemistry and microbiology; II. Environmental requirements and control; III. Toxic materials and their control. *Public Works*, **95**, 91–94; 107–12; 123–26.

McCarty, P.L. (1975) Stoichiometry of biological reactions. *Progress in Water*, 7, 157–172.

Muyzer, G. and Stams, A.J.M. (2008) The ecology and biotechnology of sulphate-reducing bacteria. *Nature Reviews Microbiology*, **6**, 44–54. https://doi.org/10.1038/nrmicro1892

Nie, E., He, P., Peng, W., Hang, H. Lü, F. (2023) Microbial volatile organic compounds as novel indicators of anaerobic digestion instability: Potential and challenges. *Biotechnology Advances*, **67**, 108204. https://doi.org/10.1016/j.biotechadv.2023.108204

Ocko, I.B., Sun, T., Shindell, D., Oppenheimer, M., Hristov, A.N., Pacala, S.W., Mauzerall, D.L., Xu, Y. and Hamburg, S.P. (2021) Acting rapidly to deploy readily available methane mitigation measures by sector can immediately slow global warming. *Environmental Research Letters*, **16**, 054042. https://iopscience.iop.org/article/10.1088/1748-9326/abf9c8

Paglini, R., Gandiglio, M. and Lanzini, A. (2022) Technologies for deep biogas purification and use in zero-emission fuel cells systems — a review. *Energies*, **15**, (3551), 1–30. https://doi.org/10.3390/en15103551

PHE (2019) Methane general information. Public Health England Centre for Radiation, Chemical and Environmental Hazards, London. https://assets.publishing.service.gov.uk/media/5c34c0b240f0b6445ac3e198/Methane_PHE_general_information_070119.pdf

Rajagopal, R., Massé, D.I. and Singh, G. (2013). A critical review on inhibition of anaerobic digestion process by excess ammonia. *Bioresource Technology*, **143**, 632–641. https://doi.org/10.1016/j.biortech.2013.06.030

Rasi , S., Veijanen, A. and Rintala, J. (2006) Trace compounds of biogas from different biogas production plants. *Energy*, **32**, (8), 1375–1380. https://doi.org/10.1016/j.energy.2006.10.018

Rasi, S., Veijanen, A. and Rintala J. (2007) Trace compounds of biogas from different biogas production plants. *Energy*, **32**, 1375–1380. 10.1016/j.energy.2006.10.018

Song, S.-K., Shon, Z.-H., Kang, Y.-H., Kim, K.-H., Han, S.-B., Kang, M., Bang, J.-H. and Oh, I. (2019) Source apportionment of VOCs and their impact on air quality and health in the megacity of Seoul. *Environmental*

Pollution, **247**, 763–774. https://doi.org/10.1016/j.envpol.2019.01.102.

Strik, D.P., Domnanovich, A.M. and Holubar, P. (2006) A pH-based control of ammonia in biogas during anaerobic digestion of artificial manure and maize silage. *Process Biochemistry*, **41**, 1235–1238. https://doi.org/10.1016/j.procbio.2005.12.008

Thauer R.K. (2012) The Wolfe cycle comes full circle. *Proceedings of the National Academy of Sciences of the United States of America*. **109**, 15084–15085. https://doi.org/10.1073/pnas.1213193109

Tian, G., Xi, J., Yeung, M. and Ren, G. (2020) Characteristics and mechanisms of H_2S production in anaerobic digestion of food waste. *Science of the Total Environment*, **724**, 137977. https://doi.org/10.1016/j.scitotenv.2020.137977

UNEP (2021) Global methane assessment: benefits and costs of mitigating methane emissions. United Nations Environment Programme, Nairobi, Kenya. https://www.unep.org/resources/report/global-methane-assessment-benefits-and-costs-mitigating-methane-emissions

Wang, J. and He, Q.P. (2023) Methane removal from air: Challenges and opportunities. *Methane*, **2**, 404–414. https://doi.org/10.3390/methane2040027

Wasajja, H., Lindeboom, R.E.F., van Lier, J.B. and Aravind, P.V. (2020) Techno-economic review of biogas cleaning technologies for small scale off-grid solid oxide fuel cell applications. *Fuel Processing Technology*, **197**, 106215. https://doi.org/10.1016/j.fuproc.2019.106215

Zeikus, J.G. (1980) Microbial populations in digesters. In *Anaerobic Digestion* (eds. D.A. Stafford *et al.*). Applied Science Publishers, London, pp. 61–89.

Zhang, L., Qiu, Y.-Y., Sharma, K.R., Shi, Y., Sun, J., Liang, Z., Yuan, Z. and Jiang, F. (2023) Hydrogen sulfide control in sewer systems: A critical review of recent progress. *Water Research*, **240**, 120046. https://doi.org/10.1016/j.watres.2023.120046

Zhang, Z., Zhang, C., Yang, Y., Zhang, Z., Tang, Y., Su, P. and Lin, Z. (2022) A review of sulphate reducing bacteria: Metabolism, influencing factors and application in wastewater treatment. *Journal of Cleaner Production*, **376**, 134109. https://doi.org/10.1016/j.jclepro.2022.134109

Zheng, G., Liu, J., Shao, Z. and Chen, T. (2020) Emission characteristics and health risk assessment of VOCs from a food waste anaerobic digestion plant: A case study of Suzhou, China. *Environmental Pollution*, **357**, 113546. https://doi.org/10.1016/j.envpol.2019.113546

Section B

Flow-Through (Conventional) Anaerobic Treatment Processes

4
Combined Anaerobic Wastewater Treatment Systems

4.1 Introduction

Anaerobic treatment is a biological process in which the organic fraction of wastewater sludge, comprising proteins, carbohydrates, and lipids, is degraded in the absence of oxygen to methane and carbon dioxide by bacteria and archaea. The process occurs quite widely in natural environments, such as lake sediments, marshes, peat bogs, and even in the rumen of certain herbivorous animals.

The first direct application of anaerobic treatment of wastewater is attributed to Louis H. Mouras of Vesoul in France, who developed a flow-through cesspool (c. 1860) in which he claimed, "sewage solids are liquified by fermentation." He applied for a patent on 22 September 1881, with the design submitted very similar to the basic septic tank of today (Fig. 4.1). By 1895, digester gas, also known as biogas or methane, was being collected and used as fuel. Donald Cameron constructed a large septic tank in Exeter (United Kingdom (UK)) and was able to use the biogas for lighting the area around the treatment plant (Fig. 4.2). By the turn of the century, it had become common practice to incorporate a digestion chamber within sedimentation tanks, with the Travis hydraulic tank and the Imhoff tank the most widely adopted. The Imhoff tank, in particular, was installed at most small to moderate-sized treatment plants and is still to be found today in older wastewater treatment plants (WWTPs) in

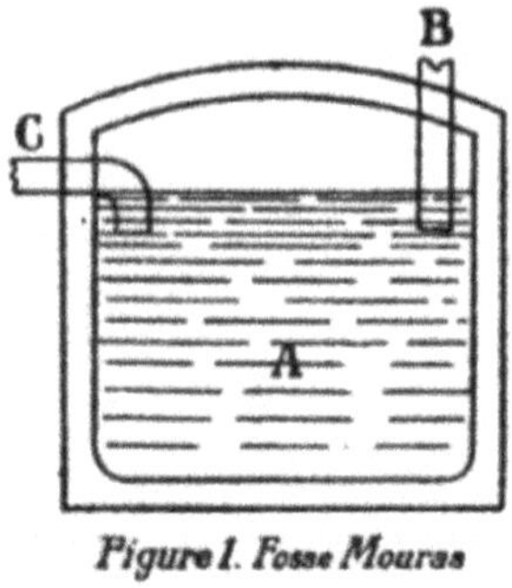

Figure 4.1. The actual design used in the patent application (1881) for the first anaerobic wastewater treatment system designed by Louis Mouras in France. It was described as a flow-through cesspool, where A is the fermentation tank, B is the inlet pipe, and C is the outlet pipe to the sewer.

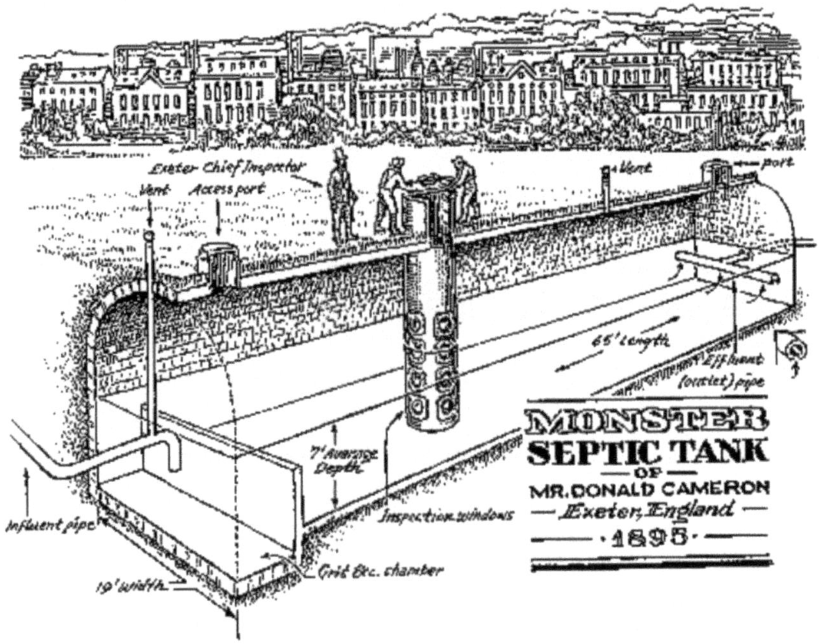

Figure 4.2. It was the English engineer Donald Cameron who coined the term "septic tank," with his patented system used to collect biogas. In 1895, he was able to produce enough biogas from his community septic tank to power the gas street-lights in Exeter.

Ireland. While few have been constructed *in-situ* since 1950, they are still commercially available as packaged units. The development of specific digesters for the anaerobic breakdown of sludges began in the early 1920s, with their use restricted to large cities. During World War II, there was renewed interest in biogas production from wastewater sludges due to fuel shortages for transport, but very little development occurred after the war. However, there has been a resurgence of interest in anaerobic digestion in the past 20–25 years, especially in France and China, with much fundamental research and development work being done, leading to the creation of new, more robust and reliable, digester, and anaerobic reactor designs. The basic flow-through system can be separated into two categories: systems that combine settlement with digestion and separate systems built for digestion only (Fig. 4.3).

Combined systems of settlement and digestion are restricted to situations where there is only a small volume of wastewater to be treated. The combined system provides primary settlement and then sludge stabilization by digestion, all within a single unit. By reducing

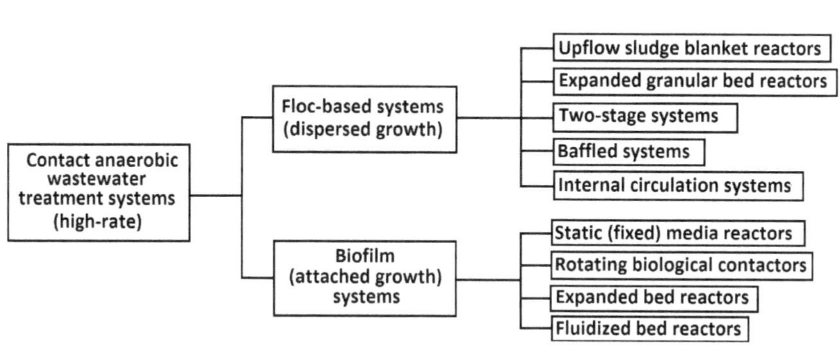

Figure 4.3. The principle conventional and high-rate anaerobic wastewater treatment systems currently in use.

the volume of sludge produced and storing it in such a way so as not to impede the settlement process, such systems are ideal for single houses, hotels, restaurants, or small communities. Anaerobic lagoons can also be categorized as combined systems, although their use in Europe is largely restricted to strong organic wastes from the food processing and agricultural industries. The selection, installation, and maintenance of small wastewater treatment systems, including combined anaerobic processes, are covered by a series of standards (EN 12566) by the Comité Européen de Normalisation (CEN) (Table 4.1).

Table 4.1. The European Committee for Standardization (CEN) Standards for septic tanks and small treatment systems. CEN is the provider of standards for all 30 Member States of the EU, including EN 12566, which is a series of standards on small wastewater treatment systems. https://standards.cen.eu/

Standard	Title	Function	Reference
EN 12566-1	Prefabricated septic tanks	Design, installation and testing of septic tanks	CEN 2000; 2003
EN 12566-2	Soil infiltration systems	Design, installation of percolation areas	CEN 2005
EN 12566-3	Packaged and/or site assembled domestic wastewater treatment plants	Requirements and test methods for packaged secondary treatment units	CEN 2009
EN 12566-4	Septic tanks assembled *in situ* from pre-fabricated kits	Assembly, installation and testing of site-assembled septic tanks	CEN 2007
EN 12566-5	Pre-treated effluent filtration systems	Design, construction, installation and testing of secondary treatment filtration units for septic tanks	CEN 2008
EN 12566-6	Pre-fabricated treatment units for septic tank effluent	Test methods and evaluation for conformity for secondary treatment units used with septic tanks	CEN 2013a
EN 12566-7	Pre-fabricated tertiary treatment units	Treatment requirements and test methods for packaged and site assembled tertiary treatment units	CEN 2013b

4.2 Septic Tanks

Septic tanks are often confused with cesspools or cesspits. A cesspool is an underground chamber constructed solely for the reception and storage of wastewater, with no treatment taking place. A cesspool is a storage tank that requires periodic emptying and is not intended as a septic tank in which decomposition of the settled material occurs. It is important to differentiate between the two, as these terms are used interchangeably in many areas. A cesspit is the predecessor of the cesspool, which allowed the liquid fraction to soakaway into the soil layer. Made of brick or stone with gaps to allow liquid to percolate into the surrounding soil, they were generally circular in design and approximately 2 m in depth. They were common in rural areas but were frequently associated with contamination of nearby wells and boreholes. They were made illegal in the UK by the Public Health Act of 1936. Where a septic tank or cesspool is not watertight, it acts as a cesspit and is thus illegal.

4.2.1 *Cesspools*

Cesspools can be constructed out of concrete, plastic, or fiberglass, and according to the British Code of Practice (BSI, 1972; 1983), must be: (a) impervious (i.e., watertight), (b) not be able to overflow, (c) have a minimum capacity of 18 m^3 or 45 days retention for two people assuming a per capita water usage of 180 L d^{-1}, (d) be constructed so that it can be completely emptied, (e) adequately ventilated, and, most importantly, (f) adequately covered to ensure safety. Clearly, cesspools are only used where no other form of treatment is possible, and the need to have them regularly emptied means that they are the most expensive form of treatment for domestic dwellings in terms of both capital and operational costs. The only improvement in this system can be achieved by adopting water-saving improvements that reduce the volume of wastewater discharged. There are advantages, however, such as no power requirement, no quality control required, no mechanism that can go wrong, the process is not injured by intermittent use, and as there is no effluent discharge, there is no immediate environmental impact. The major limitation on the use of cesspools is the cost of emptying, although the construction of large underground storage tanks can be both difficult and expensive. Sealed prefabricated units are available up to 54 m^3 of capacity and are cylindrical in design (12,000 mm × 2,740 mm) and must be embedded in a minimum of 150 mm of concrete. Several units can be used in series to increase capacity. The advantages and disadvantages of cesspools have been examined by Mann (1979).

One area where storage tanks are still widely used is on farms. Storage of farm effluents is expensive, and ideally, raw effluent should be spread immediately onto the land. However, this is impossible because of seasonal crop cycles and, at certain times of the year, the risk of surface runoff causing pollution. The use of slurry tanks, both of underground and above-ground construction, is widespread, and the design of such systems is discussed in (DAFM, 2022).

4.2.2 *Septic Tanks*

Septic tanks are essentially a chamber in which settleable solids settle out of suspension to form a sludge, which undergoes anaerobic breakdown. The process provides partial treatment only and cannot produce an effluent equivalent to a normal WWTP but is similar in composition to settled sewage (Table 4.2). Therefore, further treatment is required either by a secondary treatment process or by the provision of a percolation (i.e., irrigation) area in which the sewage percolates into the soil via a system of underground distribution pipes

Table 4.2. Variation in the strength of the effluent from 28 Irish septic tanks ($n = 84$)

	Mean	Standard deviation	Minimum	Maximum
pH	7.0	0.24	6.3	7.5
BOD (mg L^{-1})	264	94.1	110	510
COD (mg L^{-1})	500	175.7	225	900
Suspended solids (mg L^{-1})	124	53.4	40	260
NH_4–N (mg L^{-1})	55	21.4	16	106
Total phosphorus (mg L^{-1})	16.2	6.4	2.0	29.0
COD:BOD	1.9	0.21	1.3	2.4

BOD: biochemical oxygen demand; COD: chemical oxygen demand.

and undergoes treatment by the formation of a biomat (Gray, 2022). If the effluent from a septic tank is discharged directly to surface waters, then a minimum dilution factor of 300–400 is required. Although most commonly used for individual houses, they can be used for small communities of up to 500 and are commonly used in combination with a percolating filter, rotating biological contactor, reed bed, or a submerged aerated filter system for small rural villages. Septic tanks themselves require no power, and where subsurface drainage is used, quality control can be confined to suspended solids removal only to prevent blockage of the percolation pipework. They are not adversely affected by intermittent use, have a very small head loss, and can achieve 40–50% BOD (biochemical oxygen demand) and 80% suspended solids removal.

With 50% of the global population living in rural areas, a large portion of these households employ septic tanks or small-packaged systems incorporating a septic tank section for the treatment of their wastewater (Singh *et al.*, 2019). In its simplest form, the septic tank consists of a single chamber (Fig. 4.4a) with a single input and output. They can be of any shape, although they are traditionally rectangular and made from concrete comprising two or, frequently in Scandinavia and China, three chambers (Fig. 4.4b). From 2015 to 2020, 10 million three-chambered septic tanks were installed in China alone, with the final effluent used for irrigation (Tan *et al.*, 2021). New prefabricated units in plastic or fiberglass come in a variety of shapes, designs, and even colors. The tank consists of three separate zones, a scum layer made up largely of emulsified fats that floats on top of the clarified liquid, with a sludge layer in the base. Wastewater enters and leaves the tank via T-shaped pipes that prevent disturbing the scum layer or allowing solids or floatable material to be carried out of the tank. As the wastewater moves through the chamber, settleable solids gradually settle forming a sludge, and any fats or buoyant material floats to the surface where they are retained by the baffle edge of the T-pipe and form the scum layer. The scum layer, although not vital to the successful operation of the tank, helps its operation in three ways: it prevents oxygen transfer through the air-water interface, insulates the anaerobic chamber by preventing heat

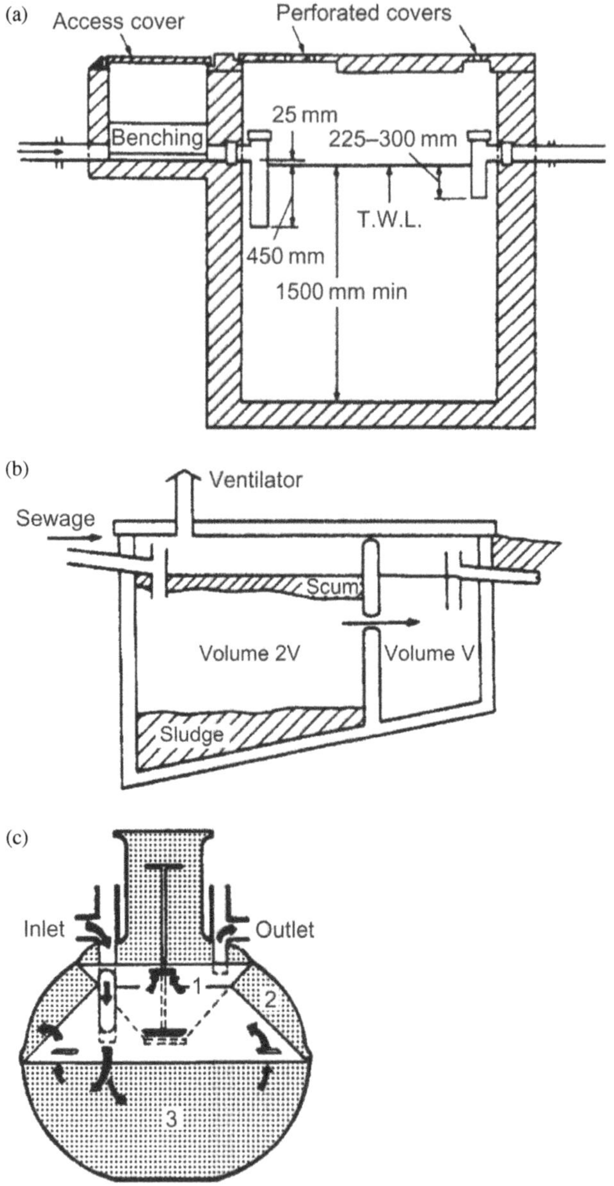

Figure 4.4. Examples of manufactured septic tanks: (a) Original single-chamber system, (b) more common two-chamber system, and (c) GRP septic tank.

loss, and by attracting and retaining fats, floatable material, including raised solids from the sludge layer. It is in the sludge layer where anaerobic decomposition takes place, with the organic fraction slowly being degraded. Often, decomposition is incomplete, and the settled material is only hydrolyzed and not broken down to methane. This results in intermediate products of anaerobic digestion, such as short chain fatty acids, being produced and slowly diffused back into the clarified liquid to be discharged from the tank. This incomplete anaerobic activity results in unpleasant odors, which the scum layer helps to retain within the tank.

Methane production is inhibited by a number of factors in septic tanks, most notably, low temperatures and insufficient sludge storage capacity. Methane production in septic tanks is low, even in the summer, whereas in winter, it may cease completely, resulting in an overall increase in sludge accumulation rate. However, gas formation occurs even in the absence of methane formation, which can carry solids back into the liquid phase and cause an increase in the rate of scum accumulation. Most wastewaters contain a significant proportion of non-degradable solids, and even if anaerobic digestion is highly efficient, there will be a gradual increase in solids in the tank. As these build up, the volume of the liquid zone in the chamber is reduced, thus reducing the retention time of the wastewater and the degree of settlement that is possible. If the discharge of solids to the next treatment phase is to be minimized, then the septic tank must be regularly desludged. Ideally, the floor of the chamber is sloped toward the inlet end (Fig 4.4b), above which a manhole is situated into which a suction pipe can be lowered for sludge removal. Normally, they have a flat bottom so that sludge accumulation tends to be deeper at the outlet end of the tank (Fig. 4.4a). It is recommended septic tanks should be desludged every 12 months, leaving a small quantity of the sludge to re-seed the tank, ensuring rapid recommencement of anaerobic activity. In practice, it is easier for operators to completely empty the tank, and it is felt that enough sludge will be left behind in the corners of the chamber and adhering to the walls to ensure seeding. If possible, the tank, which is designed to operate full, should be refilled with water as soon as possible after desludging. This also prevents the tank from rising up out of the ground if the soil is saturated with water. In practice, the scum collapses as the sludge is pumped out below. If this occurs, then the scum must also be pumped out to prevent the overflow pipe becoming blocked as the tank refills. Also, an excessively thick scum layer should be removed to prevent the inlet and outlet pipes from becoming blocked. Where excessive scum formation occurs, a grease trap should be installed.

When commissioning septic tanks, they should be filled with water before use and tested for leaks by checking the water level over a period of a few days. This is a common problem with prefabricated concrete tanks, which come in two sections — a top and bottom, often with a rubber seal that is placed between the two sections as they are assembled. Although not strictly necessary, it is advantageous if tanks can be seeded with a few liters of sludge when first commissioned or a kilogram of cow or sheep dung added to help fermentation to begin. Commercial seeds are now available, which are easier to handle and more convenient. There are a number of chemicals and biological bacterial seeds advertised for septic tank problems. There is no shortcut to effective tank care and the addition of yeast, enzymes, and bacteria are not necessary for digestion within the tank. It is a common fallacy that odors produced by septic tanks can be reduced by cleaning the chamber using a proprietary cleaner. Under no circumstances should a tank be disinfected, as this will only result in even worse odors being produced on recovery and inhibiting the degradation processes already taking place in the chamber. If odors are a problem, then the tank should be desludged.

Septic tank design

The design and construction details of septic tanks are given by the European standard EN12566 (Table 4.1), British Standard Institution BS 6297 (BSI, 1983; 2007), and the Irish Environmental Protection Agency (2009; 2021), although most countries have their own standards. The two most important design criteria for septic tanks are the suitability of the land and the capacity of the tank. Many by-laws insist on minimum distances for the siting of tanks and percolation areas from houses, wells, streams, and boundaries. These vary, but in Ireland, no part of the percolation area should be closer than 10 m from the nearest habitable building (or 4 m from the nearest road) and 10 m from a ditch or stream (or 3 m from the boundary of the adjoining site). Distances from wells vary from 30 to 60 m, depending on percolation characteristics, the minimum depth of subsoil above rock, and the minimum depth of the water table (EPA, 2009). The capacity and hence the retention time for settlement and sludge digestion is calculated according to the number of people discharging. The minimum capacity of a septic tank must not be less than 2720 L, with the actual capacity calculated as:

$$C = (180P + 2000) \text{ liters} \tag{4.1}$$

where C is the capacity in liters, and P is the number of people discharging into the system. It is quite common for people to purchase a country cottage in which an elderly couple had lived for many years, only to find that once modernized, with mains water, a new bathroom, and a kitchen full of labor-saving devices, that the septic tank system can no longer cope. For example, a disposal unit for household kitchen waste (garbage grinder) results in a 30% increase in the BOD and a 60% increase in suspended solids discharged to the tank. Where these are installed, the capacity of the septic tank must be calculated as:

$$C = (250P + 2000) \text{ liters} \tag{4.2}$$

Therefore, in the design of a septic tank, the future size of the family, number of bathrooms, visitors, and other potential developments must be included in the design. It is unacceptable to use half values for children in such calculations as their water demands are just as great as adults. In Ireland, for example, the capacity of the septic tank is based on the number of people served, with two sizes recommended. For 2–5 people, a tank of nominal capacity of 3 m^3 is required, or 4 m^3 for 6–10 people. However, in practice, larger tanks may be needed if water consumption is high or if the household works from home. Some wastewaters are not suitable for discharge to septic tanks, for example, inert, dense, or highly compactable solids. Septic tanks are generally unsuitable for industrial wastewaters or agricultural use.

While the majority of septic tanks are made from precast concrete or are cast *in situ*, new prefabricated units in plastic or fiberglass are now widely available and come in a variety of shapes, designs, and even colors. The Klargester septic tank has a characteristic onion shape (Fig. 4.4c). It is a three-chamber design that must be positioned vertically in the ground. To prevent misalignment, it must be set in concrete, and the immediate site protected from traffic. The wastewater enters at the base (zone 3), where settlement and sludge storage occur. Settled wastewater is displaced upward into the second chamber (zone 2), where the inclined base encourages further settlement, with solids falling into the lower chamber through the peripheral slots. The wastewater passes into the final chamber (zone 1), from where it is discharged via the outlet to the percolation area. These plastic or fiberglass tanks are ideal where the location of the site does not permit access by heavy-lifting equipment, such as a JCB machine, required for moving concrete tanks. A standard plastic or fiberglass septic tank can be easily carried and installed by two people.

Septic tanks are easy to uprate by adding extra chambers in series. This reduces the effects of surge flows and excessive sludge accumulation in the first chamber. Existing multi-chambered tanks can be uprated by adding new septic tank units in parallel. The commonest faults associated with septic tanks include: (i) Leaking joints when tanks have been constructed from concrete sections or concrete rings. It is important that septic tanks should be watertight to prevent contamination of the groundwater and should be constructed, if possible, without joints. (ii) Non-desludging is the commonest fault, resulting in a reduced retention time and a stronger effluent due to less settlement. The loss of solids under these circumstances can block pipes, percolation areas, or increase loadings to secondary treatment units. (iii) Blocked outlet pipes are frequently a problem because of the scum layer becoming too thick or sludge physically blocking the outlet pipe, which causes the tank to overflow. Tampons tend to become suspended in the liquid phase of the tank and can easily be transferred into the final chamber and eventually into the percolation area and so, if possible, should not be flushed along with nappies, sanitary and incontinent pads, wipes, and cotton wool balls or pads. The insertion of a circular mesh sieve into the final outlet T-piece can reduce the loss of solids from the septic tank but must be regularly cleaned to prevent blockage and overflow through the access manhole. In Ireland, many systems have inadequate percolation areas or suitable access for the sludge tanker, with desludging pipes occasionally having to pass through the owners' houses to reach the tank itself. The design of percolation (sub-surface irrigation) areas and problems associated with soil percolation have been reviewed by Gray (2022).

Operation

Little information is available on the operation of septic tank systems, but on the whole, they are very robust. However, care should be taken when using the following: *Disinfectants* should be used moderately, as their bactericidal properties inhibit or kill the anaerobic bacteria, which can result in unpleasant odors being produced during recovery. Phenol-based disinfectants can cause significant odors. It is best to use disinfectants having free chlorine, as it reacts with the organic matter in the sewage, rendering it harmless by the time it reaches the tank. *Caustic soda*, which is often used to remove grease from drains, can cause the sludge to flocculate and rise. Excessive use can result in sludge passing out of the tank and blocking the percolation area. Small amounts of acidic or alkaline cleaners do no lasting harm and should be used in preference. Some strong cleaners can upset the pH of the tank, which should be as near to neutral as possible. Under good operational practice, the tank will buffer itself. *High sodium* concentrations in the water do not affect the septic tank system directly but can impair the drainage properties of the soil. Those who have water softeners and use soil percolation as a secondary treatment process of septic tank effluents should take advice. *Detergents*, especially alkyl benzene sulfonate, are known to inhibit the digestion process, although, providing the tank capacity is sufficient, if normal concentrations of detergents are used, the performance of the system will not be impaired. Enzyme-based washing powders have no effect on septic tank systems. *Large flushes of water* to the tank should be avoided if possible to prevent scouring of the sludge unless the tank is large enough to withstand them or a three-chambered system has been installed. Wherever legally permissible, bathwater should be diverted to a soakaway, as should rainwater and melted snow from roofs and paved areas. *Solid materials* such as disposable nappies, coffee grinds, bones, cigarette ends, and cat litter will not degrade in the tank and should not be discharged to the septic tank as they will reduce the volume of the settlement chamber very quickly and can be difficult to remove due to compaction. Excessive amounts of fats, oils, and greases should also be disposed of separately whenever possible, or a grease trap installed before the septic tank.

If laundry wastewater exceeds 10% of the total flow to the septic tank, then this should be treated separately using a treatment system with a separate percolation area. Alternatively, it can be stored and sprayed onto lawns during dry periods. Catering facilities such as restaurants produce a large quantity of grease that cannot be dealt with by the system, so fats must be kept to a minimum, especially cooking oils. A grease trap should be installed on the kitchen waste line as close to the source as possible and before any foul water enters. Any settleable solids will also be retained in the trap, which reduces its efficiency and rapidly blocks it. So, where necessary, a dedicated disposal line should be installed for grease and oils, or solids should not be allowed to enter the drain. It is important that oils must be cooled before entering the trap. High-protein wastewaters can also lead to excessive hydrogen sulfide formation (Sec. 3.3).

Petrol and oil from car parking areas, garages, and petrol station forecourts must be removed prior to septic tank or secondary treatment. This is best done by the use of oil interceptors, which are installed before the foul water from paved areas enters the system. They are designed to ensure maximum separation with retention times for adequate separation ranging from 6 to 20 minutes, depending on the density of the oil or petrol. Units are normally prefabricated in plastic or glass fiber and vary in size from 2,000 to 54,000 L. The design of the unit depends on the area served and the maximum expected rainfall per hour. They are also useful for containing spillages of light-density liquids.

Maintenance

Maintenance of the septic tank, like all sewage treatment units, is important in maximizing treatment efficiency and preventing pollution. The scum and sludge accumulation should be inspected twice a year in systems serving more than one household, depending on its volume. In many countries, it is a requirement to keep an up-to-date record book noting inspection dates, observations, and details of desludging. In the United States (US), the scum and sludge depth should be measured twice a year. The depth of the sludge and the thickness of the scum can be measured in the vicinity of the outlet pipe in a single-chambered tank, which can help to predict when desludging is necessary. The tank should be cleaned whenever the bottom of the scum layer is within 7.5 cm of the bottom of the outlet pipe or the sludge level is within 25–30 cm of the bottom of the outlet pipe in a single-chambered tank. In practice, there should be as little accumulated sludge as possible in the final chamber of a two or three-chambered septic tank to ensure minimal solids loss from the tank. Scum thickness is measured using a hinged flap device, which is a weighted flap attached to a long rod (Fig. 4.5a). Any device can be used, which allows the bottom of the scum mat to be felt. The measuring device is pushed through the scum layer until the hinged flap falls into the horizontal position. It is then gently pulled upward until the flap engages against the bottom of the scum layer. The handle is marked to correspond to a reference point on top of the tank. The same procedure is used to locate the lower end of the outlet pipe. The difference in height on the handle corresponds to the distance the scum is from the outlet. Sludge level detectors, using a light-sensitive cell or ultrasonic immersion sensors, are now widely used by professionals to monitor sludge levels (Fig 4.5b). Ultrasound is not only used to measure the blanket surface but to actually measure the TSS (total suspended solids), thereby giving a profile of sludge density with depth. A transducer sends an ultra-sonic pulse, which is

(a)

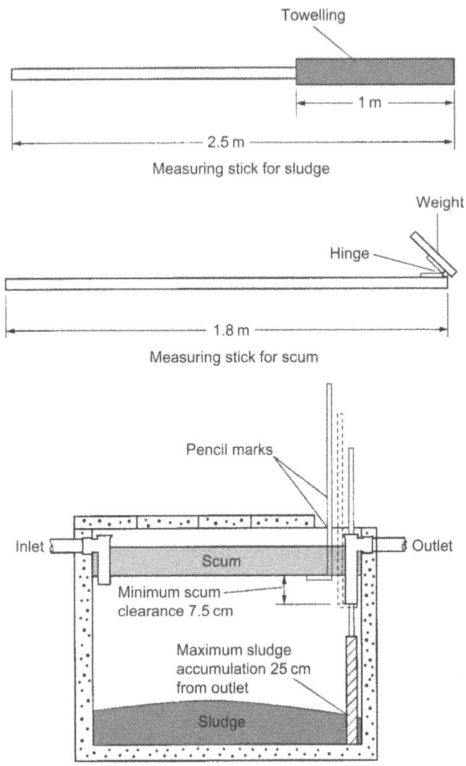
Towelling
1 m
2.5 m
Measuring stick for sludge
Weight
Hinge
1.8 m
Measuring stick for scum
Pencil marks
Inlet
Outlet
Scum
Minimum scum clearance 7.5 cm
Maximum sludge accumulation 25 cm from outlet
Sludge

(b)

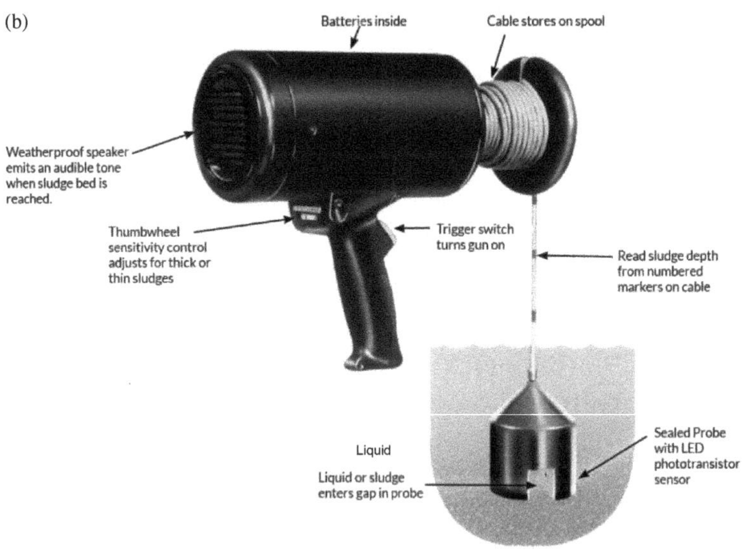
Batteries inside
Cable stores on spool
Weatherproof speaker emits an audible tone when sludge bed is reached.
Thumbwheel sensitivity control adjusts for thick or thin sludges
Trigger switch turns gun on
Read sludge depth from numbered markers on cable
Liquid
Liquid or sludge enters gap in probe
Sealed Probe with LED phototransistor sensor

reflected back to a second transducer. However, if not available, the depth of sludge can be measured by wrapping a long stick in rough light-colored toweling, which is tied securely. The stick is slowly lowered into the tank through the vertical piece of the outlet pipe to the bottom of the tank to avoid the scum. It is left for a few minutes and then slowly removed. The depth of the sludge can be distinguished on the toweling by black particles clinging to it (Fig. 4.5a). If the depth of the sludge is more than one-third of the total liquid and sludge depth at this point, desludging should be arranged. In practice, measuring sludge depth in septic tanks can be quite difficult and messy, requiring experimentation to get it right, making annual desludging a simpler option.

In theory, the rate of sludge accumulation determines the rate of desludging required, although in the UK and Ireland, desludging every 12–18 months is recommended. Twenty-eight septic tank systems of various sizes and configurations were studied in Ireland, with sludge accumulation rates observed to decline steadily over time, with the rate significantly less after 12 months (Gray, 1995). Specific sludge accumulation per capita per day fell from 0.254 L $ca^{-1}d^{-1}$ (equivalent to 92.7 L $ca^{-1}yr^{-1}$) over the first 6 months of operation to 0.178 L $ca^{-1}d^{-1}$ (64.9 L $ca^{-1}yr^{-1}$) after 60 months. The mean sludge accumulation rate, irrespective of sludge age, was 0.234 L $ca^{-1}d^{-1}$ (85.3 L $ca^{-1}yr^{-1}$). Increasing the desludging interval reduces the volume of sludge produced, and with the average cost of desludging in Ireland between €300 and €400 (2025), this could amount to a significant saving over time. There is also an increasing problem in finding WWTPs willing to take septic tank sludge due to reported

Figure 4.5. (*Continued on facing page*) Sludge level detection. (a) Simple equipment and procedures for measuring the accumulation of sludge and scum in a septic tank. (b) The Sludge Gun™ portable sludge blanket level detector. A pulse of infrared light is emitted and received by a phototransistor. In clear water, nothing happens, but as the solids content of the water increases, a sound is emitted, which increases in pitch and volume until the thickest sludge is located. This allows the exact location of the sludge blanket and the nature of the overlaying liquid by reading off the depth marked on the cable. Reproduced with permission of Markland Speciality Engineering Ltd, Georgetown, Canada. https://sludgecontrols.com/

problems of septicity causing bulking problems when treated by the activated sludge process (Gray, 2023). In theory, the longer the sludge age, the more stabilized the sludge, which in warmer areas would mean that sludges would not necessarily have to be disposed to a sewage treatment works. Therefore, a standard 3,000 L septic tank serving a population of 4 should require desludging every 3–5 years if operating and loaded correctly while retaining a minimum hydraulic retention time of 2.0 days and a liquid settlement volume of 48% overall. A disadvantage is that there is a discernible increase in the carryover of solids from the primary chamber of some septic tanks with sludge age. This suggests a potential for better-designed systems to take advantage of longer sludge ages, with mutual benefits for owners and the environment. An earlier study conducted by Philip *et al.* (1993) reached similar conclusions, suggesting that desludging intervals of up to 5 years were possible. They found that over the first two years of operation, hydrolysis led to a steady build-up of volatile fatty acids, with methane activity only recorded after 2 years and optimal sludge reduction seen after 2.5 years, with a *n* average sludge accumulation rate of 82 L $ca^{-1}yr^{-1}$ after 3 years falling to 60 L $ca^{-1}yr^{-1}$ after 3.5 years (Fig. 4.6). Elmitwalli (2013) recorded similar sludge

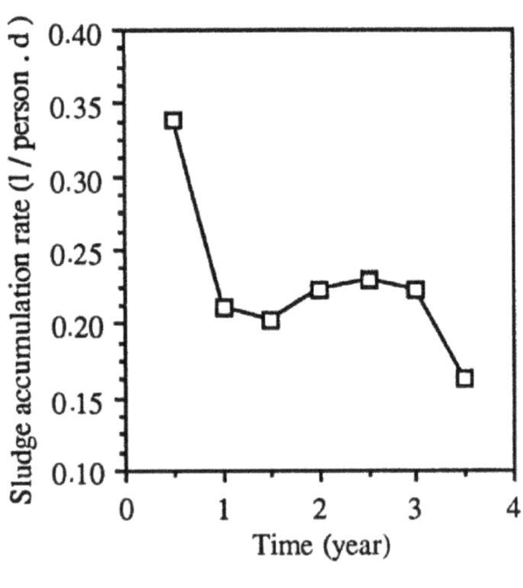

Figure 4.6. The average sludge accumulation rate over time in septic tanks in France expressed as liters per capita per day (Philip *et al.*, 1993). Reproduced with permission of IWA Publishing.

accumulation rates of between 0.19 and 0.22 L $ca^{-1}d^{-1}$ for domestic wastewater and 0.13–0.15 L $ca^{-1}d^{-1}$ for black water. He concluded that conversion rates of organic matter to methane were 50% and 60% for the different wastewaters, respectively, but that to obtain a high and stable conversion rate, desludging intervals had to be greater than one year.

A detailed study of 27 two-chambered septic tanks by MacMahon *et al.* (2022) recorded significantly higher accumulation rates of 250 L $ca^{-1}yr^{-1}$ over the first 12 months, falling to 150 L $ca^{-1}yr^{-1}$. This represents a much higher rate of accumulation than previously recorded but could reflect a change in lifestyle and disposal practices over the past decade or the effects of working from home, which became prevalent due to COVID-19. Figure 4.7 uniquely shows the sludge depth in both the primary and secondary chambers from the study by MacMahon *et al.* (2022). This is interesting as it illustrates that it cannot be assumed that septic tanks are all behaving in a predictive manner, as would be expected in a normal WWTP primary settlement tank, so each septic tank must be considered unique due largely to behavior differences between households, water consumption, and wastewater characteristics, as well as tank design and installation. So, as a general rule of thumb, if the solids accumulation in the primary chamber is >40% or >25% in the secondary tank, then desludging should be considered. If these increase to >50 and >35%, respectively, then the tank should be desludged immediately for the tank to operate effectively. Measuring sludge depth is very difficult in the primary chamber, but measuring the sludge level in the secondary chamber is much simpler. So, routine monitoring could be based on sludge carryover into the secondary tank, as the primary function is to protect the percolation area. MacMahon *et al.* (2022) recommend that desludging intervals could be extended in a temperate region, such as Ireland and the UK, to at least 3–5 years, so long as the accumulated sludge volume does not exceed 50%. They also agree with Gray's (1995) conclusions that the volume of the primary chamber could be enlarged to facilitate longer desludging intervals and that this should be considered as a possible change to current guidelines. What is clear from their study is that more information is needed in relation to sludge production in septic tanks, especially as work practices change.

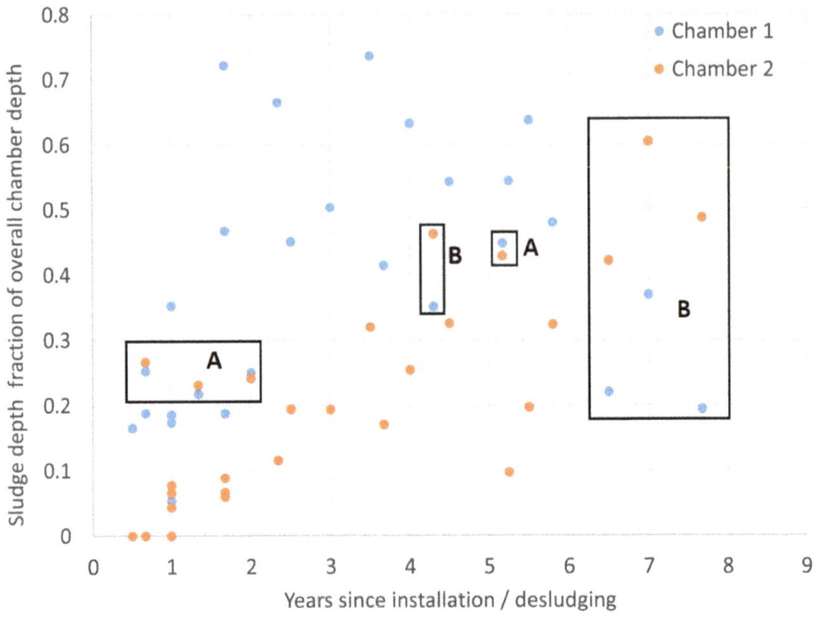

Figure 4.7. Fractional sludge depth in the first and second chambers of two-chambered septic tanks sampled in Ireland for different sludge ages. Of the 27 tanks sampled, only 12 appear to be operating within guidelines, with 7 with high sludge accumulations (>50% sludge). A further 8 septic tank systems are showing operational problems. Those marked (A) show equal sludge accumulation in both chambers and appear to be operating as a single chamber unit, and those marked (B) indicate significant carryover of sludge to the secondary chamber, which could result in solids loss. Both of these scenarios are indicative of design, installation, or operational problems. Tanks with ≥ 50% or ≥ 35% in the first and second chambers, respectively, require immediate desludging. Data from MacMahon *et al.* (2022). Reproduced under the Creative Commons Attribution (CC-BY) license 4.0.

However, in practice, if the sludge levels cannot be checked, then all septic tanks should be desludged at least every two years, if not annually.

4.2.3 *Septic Tank Modifications*

With septic tanks, the world's most common treatment system, research into making them more effective and sustainable is an exciting research area that encompasses all aspects of anaerobic treatment technologies. Conventional septic tanks offer limited treatment to wastewaters, being only able to remove settleable solids and their associated BOD. This means that a secondary treatment process is required, and when this is subsurface irrigation, this often results in ground or surface water pollution. A number of modifications to septic tanks have been made, including the inclusion of media for biofilm development or passing the wastewater through the sludge to enhance anaerobic microbial activity (Fig. 4.8). Early modifications were in the design of septic tanks, with longer narrower tanks improving effluent quality by improving laminar flow, reducing turbulence and mixing, and ensuring all the tank volume was used. Small orifices between chambers also caused short-circuiting of the wastewater and turbulence, both affecting effluent quality, especially TSS.

Anaerobic-baffled septic tanks use a series of vertical plates within the tank to ensure that the wastewater flows over and under each plate as it flows from the inlet to the outlet. Within each zone settlement, anaerobic bacteria and archaea develop, forcing the wastewater to pass through the sludge as it passes under the vertical plates or baffles, which increases the wastewater microbial biomass contact time, whereas, in a standard septic tank, the wastewater passes over the sludge biomass with little or no contact. These systems have higher BOD and COD removal rates and enhanced specific methane production rates (Chen *et al.*, 2014) and are considered in more detail in Sec. 7.6, where they can be operated using granular sludge or by the inclusion of a carrier medium.

In up-flow septic tanks (USTs), also known as UASB-septic tanks, the flow direction is changed from being horizontal to vertical, with the influent wastewater passing upward through the sludge blanket, with solids separation occurring as the wastewater rises to the overflow at the top of the tank. This increases the contact between the solids and the microbial community within the sludge blanket, improving removal efficiency, and also improves solids removal partially by enmeshment within the sludge blanket but also due to the flocculation of smaller solids as the wastewater moves upward, with the solids slowly settling within the vertical flow tank (Coelho *et al.*, 2003; Kujawa-Roeleveld *et al.*, 2005; Luostarinen *et al.*, 2007).

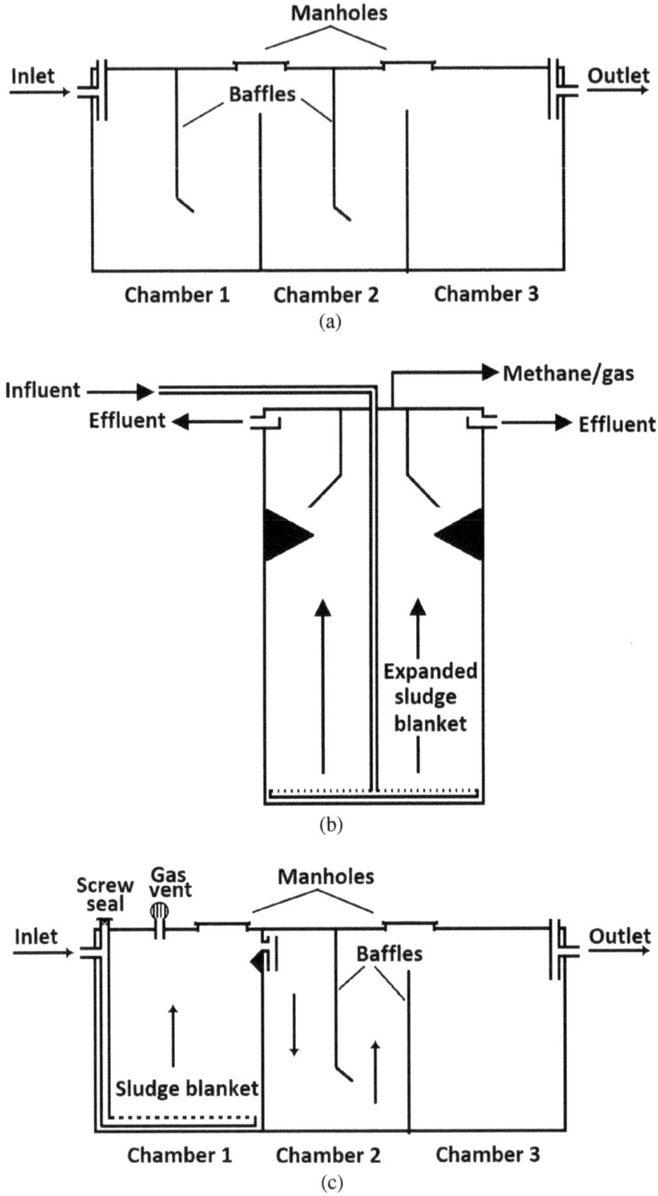

Figure 4.8. Schematic septic tank modifications: (a) baffled septic tank, (b) up-flow septic tank, and (c) integrated up-flow baffled system.

Moussavi *et al.* (2010), using a pilot UST treating sewage, observed that treatment efficiency was related to hydraulic retention time (HRT), with 85, 77, and 86% BOD, COD, and TSS removal achieved at an HRT of 24 hours. At 12 hours of removal, this fell to 71, 67, and 67%, and at an HRT of 6 hours, this fell further to 33, 31, and 33% respectively. The system had a high specific methanogenic activity at 15.2 mL CH_4 g^{-1} VSS d^{-1}, with a significant stabilization of sludge with a VSS/TSS ratio of 0.57.

Filter media has been used in anaerobic reactors to support an active biofilm to enhance carbonaceous removal. A full-scale anaerobic packaged plant was developed by Sharma and Kazmi (2015) in India, where they used the system to treat the wastewater from a toilet block at a residential school. The single low-density polyethylene tank comprised a septic tank (950 L), followed by an up-flow anaerobic filter (380 L) containing plastic pall rings with a specific surface area of 100 m^2m^{-3} (Fig. 4.9). Enhanced removal efficiency was achieved with average percentage removals in the final effluent of 73% for total

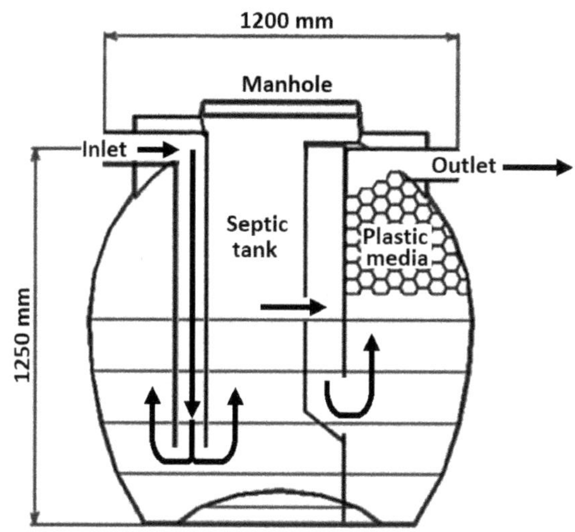

Figure 4.9. Combined septic tank and anaerobic biofilter (Sharma and Kazmi, 2015). Reproduced with permission of Elsevier Publishing.

COD, 78% BOD, and 83% TSS, with a 2.84 Log reduction in thermotolerant coliforms.

Except for the use of media for anaerobic biofilm development housed in the final chamber of a two or three-chamber septic tank, these adaptations have only been tested at the pilot level. The small size of conventional septic tanks would encounter problems if retrofitted, both due to the nature of the raw sludge and its accumulation, causing reduced flow rates into the tank, resulting in the backup of wastewater in the sewer pipe. Therefore, while these systems do work, they have to be designed as conventional anaerobic biofilm or floc-based reactors for individual houses or small communities, which would be more complex, and larger units than septic tanks and are explored in Chaps. 6 and 7, respectively. Even with modification, the final effluent will still need subsequent treatment using subsurface irrigation, a biological process, or a membrane system, before the effluent can be safely discharged. The majority of packaged plants for population equivalents of <20 employ an integrated septic tank (settlement and fermentation), with a secondary treatment option (Gray, 2017). It is likely that in warmer climates, anaerobic package plants for small communities may become the norm, coupled with treatment wetlands or ponds. Sabry (2010) describes an integrated UST and anaerobic baffle reactor in Eygpt that achieved an average COD, BOD, and TSS removal rate of 84, 88, and 89%, respectively. Once established, the baffled reactor acted as a polishing step for the UST. While TSS removal was unaffected by temperature, the difference between mean summer (35°C) and winter temperatures (22°C), both COD and BOD removal rates fell by 9%. In temperate and cold climates, septic tanks will remain primarily a settlement step prior to secondary treatment due to temperature. This has led to creating tanks with insulated walls and, even in some cases, heated ones.

4.2.4 *Microbial Community Structure of Anaerobic Sludge and Pathogen Removal*

There are remarkably few specific studies on the microbial community structure of septic tank sludges. Sharma and Kazmi (2015) produced

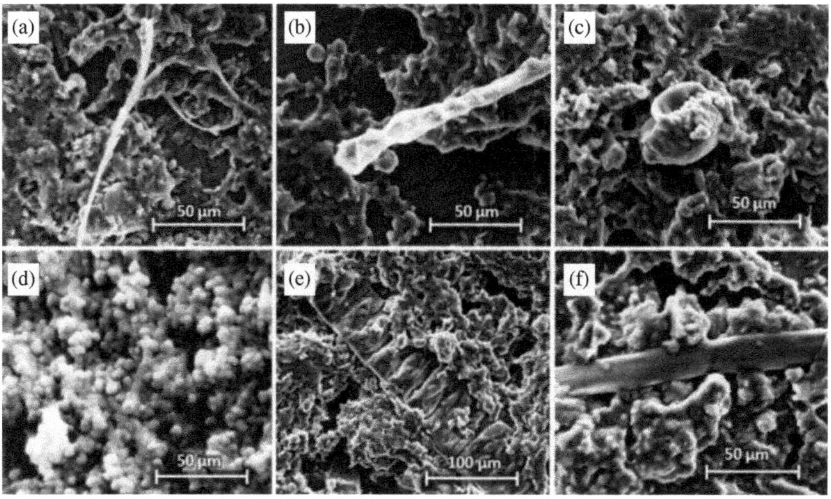

Figure 4.10. SEM images of different types of septic tank sludge bacteria identified by Sharma and Kazmi (2015): (a) inert material with aggregation of *Methanococcus*, (b) *Methanobrevibacter*-like species, (c) coccoid-shaped Archaeal species, (d) *Methanosarcina*-like species, (e) filamentous group *Methanothrix*-like species and (f) rod-shaped *Methanosaeta*-like species. Reproduced with permission of Elsevier Publishing.

some excellent SEM images of the surface sludge in their experimental full-scale septic tank in India, where the temperature varied from 22 to 30°C (Fig. 4.10). They used morphological characteristics to identify the genera and found methanogens very common, including *Methanosaeta, Methanothrix, Methanococcus, Methanospirillum, Methanobrevibacter*, and *Methanosarcina*. It is unclear whether this would be the case in temperate climates, where little methanogenic activity is observed. They also observed a strong correlation between BOD and TSS removal and the reduction in total and fecal coliforms and believed that TSS, in particular, could be used as an indicator of coliform removal.

In a comparative study of the influents and effluents from seven rural household three-chambered septic tanks in Tianjin, China, between 73.5 and 98.7% of microbes isolated were bacteria, with Archaea and viruses accounting for 0.6 and 1.5%, respectively (Tan *et al.*, 2021). All the bacteria were associated with human feces and sewage in general, specifically the phyla *Firmicutes* (28.9% ± 14.0%), *Proteobacteria* (25.2% ± 22,1%), and *Bacteroidetes* (16.4% ± 11.5%). This represented 110 microbial classes, with 59 shared by all seven households, with *Gammaproteobacteria*, *Clostridia*, and *Bacteroidia* as the most dominant, representing between 47 and 95% of the microbial community (Fig. 4.11). What was disturbing was that 104 pathogens and 83 virulence-related genes were also recorded from just the seven households, with each system having between 47 and 82 different pathogenic species. Thirty-two species were found in all samples, while 70 species were recorded in 50% of the samples tested. There was no significant difference in either diversity or abundance of pathogenic species between the influent and effluent, indicating that septic tanks are poor at removing pathogens, highlighting the high risk of contamination to ground and surface waters, especially when used for irrigation. The study highlights how vulnerable the environment is to contamination from septic tanks by a wide range of pathogens, which ultimately reflects the general health of the people served. In this study, *Enterobacteriaceae* (relative abundance of *Klebsiella pneumoniae* $(1.3 \pm 3.7) \times 10^{-3}$ % and *Escherichia coli* $6.4 (\pm 7.9) \times 10^{-4}$ %) were dominant with surprisingly *Chlamydia trachomatis*, an STI, $1.7 (+ 2.2) \times 10^{-3}$ %.The genus of *Pseudomonas* (e.g., *Pseudomonas syringae*, *P. aeruginosa*, *P. tolaasii*, *P. fuscovaginae*, *P. cichorii*, and *P. savastanoi*) was also common (Fig. 4.12). The study also showed that antibiotic-resistant genes (ARGs) proliferated in septic tanks, with a total of 441 ARGs conferring resistance to 26 antibiotic classes, allowing them to be disseminated into the environment and posing a significant risk to human and animal health when used for irrigation, which is common in China (Tan *et al.*, 2021).

4.3 Imhoff Tanks

These combined settlement and digestion units are similar to septic tanks, except they are specifically designed for sewage treatment plants rather than individual houses. They fill the design gap between septic tanks and small sedimentation tanks being used for small WWTPs. They were widely adopted throughout the world during the

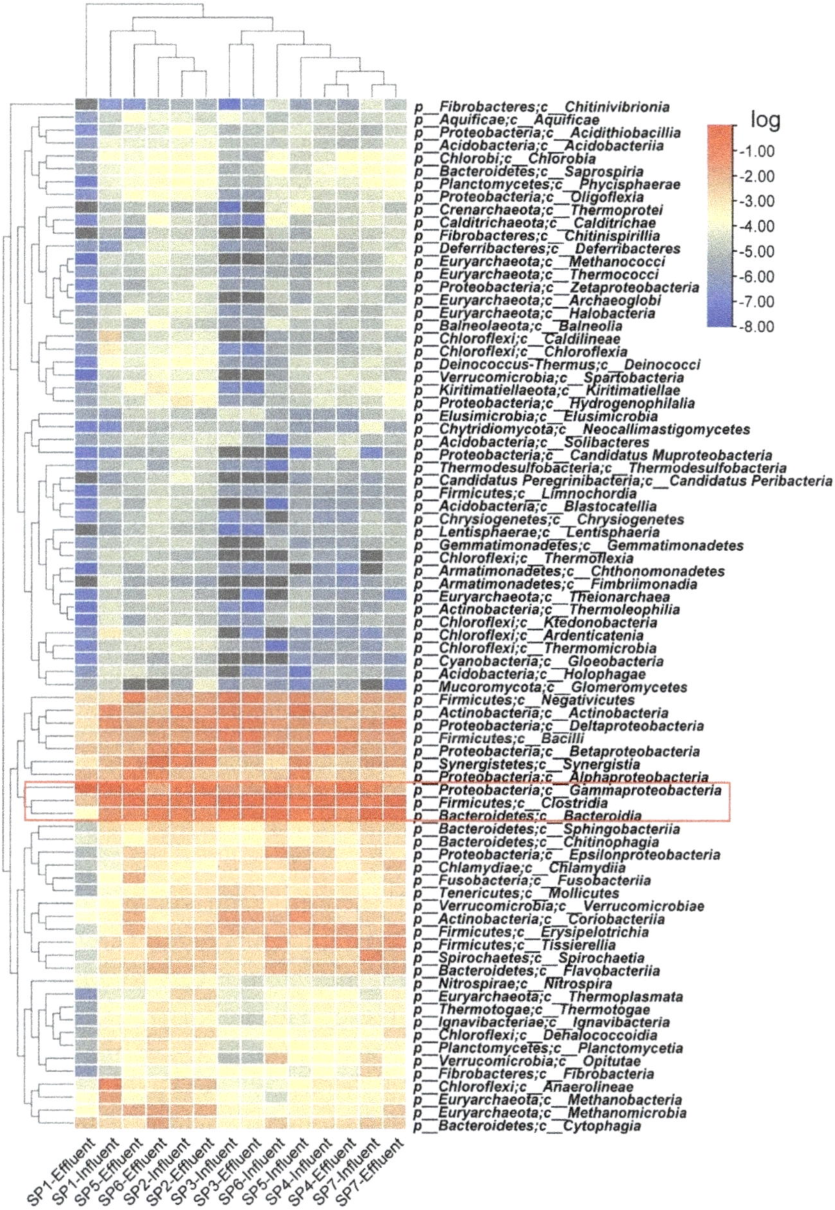

Figure 4.11. Community microbial structure at class level and clustering using the Euclidean distance of the influent and effluent from seven different septic tanks (SP1 to SP7) (Tan *et al.*, 2021). Reproduced with permission of Elsevier Publishing.

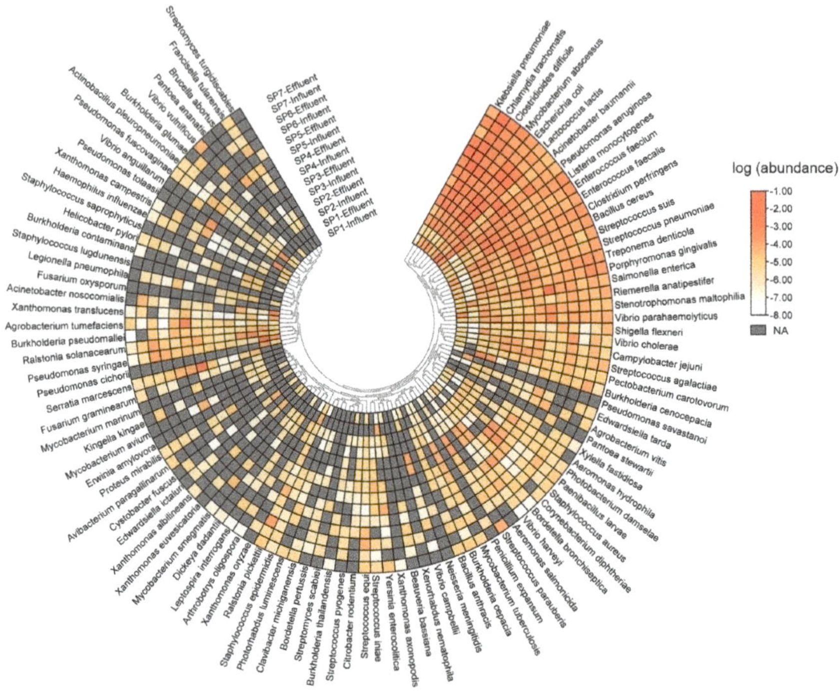

Figure 4.12. Heatmap and clustering using the Euclidean distance showing the abundance of pathogens recovered from the influent and effluent from seven different septic tanks (SP1 to SP7) (Tan *et al.*, 2021). Reproduced with permission of Elsevier Publishing.

1930s through to the 1960s for both domestic and industrial wastewaters but have been superseded by separate sedimentation and digestion units. Although simple in design, they are not easy to construct on-site in concrete, and soil stability is especially important, as Imhoff tanks are fairly deep. The cost of construction is the main reason for their selection as a primary step at small WWTPs. There has been a resurgence of interest in recent years due to new prefabricated designs using thinner reinforced concrete walls and stainless steel or aluminum internal settlement chambers allowing rapid installation, either as single units or several operated in parallel.

Imhoff tanks are rectangular in plan and comprise a settlement zone with a digestion chamber below (Fig. 4.13a). Single or double

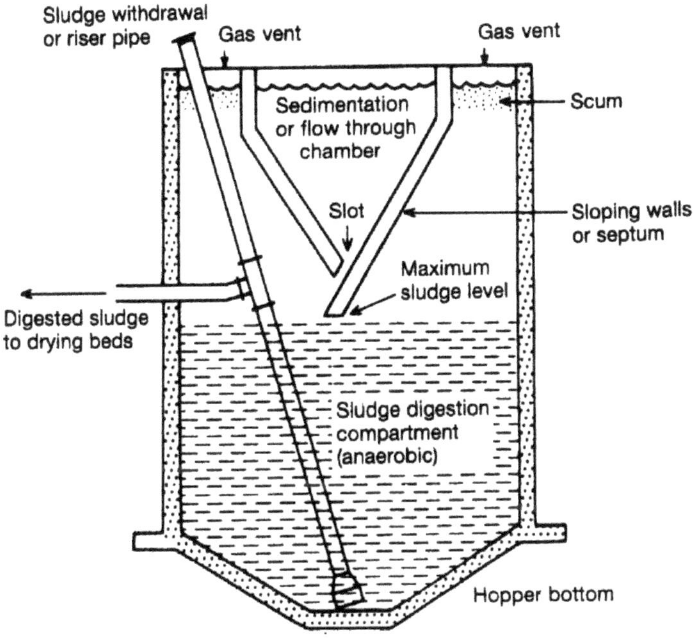

(a)

(b)

Figure 4.13. Imhoff tanks. (a) Design of a single tank and (b) double Imhoff tanks at Las Vegas, Honduras (Mikelonis *et al.*, 2010). Reproduced under the Creative Commons Attribution (CC-BY) license 4.0.

units were made in a variety of sizes, serving populations ranging from 300 to 10,000 (Fig. 4.13b). The advantage over conventional sedimentation tanks is that Imhoff tanks do not need to be desludged regularly (i.e., once or twice daily) to prevent anoxic conditions from developing, as the settlement chamber above is completely separated from the fermenting sludge below. Basic design and operational data are summarized in Table 4.6 (see page 131).

Screened sewage enters the V-shaped settlement chamber with a retention time of 2 to 4 hours, and as the solids settle, they fall down the inclined sides of the settlement unit and pass through the wide longitudinal slot that runs along the base of the settlement chamber and collects as a sludge in the digestion compartment below. The overlapping lip of the base of the settlement chamber prevents solids buoyed up by gas or gas bubbles themselves from entering the settlement chamber and interfering with the sedimentation process. These escape via two longitudinal vents that run alongside the outer wall of the settlement tank, where a surface scum slowly forms. The digestion compartment generally has a large storage capacity and sludge only needs to be removed at long intervals. The long sludge retention time allows digestion to take place, with any gas produced vented to the atmosphere (Fig. 4.13). In some warmer climates, digestion proceeds at a more rapid rate, allowing the gas to be collected. Imhoff tanks are more efficient than septic tanks, with enhanced settlement due to the better longitudinal flow. They are operated in the same way as a primary sedimentation tank, with a 50–70% reduction in suspended solids and a 30–50% reduction in BOD being possible. They are compact treatment units that require little maintenance, making them ideal for small communities. However, as digestion is psychrophilic, there is an incomplete breakdown of the sludge during winter, which produces odors and reduces gas production. In the summer, gas production can become excessive and interfere with settlement, causing foaming and bubbling.

The main problems associated with their operation in Ireland are: infrequent desludging, so that sludge is retained within the settlement chamber, reducing the time for settlement; hydraulic overloading again, reducing the retention time; the longitudinal slot at the base of the settlement chamber becoming blocked with large solids and grease; and while in the summer, gas accumulates under the hardened crust of scum, creating a potential safety hazard. Apart from proper operational management, the more interesting attempts to uprate Imhoff tanks have included converting them into rotating biological contactor (RBC) units (Gray, 2021). Modern prefabricated units are now available and are being used for small WWTPs and specific industrial applications (Fig. 4.14a and 4.14b)

4.4 Anaerobic Lagoons

4.4.1 *The Process*

Treatment in anaerobic lagoons relies on the development of a biologically active sludge layer. The sludge takes several months to build up before maximum biological activity is reached. Oxygen transfer through the air–water interface is undesirable and is minimized by reducing the surface area to volume ratio by increasing the depth of lagoons to between 2 and 5 m, effectively minimizing both oxygen transfer and heat loss. Construction of lagoons of this depth requires specialist engineering training and must be lined first with sand, and then several layers of welded butyl rubber or PVC sheeting to prevent any seepage to the unsaturated soil below that may ultimately result in groundwater pollution. Anaerobic lagoons need to be desludged occasionally, which may require the use of heavy machinery working from the edge of the lagoon, so simple earthen banks may not always be suitable, although smaller lagoons can be desludged using pumps. Sludge accumulation rates, sludge characteristics, and disposal are reviewed by Owusu-Twum and Sharara (2020).

As in septic tanks, three identifiable zones are observed in lagoons: the scum, the supernatant, which contains about 0.1% volatile solids, and the sludge, with 3–4% volatile solids (Fig. 14.15). The scum or grease layer, which forms on the surface of the lagoon, can become 40–60 mm thick and has a number of important functions. It insulates the lagoon, thus preventing heat loss and suppressing odors, but most importantly, maintains anaerobic conditions in the supernatant

(a)

(b)

Figure 4.14. Modern packaged Imhoff tanks. (a) View of the settlement area showing slits in the base for settleable solids to enter into the digestion chamber below and ventilation pipes for digester gas. (b) Prefabricated unit being lowered onto the concrete platform. Reproduced with permission of GAZEBO S.p.A., Gatteo, Italy. https://www.gazebo.it/

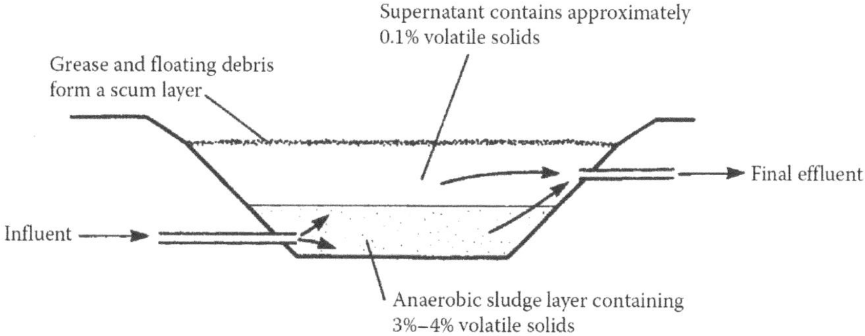

Figure 4.15. Basic design of an anaerobic lagoon used for treating strong organic wastewaters.

by eliminating oxygen transfer between the air–water interface. If the grease content of the wastewater is low (e.g., vegetable-processing wastewaters), then the lagoon surface can be covered using 50–100 mm-diameter HDPE plastic balls. To be as effective as a grease layer, at least 90% of the surface must be covered. Some animal wastewaters can also be problematic; for example, while cattle slurry rapidly forms a crust due to the high fiber content of their feces, pig slurry does not. The scum layer can become a breeding ground for a variety of flies, and in warmer climates, this can become a serious nuisance. Crust formation is quicker in summer, and in a well-constructed anaerobic lagoon, the BOD and suspended solids of the slurry will be reduced by 80–90%. During the summer, methane production can be high, and large pockets of gas will be seen breaking through the crust. Desludging of farm lagoons should be done every 3–5 years, depending on depth, total capacity, and degree of anaerobic degradation achieved, with the sludge spread directly onto the land without any of the associated pathogen problems. As most of these lagoons are between 2 and 3 m deep, they are potentially dangerous, and the crust rapidly becomes covered with grass, weeds, and even small bushes, giving it a false appearance of stability. Although in certain circumstances, they may be strong enough for an animal or even a child to walk on, they cannot support the weight of a small vehicle or possibly an adult. So, for safety, they should be securely fenced. The operation of anaerobic lagoons in series is not recommended as it is difficult to maintain an adequate scum cover on the second lagoon. The reduced loading results in the surface layer of the lagoon becoming aerobic; thus, the system becomes facultative rather than strictly anaerobic (Gray, 2022). Under-loaded lagoons will also tend to become facultative ponds.

Although the inlet pipe can be at the surface, anaerobic lagoons function most efficiently if the influent wastewater enters near the bottom of the basin. This prevents the short-circuiting of the liquid and ensures that it mixes with the microbial solids in the active anaerobic sludge layer. The flow pattern of the wastewater and the constant release of methane and carbon dioxide resulting from the degradation of organic matter ensure that some sludge particles remain in suspension within the supernatant. In this way, decomposition of organic matter continues, both in the liquid phase and within the sludge blanket; thus, a high mixing efficiency results in a high rate of BOD removal (Parker *et al.*, 1950). The discharge pipe is located just below the scum layer at the opposite end of the basin to ensure an upward flow pattern. This encourages the bacterial flocs to settle out of suspension, ensuring a clarified effluent and also eliminating the need for sludge return by retaining the microbial biomass in the lagoon. Full design details are given by the USEPA (1983; 2011).

4.4.2 *Microbiology*

Anaerobic lagoons have a widely diverse population of bacteria and archaea. In a study of four anaerobic lagoons treating swine wastewater, Ducey and Hunt (2013) identified 15,682 phylotypes that could be classified into 839 individual genera, with 223 genera common to all four lagoons. These belonged primarily to four major phyla: the Firmicutes (54.1%), Proteobacteria (15.8%), Actinobacteria (7.2%), and the Bacteroidetes (4.1%) (Fig. 14.16). These anaerobes are responsible for a wide range of functions, including fermentation, sulfate reduction, odor production, and both nitrogen and sulfur cycling. Twenty-five genera accounted for 20% of the operational taxonomic units (OTUs) measured in the four lagoons (Table 4.3). Each lagoon had a unique community structure controlled by key environmental

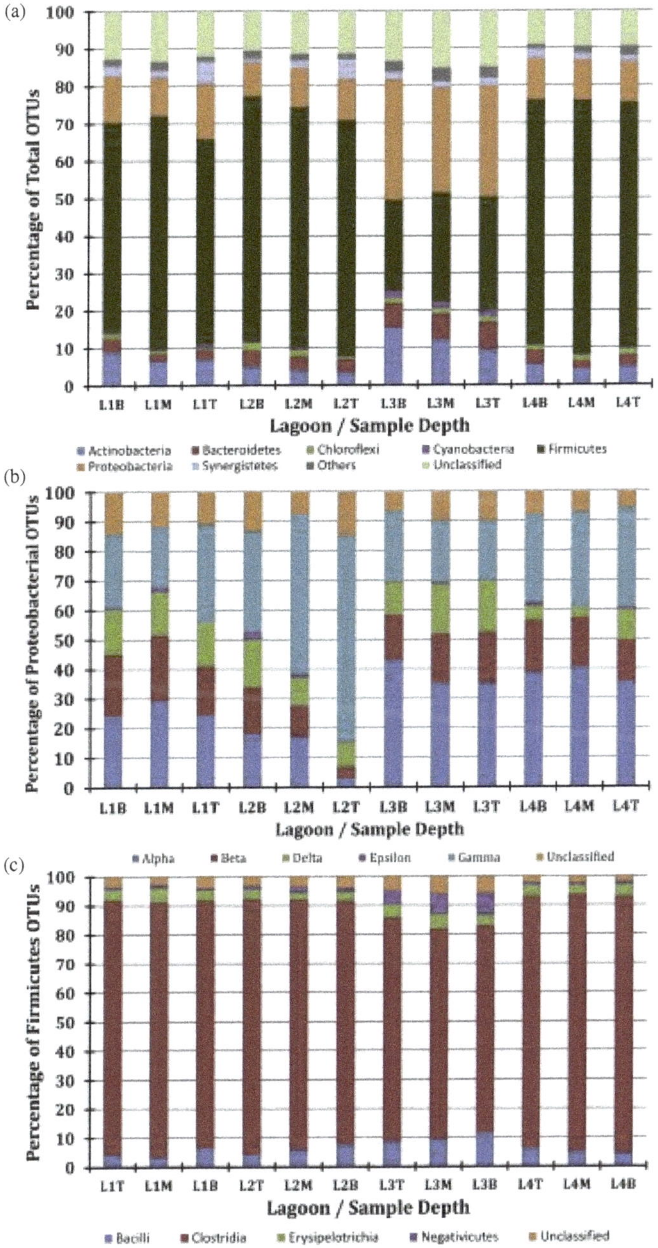

Figure 4.16. Relative abundance of bacterial 16S rDNA genes from swine wastewater anaerobic lagoons expressed as operational taxonomic units (OTUs) at (a) the phylum level, (b) breakdown of proteobacteria, and (c) Firmicutes (Ducey and Hunt, 2013). Reproduced with permission of Elsevier Publishing.

Table 4.3. The relative abundance (Ra) (%) and the number of operational taxonomic units (#OTUs) of the commonest 25 genera isolated from four anaerobic lagoons treating piggery wastewater (Ducey and Hunt, 2013). Reproduced with permission of Elsevier Publishing.

		Lagoon #1		Lagoon #2		Lagoon #3		Lagoon #4	
Phylum	**Genus**	**Ra**	**#OTUs**	**Ra**	**#OTUs**	**Ra**	**#OTUs**	**Ra**	**# OTU**
Firmicutes	*Anaerovorax*	2.58%	96	2.45%	126	1.94%	73	2.78%	85
Proteobacteria	*Thiolamprovum*	1.59%	59	3.81%	196	1.38%	52	1.14%	35
Firmicutes	*Clostridium* sp. cluster XI	3.14%	117	0.70%	36	0.40%	15	2.16%	66
TM7	TM7	0.89%	33	1.24%	64	1.04%	39	1.34%	41
Firmicutes	*Clostridium* (sensu stricto)	1.53%	57	0.64%	33	0.48%	18	1.63%	50
Firmicutes	*Oscillibacter*	0.43%	16	1.13%	58	0.85%	32	0.98%	30
Actinobacteria	*Leucobacter*	1.67%	62	0.45%	23	0.37%	14	0.78%	24
Proteobacteria	*Methylocystis*	0.11%	4	0.10%	5	2.77%	104	0.00%	0
Firmicutes	*Tissierella*	0.67%	25	0.86%	44	0.05%	2	1.18%	36
Firmicutes	*Acetanaerobacterium*	1.05%	39	0.56%	29	0.16%	6	1.05%	32
Firmicutes	*Turicibacter*	1.42%	53	0.40%	19	0.13%	5	0.85%	26
Firmicutes	*Saccharofermentans*	0.67%	25	0.70%	36	0.67%	25	0.33%	10
Firmicutes	*Lachnospiracea*	0.38%	14	0.80%	41	0.56%	21	0.62%	19
Actinobacteria	*Corynebacterium*	0.73%	27	0.52%	27	0.59%	22	0.56%	17
Firmicutes	*Sporobacter*	0.54%	20	0.76%	39	0.05%	2	0.75%	23

Synergistetes	*Aminobacterium*	0.73%	27	0.51%	26	0.24%	9	0.26%	8
Firmicutes	*Lactobacillus*	0.38%	14	0.45%	23	0.64%	24	0.29%	9
Proteobacteria	*Desulfomonile*	0.38%	14	0.06%	3	1.23%	46	0.20%	6
Proteobacteria	*Acinetobacter*	0.40%	15	0.45%	23	0.19%	7	0.59%	18
Actinobacteria	*Mycobacterium*	0.13%	5	0.08%	4	1.25%	47	0.03%	1
Cyanobacteria	Group IIa	0.00%	0	0.00%	0	1.52%	57	0.00%	0
Actinobacteria	*Actinomyces*	0.70%	26	0.25%	13	0.11%	4	0.39%	12
Proteobacteria	*Thioflavicoccus*	0.00%	0	0.00%	0	1.46%	55	0.00%	0
Bacteroidetes	*Prevotella*	0.16%	6	0.27%	14	0.53%	20	0.42%	13
Actinobacteria	*Klugiella*	0.02%	1	0.00%	0	1.09%	41	0.07%	2
Total		20.30%	755	17.19%	882	19.70%	740	18.40%	563

parameters, such as Kjeldhal nitrogen, COD, orthophosphate, total suspended solids, and dissolved oxygen (DO) (Ducey and Hunt, 2013). The microbial community structure changes seasonally, which may also affect odor production (Cook *et al.*, 2010).

In anaerobic lagoons, the organic substrate is degraded in the absence of oxygen to carbon dioxide and methane, with only a small amount of bacterial growth. Approximately 90% of the available chemical energy, in the form of organic material, is retained as methane production (McInerney *et al.*, 1980). Apart from the economic value of the methane gas produced (Sec. 4.4.8), anaerobic treatment has many advantages over aerobic treatment processes, such as less biomass produced per unit of substrate utilized (the lower biomass production means a lower requirement for nitrogen, phosphorus, and other nutrient and growth factors), higher organic loadings, as anaerobic processes are not limited by oxygen transfer rates, and lower constructional and operational costs. However, the major disadvantage in temperate and colder climates is the elevated temperatures required to maintain microbial activity at a reasonable level.

When operating under ideal conditions, the rate of volatile acid production equals the rate of conversion to methane. However, methanogenic activity is adversely affected by fluctuations in organic loading, temperature, and pH. So, if the methanogens are inhibited, then there is an immediate reduction in the conversion of volatile organic acids, leading to an increase in their concentration and a rapid reduction in pH, causing further inhibition of methanogenic activity, leading to process failure. Optimum (and maximum) operating conditions for methane formation are temperature 30–35°C (25–40°C), pH 6.8–7.4 (6.2–7.8), redox (or ORP) –520 to –530 mV (–490 to –550 mV), volatile organic acids 50–500 mg acetic acid L^{-1} (50–2,000 mg L^{-1}), and alkalinity 2,000–3,000 mg $CaCO_3$ L^{-1} (1,000–5,000 mg L^{-1}) (USEPA, 2011).

4.4.3 *Application*

Anaerobic lagoons are generally used as a preliminary treatment for strong organic wastes, with an 80% reduction in BOD and a 90%

reduction in suspended solids possible. This means that the lagoon only partially stabilizes the wastewater, and secondary treatment is required before the effluent can be discharged to natural waters — usually, this is achieved using an aerobic process either in the form of facultative ponds in series or activated sludge, although recent developments in granular anaerobic processes, in particular, those employing membranes, means that a full anaerobic treatment process could be feasible, enhancing carbon recovery through extra methane production and reducing energy use. However, anaerobic treatment can considerably reduce the organic loading to secondary treatment units, thus reducing the secondary treatment capacity required. If the subsequent aerobic treatment process is a facultative pond, then the sludge will be less likely to rise to the surface in warm weather as the sludge has already undergone stabilization by anaerobic breakdown. Most industrial ponds operate as anaerobic lagoons, and the process is widely used for the treatment of slaughterhouse waste, animal production, sugar beet, paper pulp, potato, and other food processing wastewaters (Porges and Brackney, 1962; Barnes *et al.*, 1984; Vuillot and Boutin, 1987). Wastewaters most amenable to stabilization by anaerobic breakdown in lagoons have a high organic strength, are rich in fats and protein, have a relatively high temperature, are free from toxic materials, especially heavy metals, and contain sufficient biological nutrients.

Slaughterhouse and meat-processing wastewaters are particularly suitable for treatment by anaerobic lagoons as they have a high BOD (~1,400 mg L^{-1}), grease content (~500 mg L^{-1}), temperature (~28°C), and a neutral pH. Minimum pre-treatment is required prior to discharge to an anaerobic lagoon, which should normally be limited to blood recovery, screening to remove coarse solids, and dissolved air flotation or skimming to remove excessive grease. If pre-treatment is too extensive, then insufficient grease may remain in the wastewater to form a scum when discharged to the lagoon. Domestic sewage is normally not suitable for treatment by this method due to a relatively low BOD concentration and grease content, although facultative ponds work well in warmer climates. When sewage has a high industrial content, anaerobic treatment often leads to serious odor problems due to the inhibition of certain microbial groups.

Anaerobic lagoons used for vegetable-processing wastewaters tend to be shallower in design and are widely used to treat sugar beet and potato wastewaters. The organic strength of sugar beet processing wastewaters can be very high, and although a 60–70% reduction of BOD and 80–90% reduction in suspended solids is possible, the final effluent is still far too strong (BOD >500 mg L^{-1}) for direct discharge to surface waters, requiring further aerobic biological treatment. These wastewaters are difficult to treat, as they are seasonal, with <100 days wastewater production per year as is the case with sugar beet processing. This can leave insufficient time for adequate active anaerobic biomass to accumulate, resulting in partial anaerobic breakdown and the release of odors. The shallow lagoons can also result in partial reaeration, while low temperatures and a high C:N ratio, common with vegetable wastewaters, all increase the inhibition of anaerobic activity.

In arid and semi-arid areas, anaerobic-lagoon-treated effluent may be used for irrigation, although it is more commonly further treated. In the US, new regulations governing wastewater irrigation have been introduced, which include the use of nutrient (N and P) management plans, which are also commonly used in Europe (Sharpley *et al.*, 2011). Microbial standards for wastewater reuse on agricultural land are not included, but the World Health Organization (WHO) has set standards based on fecal coliform counts of $\leq 1 \times 10^5$ 100 mL^{-1} (WHO, 2006). However, little attention has been paid to the transfer of fecal bacteria and zoonotic pathogens from anaerobic lagoons treating animal rearing and slaughterhouse wastewaters. In a study of water quality from an anaerobic lagoon treating piggery wastewater, McLaughlin *et al.* (2012) studied the seasonal variations in both nutrients and pathogens, recording higher nitrogen and carbon losses at temperatures above 19°C, with the N:P ratio falling from 9.7 to 2.8 during the warmer months when irrigation was used. Enterococci and zoonotic pathogens were present throughout the year but decreased at higher summer temperatures, while water-soluble metals, including Cu and Zn, increased. Temperature is a key factor in pathogen inactivation and removal, with aerobic systems more effective than anaerobic lagoons (Pandey *et al.*, 2016). A key nutrient in eutrophication is phosphorus, and the effluent from a piggery waste anaerobic lagoon is rich in this nutrient with N:P ratios normally between 2:1 and 4:1. In order to comply with crop requirements of a ratio of 4:1 to 9:1, either additional N must be added or the P concentration reduced. As both NH_4^+ and Mg^{2+} are common in the lagoon effluent, Nelson *et al.* (2003) have proposed P removal by struvite (magnesium ammonium phosphate hexahydrate $MgNH_4PO_4 \cdot 6H_2O$) precipitation, which can reduce effluent PO_4-P concentration to 2 mg L^{-1} as well also reducing the ammonium concentration and creating a useful fertilizer.

4.4.4 *Design of Anaerobic Lagoons*

The design loadings for stabilization ponds are normally measured in grams of BOD per unit area (g BOD $m^{-2}d^{-1}$), but as light is not an important factor in anaerobic lagoons, loading is expressed as BOD per unit volume (g BOD $m^{-3}d^{-1}$) or as the HRT ($m^{-3}h^{-1}$). The rate of treatment is temperature dependent (Table 4.4), so in temperate or cold climates, lagoons may have very reduced anaerobic activity in the winter, acting primarily as settlement ponds. BOD reduction is a function of both hydraulic retention time and temperature. The WHO (1987) has given different BOD reduction values at temperatures >20°C based on HRT as 1 day for 50%, 2.5 days for 60%, and 5 days for 70% reductions. Hammer (1977) gives a mean organic loading value of 320 g BOD $m^{-3}d^{-1}$ at a minimum temperature of 25°C to achieve 75% BOD removal at a minimum retention time of 4 days. Similar figures are given by Gloyna (1971), who cites examples where longer retention times of up to 19 days are needed for strong organic industrial wastewaters. The volume of the lagoon can be calculated as:

$$\lambda_v = L_i Q / V_a \text{ g m}^{-3}\text{d}^{-1} \quad (4.3)$$

where λ_v is the organic loading rate (g BOD $m^{-3}d^{-1}$), L_i is the influent wastewater BOD (mg L^{-1} or g m^{-3}), Q is the influent flow rate (m^3d^{-1}), and V_a is the volume of the lagoon (m^3) (Mara and Pearson, 1998).

Table 4.4. Design values calculated at various temperatures for the organic loading rate and percentage BOD removal in anaerobic lagoons.

Temperature (°C)	Organic loading rate (g BOD m^{-3} day^{-1}	% BOD removal
≤10	100	40
11	120	42
12	140	44
13	160	46
14	180	48
15	200	50
16	220	52
17	240	54
18	260	56
19	280	58
20	300	60
21	300	62
22	300	64
23	300	66
24	300	68
≥25	300	70

The value λ_v increases with temperature but lies between 100 and 350 g $m^{-3}d^{-1}$, the former to maintain anaerobic conditions and the latter to avoid odor release (Table 4.4). The mean temperature of the lagoon is used to select λ_v from Table 4.4, from which the lagoon volume (V_a) can be calculated using Eq. (4.3). The mean hydraulic retention time (θ_a) is calculated as:

$$\theta_a = V_a/Q \text{ days} \tag{4.4}$$

The major advantages of anaerobic systems are that less sludge is produced compared with an aerobic system, and no aeration equipment or power supply is required. The main operating limitations of the process are the temperature and strength of the wastewater. The temperature of the lagoon controls the BOD removal, and below

15°C, purification will be due to physical settlement only, with limited anaerobic breakdown of organic matter occurring. At higher temperatures, anaerobic digestion proceeds rapidly, with biogas (70% CH_4, 30% CO_2) bubbling to the surface. The same bacterial groups are involved as with other anaerobic systems. Thus, the process is inhibited by the same conditions (e.g., pH <6.2). In temperate climates, anaerobic lagoons act as settling tanks during the winter months, and the accumulated sludge is subsequently degraded during the warmer summer months. This was originally demonstrated by Parker *et al.* (1950), who recorded a 65–80% BOD removal in lagoons during the summer, with a retention period of 1.2 days, which fell to 45–65% removal in the winter, even after 5 days retention. Therefore, special attention should be paid to those factors affecting the temperature of the lagoon, such as ambient and wastewater temperature and insufficient scum development for insulation. A well-designed anaerobic lagoon will achieve, on average, 40% BOD removal at 10°C, rising to 60% at 20°C. The required hydraulic retention time is a function of temperature and wastewater strength. Low BOD and grease contents of wastewater will also reduce scum development; where no scum formation occurs, then a thin surface film of the algae *Chlamydomonas* often develops. Protection from the wind is important in preventing mixing and the resultant break-up of surface scum. Increasing concern over methane released into the atmosphere, which is a greenhouse gas, from lagoons will eventually lead to all lagoons being covered to capture the biogas (Sec. 4.4.8).

The large depth of anaerobic lagoons used to conserve heat and maintain anaerobicity can lead to the stratification of key nutrients, trace elements, and the microbial community (Lovanh *et al.,* 2009; Cook *et al.,* 2010). These effects have been shown to be both seasonal as well as temperature-dependent. Lagoons have been recorded as being up to 9 m in depth, resulting in some becoming vertically thermally stratified, which can result in the sludge layer remaining colder or warmer than the upper water phase, depending on the season. When located in exposed areas, lagoons are also affected by wind action, which causes mixing deep within the lagoon, resulting in the settled solids being mixed with the water phase. The temperature variation within lagoons is illustrated in Fig. 4.17.

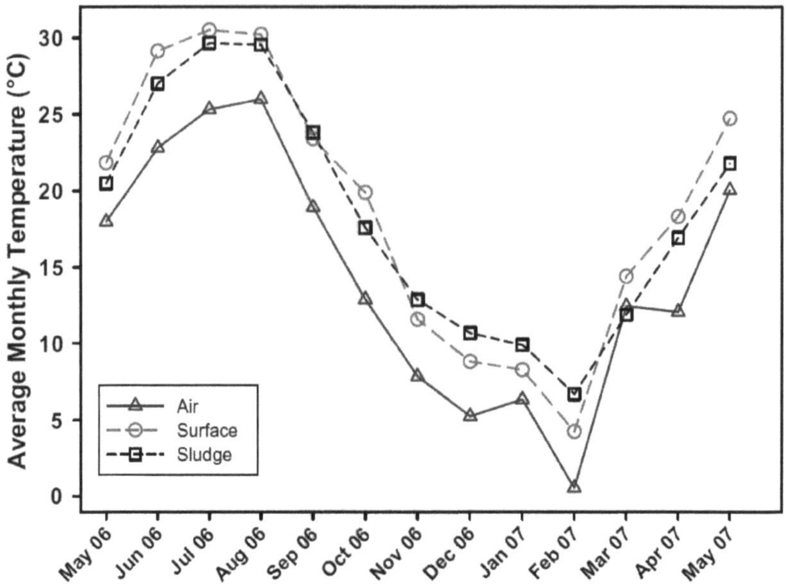

Figure 4.17. The temperature profile of an anaerobic lagoon (0.4 ha) treating piggery wastewater (approximately 2,000 sows) (Lovanh *et al.*, 2009). Reproduced with permission of Elsevier Publishing.

4.4.5 *Gas Emissions from Lagoons*

The final phase in anaerobic digestion is methane fermentation, where the end products of acid fermentation are converted to gases, mainly methane and carbon dioxide, but also ammonia and hydrogen sulfide, by several different species of obligate anaerobic archaea, see Eq. (4.5). At this stage, complete stabilization of the substrate occurs, and as the end products are only gases, it is more efficient than complete aerobic stabilization.

$$\text{Organic matter} \rightarrow CH_4 + CO_2 + H_2 + NH_3 + H_2S \qquad (4.5)$$

In anaerobic lagoons, the biogas production rate is a function of organic acid concentration and the lagoon temperature. Safley and Westerman (1988) studied a number of covered lagoons (Sec. 4.4.8) and found a minimum organic loading rate of 0·06 kg VS m^{-3} was required for consistent biogas production, with rates varying between

0·02 and 0·50 $m^3m^{-2}d^{-1}$, which is equivalent to 0·03 to 0·23 $m^3m^{-3}d^{-1}$, although the carbon dioxide content was generally less than 35%. These researchers later determined a relationship between methane production and lagoon temperature (range: 6.7 to 27.2°C) after a four-year study on an anaerobic lagoon treating dairy manure:

$$CH_4 = 89.514e^{-0.00381T} \quad (4.6)$$

where CH_4 is the percentage methane concentration and T is the lagoon temperature (°C) (Safley and Westerman, 1992).

Ammonia can be found at high concentrations in anaerobic lagoons and can, therefore, be an important local source of atmospheric ammonia. Aneja *et al.* (2001) studied a number of different anaerobic lagoons (i.e., primary, secondary, and tertiary) treating swine wastewater, with ammonia concentrations ranging from 37 to 909 mg N L^{-1}. The rate of ammonia volatilization was correlated with surface water temperature and the NH_x (i.e., NH_3 + NH_4^+) concentration in the lagoon, with the highest mean emission rate of 120.3 µg N m^{-2} min^{-1} (range 50–374 µg N m^{-2} min^{-1}) recorded in a primary lagoon and falling to the lowest rate 40.7 µg N m^{-2} min^{-1} (range 22–84 µg N m^{-2} min^{-1}) in a tertiary anaerobic lagoon over a temperature range of 10.3–23.3°C. They were able to model the emission rate from their anaerobic lagoons as:

$$\text{In}(NH_3\text{-N flux}) = 1.0788 + 0.0406\ T_L + 0.0015\ (NH_x) \quad (4.7)$$

where the NH_3-N flux is the loss of ammonia from the lagoon surface (µg N m^{-2} min^{-1}), T_L is the surface water temperature of the lagoon (°C), and NH_x is the total ammonia-nitrogen concentration (mg N L^{-1}). A similar study by Grant *et al.* (2016) expressed the annual mean ammonia emissions rate in terms of sow live weight, which ranged from 0.124 to 0.240 g N d^{-1} kg^{-1} for breeding and weaning units and 0.072 to 0.208 g N d^{-1} kg^{-1} for finishing units (i.e., growing on to market). They also produce several models: (i) incorporating wind and temperature, which accounted for 74% of emissions variability, and (ii) incorporating suspended organic concentration, pH, and nitrogen excretion rates, which accounted for 66% of the emissions

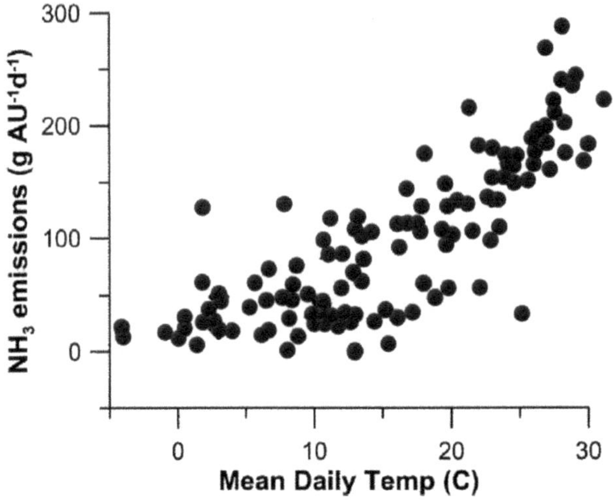

Figure 4.18. Influence of air temperature on emissions from an anaerobic lagoon. The unit AU represents 500 kg live pig weight (Grant *et al.*, 2016). Reproduced with permission of Elsevier Publishing.

variability. The solubility of NH_3 and dissociation of NH_4^+ to NH_3 in the lagoon liquid is temperature-dependent, resulting in more NH_3 volatilizing from the surface at higher temperatures, creating a seasonal variation in emissions (Fig. 14.18). There is also a significant diurnal variation in the release of ammonia into the atmosphere from lagoons, reaching a maximum during the warmest time of day (Aneja *et al.*, 2001). The temperature in the lagoon is directly related to air temperature, although the scum layer or cover, where used, does insulate the wastewater to a certain extent (Fig. 4.19). Grant *et al.* (2016) calculated that for each 1°C rise in the mean daily temperature, there is a corresponding 9.5% increase in maximum ammonia emissions and a 6.9% increase in annual emissions. Ammonia emissions from a dairy waste anaerobic lagoon were 55 kg NH_3 per cow per annum, which is 24% of the annual 180 kg N excreted per cow per annum (Rumburg *et al.*, 2008). An intetgrated physical/chemical model for anaerobic lagoons has also been produced by Wu and Chen (2011).

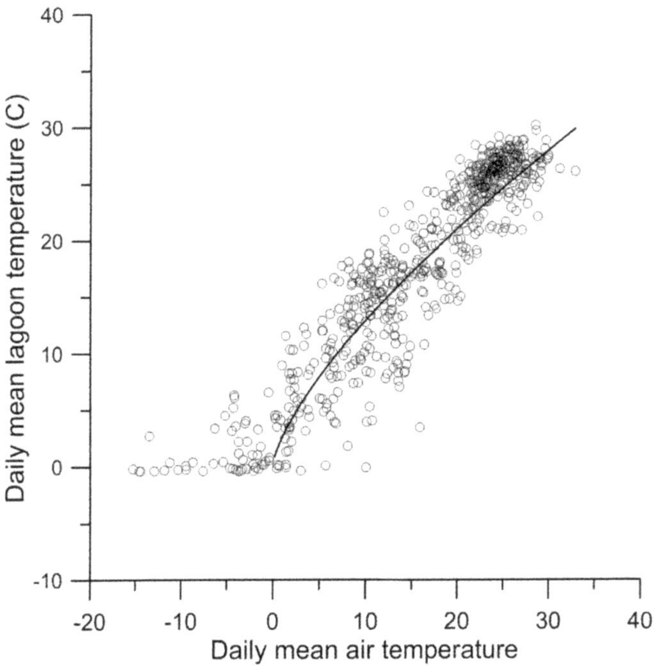

Figure 4.19. Correlation between the daily mean air temperature measured 2 m above the lagoon surface and the daily mean wastewater temperature measured 0.3 m below the lagoon surface (R^2 0.87) (Grant *et al.*, 2016). Reproduced with permission of Elsevier Publishing.

4.4.6 *Odors*

Many of the breakdown products of anaerobic fermentation in anaerobic lagoons are odorous. Odors are caused by the release of volatile or gaseous compounds into the atmosphere and can have very low odor thresholds, which is generally defined as the average concentration of the compound at which an odor is detected. The commonest odor encountered is the smell of rotten eggs caused by hydrogen sulfide, which is produced by sulfate reducing bacteria. The key organic acids produced during anaerobic breakdown are also odorous, and so when the final methanogenic phase is inhibited, these can cause odor problems.

Some food-processing wastewaters produce extremely strong odors, especially during treatment and storage (Gerick, 1984; Gray, 1988). For example, sugar beet wastewater undergoes partial anaerobic breakdown within the process and subsequently upon storage in lagoons, with the production of a variety of odors. This is due to the processing season taking place from October to January, with the cold temperatures inhibiting the full completion of the fermentation process. The major odors come from the volatile acids that comprise most of the organic fraction of the effluent. The odor threshold concentrations for the volatile organic acids produced during the treatment of sugar beet wastewater and waste are 24.3 ppm for acetic acid, 20.0 ppm for propionic acid, 0.05 ppm for iso-butyric acid, 0.24 ppm for butyric acid, 0.7 ppm for iso-valeric acid, and 3.0 ppm for valeric acid. Therefore, by measuring the volatile acid concentration and dividing it by the appropriate odor threshold concentration, a measure of the odor production or concentration, known as the odor number, can be calculated:

$$\textit{Odor number of wastewater} = \frac{\begin{array}{c}\textit{Compound concentration}\\ \textit{in wastewater}\end{array}}{\begin{array}{c}\textit{Odor threshold concentration}\\ \textit{for compound}\end{array}} \qquad (4.8)$$

Other odors associated with sugar beet processing include trimethylamine, which has a fishy odor, and organic sulfides, which produce a strong odor of rotting cabbage, as do the thiol compounds methyl mercaptan (CH_3SH) and ethyl mercaptan (C_2H_5SH), both of which have very low odor thresholds of 0.0011 ppm and 0.00019 ppm, respectively (Shore *et al.*, 1979). A detailed list of odor threshold values has been compiled by Fazzalari (1978), while a list of common odor compounds associated with anaerobic fermentation is given in Table 4.5.

Under normal operating conditions, no odors are released due to complete anaerobiosis and adequate scum cover. However, problems of odor because of sulfide formation at low temperatures or high sulfate or protein concentrations in the wastewater can occur. Sulfate can also be a problem in some paper mill wastes that use alum as a sizing agent.

Table 4.5. Some characteristic odors produced by compounds present in wastewaters. These degradation products can be categorized into two main groups, either degradation products of nitrogenous or sulfurous compounds. There are other odorous compounds, such as those associated with chlorine and phenolic wastes.

Compounds	General formulae	Odor produced
Nitrogenous		
Amines	CH_3NH_2, $(CH_3)_3N$	Fishy
Ammonia	NH_3	Ammoniacal, pungent
Diamines	$NH_2(CH_2)_4NH_2$, $NH_2(CH_2)_5NH_2$	Rotten flesh
Skatole	$C_8H_5NHCH_3$	Faecal, repulsive
Sulfurous		
Hydrogen sulfide	H_2S	Rotten eggs
Mercaptans	CH_3SH,$CH_3(CH_2)_3SH$	Strong decayed cabbage
Organic sulfides	$(CH_3)_2S$,CH_3SSCH_3	Rotten cabbage
Sulfur dioxide	SO_2	Pungent, acidic
Other		
Chlorine	Cl_2	Chlorine
Chlorophenol	$Cl.C_6H_4OH$	Medicinal, phenolic

Hydrogen sulfide is produced by the reduction of sulfate by a wide range of sulfate reducing bacteria that use the breakdown products of organic matter, in particular, acetate, butyrate, ethanol, lactate, and propionate, while using sulfate as a terminal electron acceptor. There are over 30 known genera of sulfate reducing bacteria in anaerobic lagoons, with *Desulfobacterium*, *Desulfobulbus*, and *Desulfovibrio* all commonly recorded (Zeikus, 1980; Rabus *et al.*, 2006; Cook *et al.*, 2008). Although present in relatively small numbers in wastewater, sulfate reducing bacteria rapidly grow under the ideal conditions found in anaerobic lagoons. For example, *Desulfovibrio desulfuricans* is present in low numbers in sewage (60–600 mL^{-1}) but rapidly increases on storage up to concentrations in excess of 100,000 mL^{-1} over 14 days. Ducey and Hunt (2013) recorded 26 genera of sulfate reducing bacteria in the four lagoons they studied. The genera *Desulfatirhabdium, Desulfobulbus, Desulfocurvus, Desulfomonile, Desulfovibrio, Desulfovirga, and Thermodesulfobium* were recorded in all four separate lagoons; *Desulfoglaeba, Desulfonispora, Desulfosporosinus, Desulfatiferula,* and *Desulforegula* in three lagoons; *Desulfobacca, Desulfofaba, Desulfofustis, Desulfococcus,* and *Desulfomicrobium* in two lagoons; while *Desulfarculus, Desulfonema, Desulforhabdus, Desulforhopalus, Desulfospira, Desulfovermiculus, Desulfurispora, Syntrophobacter,* and *Thermodesulforhabdus* were recorded in just one lagoon. The redox is fairly critical for sulfate reduction, with ORPs of <–50 mV normally required for hydrogen sulfide production, which normally occurs within the sludge layer. The mean concentration of hydrogen sulfide released with biogas (CH_4 80.4%; CO_2 14.7%) during a four-year study on an anaerobic lagoon treating dairy manure was 272 ppm over a lagoon temperature range of 6.7 to 27.2°C (Safley and Westerman, 1992).

With better design, it is now not always critical to have a surface scum as long as sufficiently high organic loading can be maintained to ensure anaerobic conditions. Hydrogen sulfide is present in aqueous solution as hydrogen sulfide gas, the bisulfide ion (HS^-), or the sulfide ion (S_2^-). Thus, for any given total sulfide concentration, the amount present as odorous hydrogen sulfide is pH dependent (Fig. 4.20). Mara and Pearson (1998) explain that designers have been reluctant

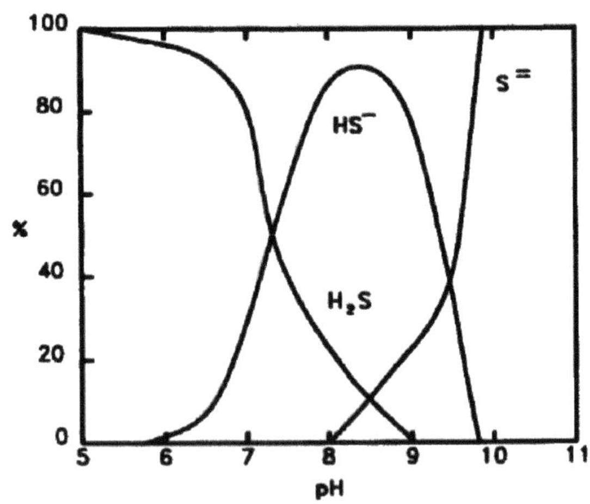

Figure 4.20. The effect of pH on the equilibrium of hydrogen sulfide–bisulfide–sulfide.

Table 4.6. Design and operational data for Imhoff tanks (Mikelonis *et al.*, 2010). Reproduced under the Creative Commons Attribution (CC-BY) license 4.0.

	Value	
Design parameter	**Range**	**Typical**
Settling compartment		
Overflow rate, peak hour $m^3/m^2/h$	1–1.75	1.36
Detention time, h	2–4	3
Length to width ratio	2:1–5:1	3:1
Slope of settling compartment ratio	1.25:1–1.75:1	1.5:1
Slope opening, cm	15–30	25
Slot overhang, cm	15–30	25
Scum baffle		
Below surface, cm	25–41	30
Above surface, cm	31	30
Freeboard, cm	45–60	60
Gas vent area		
Surface area, % total surface area	15–30	20
Width of opening, cm[1]	45–75	60
Digestion section		
Volume (unheated), storage capacity, m^3/capita		6 months sludge
Volume, m^3/6 month storage[2]	0.06–0.10	0.07
Sludge withdrawal pipe, cm	20–30	25
Depth below slot to top of sludge, m	0.30–0.90	0.60
Tank depth		
Water surface to tank bottom, m	7–10	9

[1] Minimum width of opening must be 45 cm to allow a person to enter for cleaning
[2] Based on a 6-month digestion period

to use anaerobic lagoons at treatment plants in case of odor formation. They stress that odor can be controlled by not exceeding the design loadings (Table 4.4) and ensuring that influent wastewater sulfate concentrations are <500 mg SO_4 L^{-1}. Sulfide reacts with heavy metals to form insoluble metal sulfides, but if concentrations exceed

<50 mg L^{-1}, there is a progressive inhibition of methanogenesis (Pfeffer, 1970). No odor should arise from a lagoon if the liquid temperature is above 20°C.

4.4.7 *Denitrification*

Jones *et al.* (2000) observed nitrogen gas emissions from anaerobic lagoons treating piggery wastewater. Under normal circumstances, denitrification requires the pre-step of nitrification, which is impossible under anaerobic conditions but may be possible under micro-aerobic conditions found at the air–water interface of open lagoons. A wide range of bacteria and archaea are able to denitrify under anoxic conditions, using NO_3^- and NO_2^- as terminal electron acceptors. Reduction of nitrate follows a sequence of steps, each mediated by a specific enzyme.

$$NO_3^- \text{ (aq)} \rightarrow NO_2^- \text{ (aq)} \rightarrow NO \text{ (g)} \rightarrow N_2O \text{ (g)} \rightarrow N_2 \text{ (g)} \qquad (4.9)$$

It is during incomplete denitrification that both NO and N_2O can be released into the atmosphere, causing damaging effects to the environment, both in terms of toxicity and global warming.

Denitrification

$$NO_3^- + 1.25CH_2O \rightarrow 0.5N_2 + OH^- + 0.75H_2 + 1.25CO_2 \qquad (4.10)$$

As can be seen from the equation above, denitrification releases OH^- ions into the effluent, increasing the alkalinity. However, under the strict anaerobic environment of the lagoons, nitrification would be very limited or, at best, incomplete. Jones *et al.* suggested alternative microbial-mediated denitrification processes, such as anaerobic ammonium oxidation, better known as anammox. Ammonium is used as an energy source by the anammox bacteria and carbon dioxide as a carbon source, with NO_2^- and NO_3^-, if present, used as electron acceptors (Gray, 2021). Kartel *et al.* (2011) have described the process in detail, where the enzyme nitrite oxidoreductase reduces nitrite to nitrogen oxide, which then reacts with ammonium to form hydrazine (N_2H_2), enabled by the enzyme hydrazine synthase, which is oxidized by hydrazine dehydrogenase to nitrogen. The bacteria are extremely slow-growing and strict anaerobes, unable to function if the dissolved oxygen concentration rises above 0.5 mg L^{-1}.

$$\underset{Anammox}{NH_4^+ + NO_2^- \rightarrow N_2 + 2H_2O} \quad (4.11)$$

The autotrophic bacteria responsible for the oxidation of ammonia and the simultaneous reduction of nitrite belong to the widely distributed phylum *Planctomycetes*. Not all the bacteria in this phylum are able to oxidize ammonium, this being restricted to the Anammox-*Planctomycetes*, which have a characteristic membrane-bound cell compartment called the anammoxosome (Fuerst and Sagulenko, 2011) (Sec. 6.5.3).

4.4.8 *Covered Lagoons for Biogas Collection*

In the early 1990s, with the increasing concern over the release of methane into the atmosphere, which has a high global warming potential (GWP), operators began to cover anaerobic lagoons (Green *et al.*, 1995; Hodgson and Paspaliaris, 1996). This was initially done using a sheet of geotextile membrane floated over the surface of the lagoon, with the biogas passively recovered and flared off (Fig. 4.21). Covering also means that scum formation or the use of plastic balls to act as a surface barrier is no longer required, and that odor control can be managed, opening up the process to a wider range of wastewaters that can be treated and also increasing its sustainability.

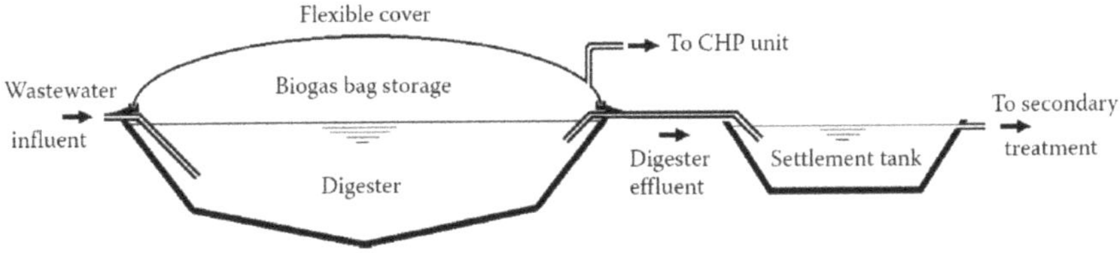

Figure 4.21. Covered anaerobic lagoons can be constructed of any size and are comparatively cheap to recover biogas from small to large-sized agricultural, industrial, and wastewater operations.

Figure 4.22. Floating scum can become compacted under covers of anaerobic lagoons due to methane gas pressure, occasionally causing covers to become damaged (Bobage *et al.*, 2022). Reproduced with permission of Elsevier Publishing.

Scum in anaerobic lagoons is primarily caused by methane bubbles bringing undigested material to the surface. The buoyancy of these solids causes surface compaction of the scum and an increase in its strength, which forces the scum against the covers (Bobage *et al.*, 2022). This can be a major problem by reducing hydrodynamic retention time, interfering with methane capture and removal, and damaging covers (Fig. 4.22).

Although mesophilic digestion using standard digesters has been used for agriculture, the cost and complexity of operation have seen its limited application. In the past decade, the use of farm-based covered anaerobic lagoons has spread throughout Europe, the US, Australia, New Zealand, and elsewhere for the treatment of animal wastewaters, primarily cattle and pig manures. This is hugely advantageous for both farmers and the environment. For the farmers, it provides safe storage, commercial biogas production, and also the onsite generation of electricity as well as heat, allowing farmers to diversify into new production areas, such as high-profit greenhouse fruit, herbs, and vegetables. Farm-based lagoons only need to be desludged every 3–5 years, depending on depth, total capacity in relation to loading, and degree of anaerobic breakdown achieved. A highly stabilized and largely odorless sludge is produced, which is generally used for land disposal (Owusu-Twum and Sharara, 2020). The effluent is rich in nutrients and can be used for surface irrigation of grassland for silage or hay.

Anaerobic lagoons are widely used in New Zealand for the treatment of predominantly dairy and also other animal wastewaters. Although originally open, these lagoons are increasingly being covered to collect the biogas to reduce greenhouse gas emissions as well as generate energy. Such ponds are operated below the optimum mesophilic temperature and so have a reduced volatile solids (VS) loading of 0.02–0.46 kg VS $m^{-3}d^{-1}$. Park and Craggs (2007) compared two anaerobic lagoons: one treating piggery wastewater (influent BOD 11,500 mg L^{-1}; BOD loading rate 0.30 kg BOD $m^{-3}d^{-1}$; VS 5,450 mg L^{-1}) and the other dairy wastewater (influent BOD 1,147 mg L^{-1}; BOD loading rate 0.01 kg BOD $m^{-3}d^{-1}$; VS 1,980 mg L^{-1}). Mean BOD and VS removal efficiency were 60.5 and 73.8% for the piggery and 81.1 and 78.5% for the dairy wastewater anaerobic lagoons, respectively. Mean biogas production 0.78 and 0.03 m^3m^{-2} d^{-1}, containing 72.0 and 80.3% methane from piggery and dairy wastewater, respectively, which is equivalent to 0.81 and 0.54 m^3CH_4 kg^{-1} VS.

Biogas production is closely linked to temperature, as is COD reduction and process stability. Optimal biogas production occurs in mesophilic systems at 38°C, which is rarely achieved under field conditions without the use of heat exchangers. In a study of covered anaerobic lagoons treating abattoir wastewater in Queensland, Australia, Schmidt *et al.* (2019) found that even the highest natural lagoon temperature of 31°C represented a reduction of potential biogas production of 30%. This increased to an 80% reduction at the lower winter temperature of 25.5°C. They concluded that the natural ambient temperatures found in covered anaerobic lagoons cannot ensure sustainable or profitable production of biogas and recommended the use of heat exchangers or waste heat, where available, to maximize biogas production. Some farms have incorporated passive solar heating to optimize the temperature. The move from low-rate anaerobic lagoons operated at ambient temperatures to high-rate lagoons maintained at 38°C maximizes the returns on investment (ROI) and overall confidence in the technology, as operating at higher temperatures also increases operational stability, improves effluent quality, and improves sludge quality.

Anaerobic lagoons are often used in series with a facultative pond, followed by at least one maturation pond. The loading to the facultative pond after the anaerobic stage is calculated using Table 4.4, which gives the expected percentage BOD reduction at the medium operating temperature.

References

Aneja, V.P., Bunton, B., Walkr, J.T. and Malik, B.P. (2001) Measurement and analysis of atmospheric ammonia emissions from anaerobic lagoons. *Atmospheric Environment*, **35**, 1949–1958.

Barnes, D., Forster, C.F. and Hrudey, S.E. (eds.) (1984) *Food and Allied Industries.* Pitman, London.

Bobage, V., Das, T., Usher, S.P., McMurrich, D., Strickland, A.D. and Eshtiaghi, N. (2022) Formation mechanism and mechanical properties of anaerobic lagoon scum. *Science of the Total Environment*, **843**, 156907. https://doi.org/10.1016/j.scitotenv.2022.156907

BSI (British Standards Institution). (1972) Small sewage treatment works. Code of Practice: CP 302. British Standards Institution, London.

BSI (British Standards Institution). (1983) Design and installation of small sewage treatment works and cesspools. Code of Practice: CP6297, British Standards Institution, London.

BSI (British Standards Institution). (2007) Code of practice for the design and installation of drainage fields for use in wastewater treatment, BS 6297:2007 +A1, British Standards Institution, London.

CEN. (2000) Small wastewater treatment systems for up to 50 PT — Part 1: Prefabricated septic tanks, Standard: EN 12566-1:2000, Comité Européen de Normalisation (CEN), Brussels.

CEN. (2003) Small wastewater treatment systems for up to 50 PT — Part 1: Prefabricated septic tanks, Standard: CEN 12566-1:2000/A1:2003, Comit é Europ é en de Normalisation (CEN), Brussels.

CEN. (2005) Small wastewater treatment systems for up to 50 PT — Part 2: Soil infiltration systems, Standard: EN 12566-2:2005, Comit é Européen de Normalisation (CEN), Brussels.

CEN. (2007) Small wastewater treatment systems for up to 50 PT — Part 4. Septic tanks assembled in situ from prefabricated kits, Standard: EN 12566-4:2007, Comit é Européen de Normalisation (CEN), Brussels.

CEN. (2008) Small wastewater treatment systems up to 50 PT — Part 5: Pre-treated effluent filtration systems, Standard: EN 12566-5:2008, Comit é Européen de Normalisation (CEN), Brussels.

CEN. (2009) Small wastewater treatment systems for up to 50 PT — Part 3. Packaged and/or site assembled domestic wastewater treatment plants, Standard: EN 12566-3:2005 +A1:2009, Comit é Europ é en de Normalisation (CEN), Brussels.

CEN. (2013a) Small wastewater treatment systems for up to 50 PT — Part 6: Prefabricated treatment units for septic tank effluent, Standard: EN 12566-6:2013, Comit é Européen de Normalisation (CEN), Brussels.

CEN. (2013b) Small wastewater treatment systems for up to 50 PT — Part 7: Prefabricated tertiary treatment units, Standard: EN 12566-7:2013, Comit é Européen de Normalisation (CEN), Brussels.

Chen, Z., Wen, Q., Guan, H., Bakke, R. and Ren, N. (2014) Anaerobic treatment of domestic sewage in modified septic tanks at low temperatures. *Environmental Technology*, **35**, 2123–2131. https://doi.org/10.1080/09593330.2014.894580

Coelho, A.L.S.S., do Nascimento, M.B.H., Cavalcanti, P.F.F. and van Haandel, A.C. (2003) The UASB reactor as an alternative for the septic tank for on-site sewage treatment. *Water Science and Technology*, **48**: 221–226.

Cook, K.L., Whitehead, T.R., Spence, C. and Cotta, M.A. (2008) Evaluation of the sulfate reducing bacterial population associated with stored swine slurry. *Anaerobe*, **14**, 172–180.

Cook, K.L., Rothrock, M.J., Lovanh, N., Sorrell, J.K. and Loughrin, J.H. (2010) Spatial and temporal changes in the microbial community in an anaerobic swine waste treatment lagoon. *Anaerobe*, **16**, 74–82. doi:10.1016/j.anaerobe.2009.06.003

DAFM. (2022) Minimum specification for bovine livestock units and reinforced tanks. Report S123, Department of Agriculture Food and the Marine, Dublin. https://assets.gov.ie/95096/ffe3a0aa-aae6-481d-ae3c-5187c8aeeaa1.pdf

Ducey, T.F. and Hunt, P.G. (2013) Microbial community analysis of swine wastewater anaerobic lagoons by next-generation DNA sequencing. *Anaerobe*, **21**, 50–57.

Elmitwalli, T. (2013) Sludge accumulation and conversion to methane in a septic tank treating domestic wastewater or black water. *Water Science and Technology*, **68**, (4), 956–964.

EPA. (2009) Code of practice for wastewater treatment and disposal systems serving single house, Environmental Protection Agency, Dublin.

EPA. (2021) Code of practice for domestic waste water treatment systems (Population equivalent ≤10). Environmental Protection Agency, Ireland.

Fazzalari, F.A. (ed.) (1978) Compilation of odor and taste threshold values data, ASTM Data Series DS 48A. American Society for Testing and Materials, Philadelphia.

Fuerst, J.A. and Sagulenko, E. (2011) Beyond the bacterium: Planctomycetes challenge our concepts of microbial structure and function. *Nature Reviews Microbiology*, **9**, 403–413.

Gerick, J.A. (1984) Land application odour study. *Journal of Water Pollution Control Federation*, **56**, 287–291.

Gloyna, E.F. (1971) Waste stabilization ponds. WHO Monograph Series, 60, World Health Organization, Geneva.

Grant, R.H., Boehm, M.T. and Heber, A.J. (2016) Ammonia emissions from anaerobic waste lagoons at pork production operations: Influence of climate. *Agricultural and Forest Meteorology*, **228–229**, 73–84. http://dx.doi.org/10.1016/j.agrformet.2013.06.006

Gray, N.F. (1988) *Sugar Beet Processing in Ireland: A Case Study*. Water Technology Press, Dublin.

Gray, N.F. (1995) The influence of sludge accumulation rate on septic tank design. *Environmental Technology*. **16**, (8), 795–800. https://doi.org/10.1080/09593331608616318

Gray, N.F. (2017) *Water Science and Technology: An Introduction*. CRC Press, Boca Raton, FL. https://doi.org/10.1201/b22082

Gray, N.F. (2021) *Fixed-Film Reactors in Wastewater Treatment*. World Scientific Publishing, Singapore.

Gray, N.F. (2022) *Natural Wastewater Treatment Systems and Sustainability*. World Scientific Publishing, Singapore.

Gray, N.F. (2023) *Activated Sludge: Developments and Sustainable Solutions*. World Scientific Publishing, Singapore.

Green, F.B., Lundquist, T.J. and Oswald, W.J. (1995) Energetics of advanced integrated wastewater pond systems. *Water Science and Technology*, **31**, (12), 9–20.

Hodgson, B. and Paspaliaris, P. (1996) Melbourne water's wastewater treatment lagoons: Design modifications to reduce odours and enhance nutrient removal. *Water Science and Technology*, **33**, (7), 157–164.

Jones, M.L., Liehr, S.K., Classon, J.J. and Robarge, W. (2000) Mechanisms of dinitrogen gas formation in anaerobic lagoons. *Advances in Environmental Research*, **4**, 133–139.

Kartel, B., Maalcke, W.J., de Almeida, N.M., Cirpus, I., Gloerich, J., Geerts, W., den Camp, H.J.M.O., Harhangi, H.R., JanssenMegens, E.M., Francoijs, K.-J., Stunnenberg, H.G., Keltjens J.T., Jetten, M.S.M. and Strousm, M. (2011) Molecular mechanism of anaerobic ammonium oxidation. *Nature*, **479**, 127–130.

Kujawa-Roeleveld, K., Fernandes, T., Wiryawan, Y., Tawfik, A., Visser, M. and Zeeman, G. (2005) Performance of UASB septic tank for treatment of concentrated black water within DESAR concept. *Water Science and Technology*, **52**, 307–313.

Lovanh, N., Loughrin, J.H., Cook, K., Rothrock, M. and Sistani, K. (2009) The effect of stratification and seasonal variability on the profile of an anaerobic swine waste treatment lagoon. *Bioresource Technology*, **100**, 3706–3712. https://doi.org/10.1016/j.biortech.2008.09.024

Luostarinen, S., Sanders, W., Kujawa-Roeleveld, K. and Zeeman, G. (2007) Effect of temperature on anaerobic treatment of black water in UASB-septic tank systems. *Bioresource Technology*, **98**, 980–986.

MacMahon, J., Knappe, J. and Gill, L.W. (2022) Sludge accumulation rates in septic tanks as part of the on-site treatment of domestic wastewater in a northern maritime temperate climate. *Journal of Environmental Management*, **304**, 114199. https://doi.org/10.1016/j.jenvman.2021.114199

Mann, H. T. (1979) Septic tanks and small sewage treatment plants. Technical Report TR107. Water Research Centre, Stevenage.

Mara, D.D. and Pearson, H.W. (1998) *Design Manual for Waste Stabilization Ponds in Mediterranean Countries.* Lagoon Technology International, Leeds, UK.

McInerney, M. L., Bryant, M. P., and Stafford, D. A. (1980) Metabolic stages and energetics of microbial anaerobic digestion. In *Anaerobic Digestion* (eds. D.A. Stafford, B.I. Wheatley and D.E. Hughes). Applied Science Publishers, London, pp. 91–98.

McLaughlin, M.R., Brooks, J.P. and Adeli, A. (2012) Temporal flux and spatial dynamics, fecal indicators, and zoonotic pathogens in anaerobic swine manure lagoon water. *Water Research*, **46**, 4949–4960. https://doi.org/10.1016/j.watres.2012.06.023

Mikelonis, A., Herrera, A.. Adams, E. and Hodge, M. (2010) Honduran Imhoff tanks: Potentials and pitfalls. *Journal of Water Management Modeling*, R236-22. https://doi.org/10.14796/JWMM.R236-22

Moussavi, G., Kazembeigi, F. and Farzadkia, M. (2010) Performance of a pilot scale up-flow septic tank for on-site decentralized treatment of residential wastewater. *Process Safety and Environmental Protection*, **88**, 47–52.

Nelson, N.O., Mikkelsen, R.L. and Hesterberg, D.L. (2003) Struvite precipitation in anaerobic swine lagoon liquid: Effect of pH and Mg:P ratio and determination of rate constant. *Bioresource Technology*, **89**, 229–236. doi:10.1016/S0960-8524(03)00076-2

Owusu-Twum, M.Y and Sharara, M.A. (2020) Sludge management in anaerobic swine lagoons: A review. *Journal of Environmental Management*, **271**, 110949. https://doi.org/10.1016/j.jenvman.2020.110949

Pandey, P.R., Cao, W., Vaddella, V., Castillo, A.R., Sousa, A. and del Rio, N.S. (2016) Simulating the effects of mesophilic anaerobic and aerobic digestions, lagoon system and composting on pathogen inactivation. *Ecological Engineering*, **97**, 633–641. http://dx.doi.org/10.1016/j.ecoleng.2016.10.047

Park, J.B.K. and Craggs, R.J. (2007) Biogas production from anaerobic waste stabilization ponds treating dairy and piggery wastewater in New Zealand. *Water Science and Technology*, **55**, (11), 257–264.

Parker, C. D., Jones, H.L. and Taylor, W.S. (1950). Purification of sewage in lagoons. *Sewage and Industrial Waste*, **22**, 760–775.

Pfeffer, J.T. (1970) Anaerobic lagoons: Theoretical considerations. In *Proceedings of the Second International Symposium on Water Treatment Lagoons* (ed. R.E. McKinney). University of Kansas, Laurence, KS, USA, pp. 310–320.

Philip, H., Maunoir, S., Rambaud, A. and Philippi, L.S. (1993) Septic tank sludges: Accumulation rate and biochemical characteristics. *Water Science and Technology*, **28**, (10), 57–64. https://doi.org/10.2166/wst.1993.0205

Porges, R. and Brackney, D. H. (1962) Industrial waste disposal by stabilization ponds. *Engineering Experimental Station Bulletin*, **68**, 131–140, Louisiana State University.

Rabus, R., Hansen, T.A. and Widdel, F. (eds.) (2006) *Dissimilatory Sulfate- and Sulfur Reducing Prokaryotes.* 3rd edition. Springer Science, New York, NY.

Rumburg, B., Mount, G.H., Yonge, D., Lamb, B., Westberg, H., Neger, M., Filipy, J., Kinacid, R. and Johnson, K. (2008) Measurements and modelling of atmospheric flux of ammonia from an anaerobic dairy waste lagoon. *Atmospheric Environment*, **42**, 3380–3393. doi:10.1016/j.atmosenv.2007.02.046

Sabry, T. (2010) Evaluation of decentralized treatment of sewage employing upflow septic tank/baffled reactor (USBR) in developing countries. *Journal of Hazardous Materials*, **174**, 500–505. https://doi.org/10.1016/j.jhazmat.2009.09.080

Safley, L.M. and Westerman, P.W. (1988) Biogas production from anaerobic lagoons. *Biological Wastes*, **23**, (3), 181–193. https://doi.org/10.1016/0269-7483(88)90033-X

Safley, L.M. and Westerman, P.W. (1992) Performance of a dairy manure anaerobic lagoon. *Bioresource Technology*, **42**, 43–52. https://doi.org/10.1016/0960-8524(92)90086-D

Schmidt, T., Harris, P., Lee, S., and McCabe, B.K. (2019) Investigating the impact of seasonal temperature variation on biogas production from covered anaerobic lagoons treating slaughterhouse wastewater using lab scale studies. *Journal of Environmental Chemical Engineering*, 7, 103077. https://doi.org/10.1016/j.jece.2019.103077

Sharma, M.K. and Kazmi, A.A. (2015) Anaerobic onsite treatment of black water using filter-based packaged system as an alternative of conventional septic tank. *Ecological Engineering*, **75**, 457–461.

Sharpley, A., Beegle, D., Bolster, C., Good, L., Joern, B., Ketterings, Q., Lory, J., Mikkelsen, R., Osmond, D. and Vadas, P. (2011) Revision of the 590 nutrient management standard: SERA-17 recommendations. *Southern Cooperative Series Bulletin*. **412**. https://sera17dotorg.files.wordpress.com/2015/02/590-sera-17-recommendations.pdf

Shore, M., Dutton, J.V., Broughton, N.W. and Bumstead, N. (1979) The effect of pH on the odours of transport and storage waters. British Sugar Corporation Research Laboratories Report 79/36. Annual Works Managers Meeting, British Sugar Corporation Ltd, Norwich.

Singh, R.P., Kun, W. and Fu, D. (2019) Designing process and operational effect of modified septic tank for the pre-treatment of rural domestic sewage. *Journal of Environmental Management*, **251**, 109552. https://doi.org/10.1016/j.jenvman.2019.109552

Tan, L., Zhang, C., Liu, F., Chen, P., Wei, X., Li, H., Yi, G., Xu, Y. and Zheng, X. (2021) Three-compartment septic tanks as suitable on-site treatment facilities? Watch out for the potential dissemination of human-associated pathogens and antibiotic resistance. *Journal of Environmental Management*, **300**, 113709. https://doi.org/10.1016/j.jenvman.2021.113709

USEPA (1983) Design manual: Municipal wastewater stabilization ponds. EPA-625/1-83-015, Office of Water, U.S. Environmental Protection Agency, Washington, DC., USA.

USEPA (2011) Principles of design and operation of wastewater treatment pond systems for plant operators, engineers and managers. EPA/600/R-10/088, Office of Research and Development, U.S. Environmental Protection Agency, Cincinnati, Ohio, USA.

Vuillot, M. and Boutin, C. (1987) Waste stabilization ponds in Europe: A state of the art review. *Water Science and Technology*, **19,** (12), 1–6. https://doi.org/10.2166/wst.1987.0119

WHO (1987) *Wastewater Stabilization Ponds: Principles of Planning and Practice*, Technical Publication: 10, Regional Office for the Eastern Mediterranean, World Health Organization, Alexandria.

WHO (2006) *Wastewater Use in Agriculture: WHO Guidelines for the Safe Use of Wastewater, Excreta and Greywater*, Vol. 2, World Health Organization, Geneva, Switzerland. https://www.who.int/water_sanitation_health/publications/gsuweg2/en/

Wu, B. and Chen, Z. (2011) An integrated physical and biological model for anaerobic lagoons. *Bioresource Technology*, **102**, 5032–5038. doi: 10.1016/j.biortech.2011.01.076

Zeikus, J.G. (1980) Microbial populations in digesters. In *Anaerobic digestion* (eds. D.A. Stafford, B.I. Wheatley, and D.E. Hughes). Applied Science Publishers, London, pp. 61–89.

5

Sludge Digesters

5.1 Introduction

Modern wastewater treatment practice separates primary settlement from digestion, with septic tanks and Imhoff tanks now superceeded with larger and more efficient primary settlement tanks, with the settled sludge then being mixed with the settled waste biomass from the secondary biological unit process, which is then thickened before being anaerobically digested in a stand-alone reactor. Digestion is carried out either in large open tanks or lagoons (Sec. 4.4) at ambient temperature (psychrophilic digestion) or more rapidly in covered tanks heated to between 30 and 35°C (mesophilic digestion) or at 50–70°C (thermophilic digestion). Psychrophilic digesters are usually unmixed, and, as they are often uncovered or vented, any gas produced is dispersed to the atmosphere as it is uneconomic to collect it. In winter, the rate of digestion will be extremely low or even zero, whereas in the summer, digestion may become rapid, inducing mixing as the gas rises. Therefore, long retention times are required of between 6 and 12 months to stabilize sludges under these conditions and to balance between sludge accumulation and sludge degradation. Psychrophilic digesters are restricted to smaller works, where the output of sludge is low, and land for sludge disposal is readily available. Both septic tanks and Imhoff tanks fall within the psychrophilic range, which operates from 5 to 25°C.

Heated digesters are much more cost-effective at larger treatment plants, where there is sufficient sludge available to ensure a continuous

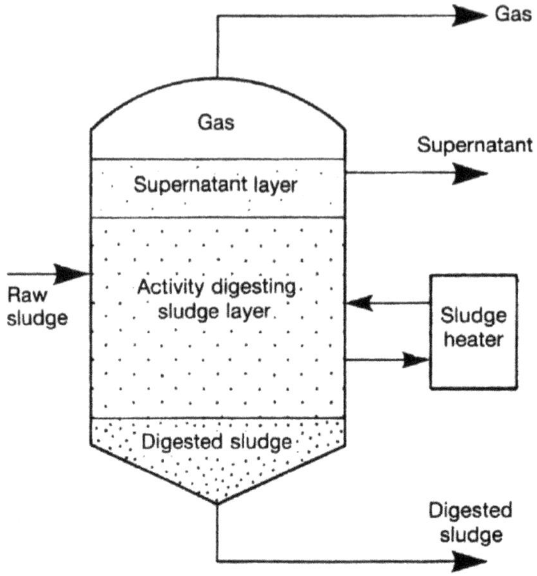

Figure 5.1. Single-stage anaerobic sludge digester.

operation, with gas collected and used to either directly or indirectly heat the digesters. In their simplest form, digesters are intermittently fed with no recycling. They can be a single-stage reactor (Fig. 5.1), but in the United Kingdom (UK) and Ireland, conventional digestion is usually a two-stage process, with the primary digester heated to the desired temperature to allow for optimum anaerobic activity, with acid formation and biogas production occurring simultaneously (Fig. 5.2). The primary reactor is continuously stirred, unlike the earlier stratified digesters (Fig. 5.3), which reduces the retention time from 30 to 60 d to <15 d. The secondary digester is unheated and can be used for two functions. Either to continue digestion under psychrophilic conditions, in which case it will be stirred to ensure complete mixing as in the primary digester, or it can be used for sludge separation. However, as the evolution of gas interferes with settlement, the second reactor cannot be used for both functions. When used for sludge separation, the secondary digester is not stirred and is generally uncovered (Fig. 5.4). Under quiescent conditions, the solids separate, leaving a strong supernatant liquor on top of the stabilized sludge. The liquor

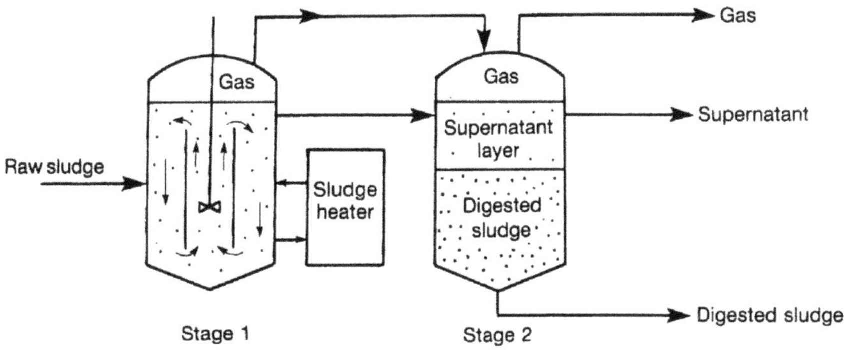

Figure 5.2. Two-stage anaerobic digester.

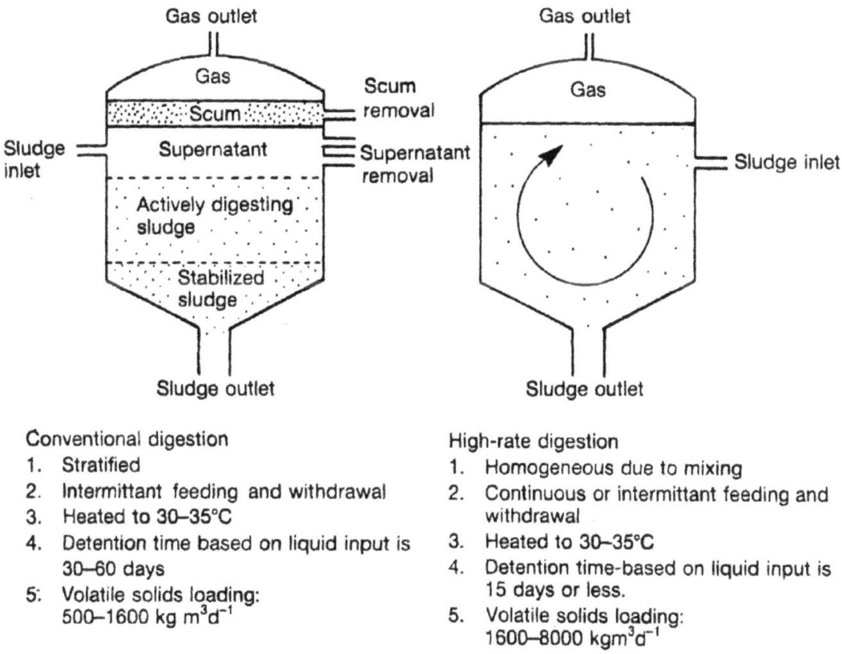

Figure 5.3. Comparison of conventional and high-rate digestion of sludge.

has a high soluble organic content with a high biochemical oxygen demand (BOD), which needs to be recirculated back to the inlet of the treatment plant. The settled solids are removed, dewatered, and disposed. When the secondary digester is used for digestion, a third

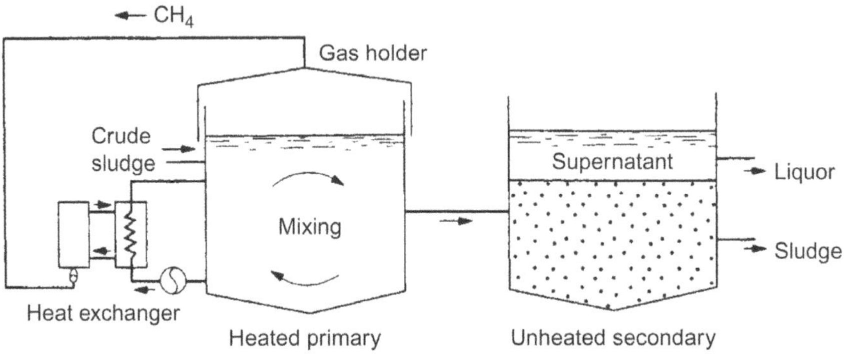

Figure 5.4. Conventional sludge digestion plant.

digester may be employed if thickening is required. This type of frequently fed reactor with no recycling of solids is used for the treatment of sludges with high solids content, such as animal slurries and sewage sludge. For example, the waste from a 300-pig unit was sucessfully treated in such a reactor with a hydraulic retention time (HRT) of 10 days, with a gas yield of 0.3 m^3 kg total solids d^{-1} (Hobson, 1984). Sludges and liquors with a lower solids concentration can only be digested in a conventional reactor where there is some form of solids recycle to prevent wash-out of bacteria. Generally, such systems have a short HRT of 0.5–5.0 days, and an efficient secondary settlement system is vital to ensure an adequate SRT in the primary reactor. The other type of flow-through reactor is the plug-flow type (Fig. 5.5), which is used mainly for animal manure. However, adequate sludge recycle is required for seeding purposes, while mixing can be achieved by gas recycling. The major advantage of this system is that it allows simple tank configurations to be used, resulting in significant savings on capital cost (Casey, 1981).

Digestion is used to stabilize both primary and secondary sludges having a solids content of between 20,000 and 60,000 mg L^{-1} (2–6%). About 70% of the sludge is degradable, and up to 80% of this will be digested, reducing the solids content by about 50%. The remaining solids form a relatively stable sludge, which has a characteristic tarry odor and can be difficult to dewater. The mixture

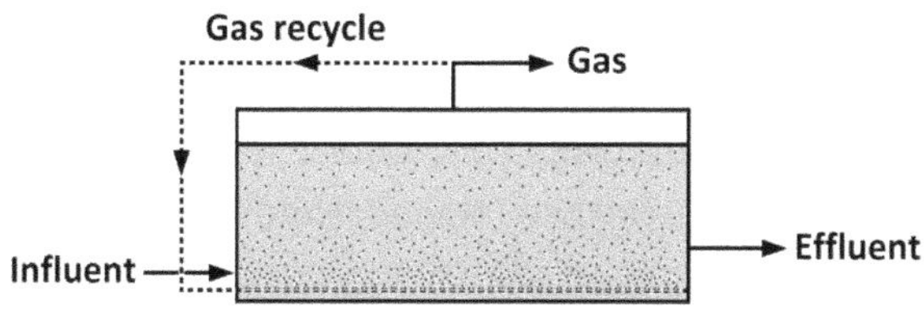

Figure 5.5. Schematic representation of a flow-through plug-flow type anaerobic reactor.

of primary and secondary domestic wastewater sludges fed into a digester is rich in carbohydrates, lipids, and proteins, which are ideal substrates for microbial degradation, and although the sludge is already rich in a variety of anaerobic bacteria, in order to obtain the correct balance of hydrolytic acid producers and methanogens, the raw sludge needs to be initially seeded. This is particularly important as the main bacterial and archaea groups responsible for digestion are dependent on the end-products of the other (Sec. 2.2; Fig. 1.3). Hydolysis and fermentation are the first two stages where particulate and high molecular weight organic matter is broken down to lower fatty acids, hydrogen, and carbon dioxide. The end-products are converted to acetic acid by acetogenic bacteria (third stage), which is subsequently converted in the final and fourth stage by methanogenic archaea to methane and carbon dioxide. Obviously, anaerobic digestion will proceed most efficiently when the rates of reaction at each stage are equal. However, if either stage one or two is limited, then the nutrient supply to the other stages will be reduced with the overall effect of suppressing the rate of digestion and biogas formation but not inhibiting it (Fig. 1.3). If either of the subsequent stages is restricted, then the first-stage products will accumulate, causing a gradual rise in the carbon dioxide fraction in the biogas (>30%) and a gradual fall in the pH as volatile fatty acids (VFAs) accumulate until the pH falls below 7.0 and the whole process becomes stressed. All stages proceed simultaneously within the same reactor, with the microbial groups in close proximity to each other. However, as the anaerobic chain of reaction proceeds, the microbial groups become progressively more sensitive to their environmental conditions. Hydrolysis and acid formation are carried out by a diverse and large group of facultative bacteria that can tolerate a wide variation in temperature, pH, and a range of inhibitory substances (Sec. 5.4.1). In contrast, the acetogenic bacteria and methanogenic archaea are far more sensitive, being highly specialized and severely inhibited by even minor changes in operating conditions. Thus, it is more common to have operational problems involving the third and fourth stages of digestion (Secs. 2.1.3 and 2.1.4).

5.2 Design

Primary digesters are normally covered with a fixed or floating top for gas collection. Floating covers rise and fall according to the volume of gas and sludge, as they are in the same design as conventional gas holders, with the weight of the floating cover creating the gas pressure (Fig. 5.6). Methane is biochemically inert and can be stored

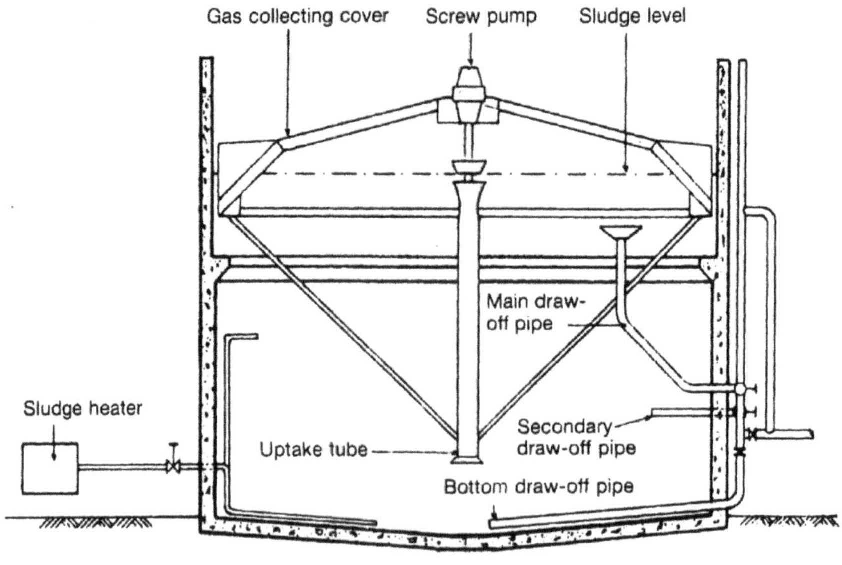

Figure 5.6. Primary digestion tank with a screw mixing pump and external heater (Institute of Water Pollution Control, 1979). Reproduced with permission of the Chartered Institution of Water and Environmental Management.

above the sludge within the reactor without affecting or inhibiting the process. This allows flexible covers to be used to collect and store gas above large reactors such as anaerobic lagoons (Fig. 4.21). Modern digesters generally use fixed covers, which require a separate gas holder that allows the gas to move freely in both directions (Figs. 5.7 and 5.8). This is particularly important if a vacuum is not to be produced when sludge is removed from the digester, which would not only affect operation but could affect the structural stability of a reactor with a fixed cover.

Digesters are usually circular in plan, with maximum diameters of approximately 25 m, and to ensure the minimum depth, the ratios of the reactor's height to diameter is normally between 1:3 and 1:2, although smaller digesters tend toward a ratio of 1:1. Conventional

Figure 5.7. Two-stage anaerobic digestion with a heated mesophilic digester on the right and an unheated settlement stage on the left. Excess supernatant (sludge liquor) is stored in the smaller tank on the extreme left prior to being returned to the inlet for treatment.

Figure 5.8. Gas storage bag with gas flare in the background where excess gas is burnt to prevent methane, which is a greenhouse gas, from escaping to the atmosphere.

tank designs are susceptible to solids settlement and scum formation, which is exacerbated when the width to height ratio is increased. The floor of the digester is sloped between 12 and 30°, with facilities to withdraw heavier solids and grit from the base. Large tanks are built above ground, and smaller units can be constructed below ground level or surrounded by earth embankments to provide additional insulation. Modern digesters are based on an egg-shaped design, which prevents any settlement or scum formation, ensuring the entire tank is always fully mixed and available for treatment (Ditchtl, 1997) (Fig. 5.9).

The entire contents of the tank need to be turned over once every 3–4 hours, with the sludge warmed by heat exchangers using hot water from gas boilers or the cooling systems of gas engines. Heating and mixing of digester contents can be achieved by a number of techniques, the most common being a central screw mixing pump with an

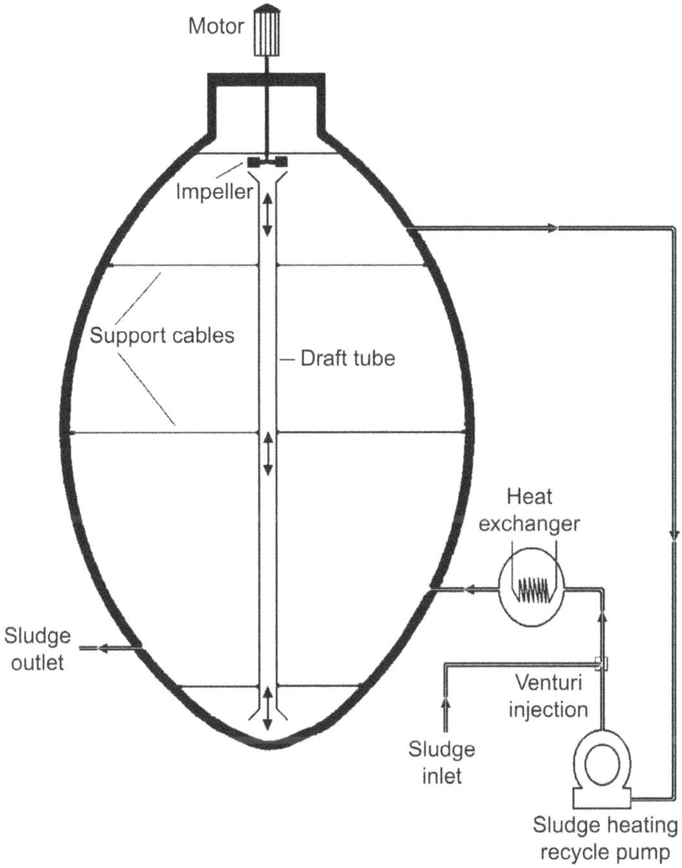

Figure 5.9. Design of a modern egg-shaped anaerobic digester that prevents both settlement of solids and scum formation.

external heat exchanger, or a circulating pump and heat exchanger housed in a projecting chamber, or internal gas-lift pumps. All these methods are fully reviewed by the Institute of Water Pollution Control (1979). The system shown in Fig. 5.6 has a screw pump mounted on the top of a vertical uptake tube. Known as the "Simplex System," it is designed and manufactured by Ames Crosta Ltd. The pump circulates the contents of the tank by drawing sludge from the bottom of the reactor and then spraying it out over the surface of the main body of sludge, providing surface scum control as well as efficient mixing. At regular intervals, the pump is reversed for a short period to prevent blockages and improve mixing. The heat exchanger is external to the digester, the sludge being drawn off from near the base, heated, and then returned toward the surface so that it is not immediately drawn up the central updraught tube, thus ensuring maximum utilization of the heat within the reactor. A simpler impellor system is used in the design (Fig 5.9), which pulls or drives the sludge up and down a draft tube.

The amount of heat required to raise the temperature of the incoming sludge to the optimum temperature and maintain it is calculated as:

$$H = WC\,\Delta T + UA\,\Delta T \qquad (5.1)$$

where H is the heat required to heat the incoming sludge and compensate for heat losses (kg cal h^{-1}), W is the mass rate of influent sludge (kg h^{-1}), ΔT is the difference between the digester temperature and influent sludge temperature, U is the heat transfer through the digester walls (kg cal $m^{-2}h^{-1}$ °C), A is the surface area of digester losing heat, and C is the mean specific heat of the feed sludge (1 kg cal kg^{-1} °C). The basic design criterion is the provision of sufficient capacity to ensure an adequate sludge retention time. It is common practice to use 25–30 days as the design retention period to allow for variation in daily sludge loading. However, as the theoretical retention time can be as low as 7–10 days, many feel that with the improvements in mixing and heating digesters now available, 15 days allows an adequate safety margin to cover most operational difficulties. Digester capacity can also be calculated on a population equivalent (PE) basis. Assuming a per capita sludge production of 1.8 L of raw sludge containing 4.5% dry solids, the capacity of a digester with a 25-day retention time is:

$$\frac{1.8 \times 2.5 \times \mathrm{PE\,m^3}}{1000} \qquad (5.2)$$

The volume of a batch digester can be more accurately calculated using the equation:

$$V_s = [V_1 - 0.66\,(V_1 - V_2)]t \qquad (5.3)$$

where V_s is the digester volume (m^3), V_1 is the initial sludge volume (m^3), V_2 is the final sludge volume (m^3), and t is the retention time (days).

For example, the digestion tank volume required to treat waste-activated sludge from a treatment plant treating a population equivalent of 35,000 is calculated as follows where the sludge production rate is 0.1 kg dry solids (DS) ca^{-1} d^{-1}, the solids content is 3.5% with a volatile content of 78% and a specific gravity of wet sludge of 1.017. After digestion, the sludge is 6% DS with a specific gravity of 1.025 and with 65% of the volatile solids destroyed.

Influent sludge production = 35000 × 0.1 = 3500 kg d^{-1}
Volatile suspended solids (VSS) = 3500 × 0.78 = 2730 kg d^{-1}
Fixed suspended solids (FSS) = 3500 × 0.22 = 770 kg d^{-1}
VSS destroyed = 2730 × 0.65 = 1774.5 kg d^{-1}

Remaining VSS in sludge after digestion = 2730 – 1774.5 = 955.5 kg d^{-1}
Remaining FSS in sludge after digestion = 770 kg d^{-1}
Total solids in sludge after digestion = 955.5 + 770 = 1725.5 kg d^{-1}

Influent sludge volume = 3500 × (100/3.5) × (1/1.017) = 98.3 m^3d^{-1}
Digested sludge volume = 1725.5 × (100/6) × (1/1.025) = 28.1 m^3d^{-1}

Using Eq. (5.3) above, the volume of digested sludge (V_s) is calculated using a sludge residence time of 25 days:

$$V_s = [98.3 - 0.66\ (98.3 - 28.1)]25 = 1299.2\ m^3$$

As the sludge in the digester only occupies two-thirds of the tank, the total tank volume required is:

$$1299.2 \times 1.5 = 1948.8\ m^3$$

Although often quoted, organic loadings expressed as kg organic or volatile matter m^{-3} are not appropriate parameters for digester design. With the standard 25-day retention time, the typical sludge feed of 4.5% total solids with an 85% organic (volatile) matter content results in an organic loading equivalent to 1.5 kg $m^{-3}d^{-1}$. However, in the UK, organic loading ranges from 0.3 to 2.8 kg $m^{-3}d^{-1}$, indicating the varying water contents of the sludges being digested. So, in order to achieve the optimum organic loading of 1.5 kg $m^{-3}d^{-1}$ with a thin sludge, a much shorter retention time will result in incomplete digestion. For this reason, sludges should be characterized and loadings specified as a concentration of total (dry) solids (%).

Secondary digestion tanks are generally used for storage and separation, although they can be used for further digestion and gas collection. Sludge is passed as frequently as possible from the primary digester into the uncovered tank, where it cools and allows the liquor to separate from the solids so that each can be withdrawn separately. Temperature differences between the cool and warm sludge can cause convection currents within the digester that will hinder settlement, as will gas production so that it becomes difficult to obtain a solids-free liquor. To overcome this, secondary digesters are relatively shallow, with a maximum depth of 3.5 m. The capacity of older tanks was between 50 and 70% of the primary digester, providing 15–20 days of retention. Newer tanks are approximately the same size as the primary tanks and provide similar retention times for the sludge. Tank volume can be estimated on a population equivalent (PE) basis using the equation:

$$\text{Secondary digester volume} = 0.035 \times \text{PE m}^3 \qquad (5.4)$$

In order to ensure adequate operating conditions for digestion, the pH, concentration of carbon dioxide in the biogas, and volatile acid concentrations should all be continuously monitored. The normal operating conditions are a pH of 7.0–7.2, alkalinity (as $CaCO_3$) of 4,000–5,000 mg L^{-1}, and a concentration of volatile acids (as acetic acid) of <1,800 mg L^{-1}. The carbon dioxide content of the biogas should not exceed 30%, and once any of these values are exceeded, then immediate remedial action is required.

5.3 Hydrolysis

In order to increase digestibility of sludges, they are often hydrolyzed prior to entering the digester. This breaks down long-chain organic molecules into water-soluble and readily degradable short-chain molecules. The cells rupture, releasing enzymes and increasing the removal of volatile solids, thus reducing the final volume of sludge requiring disposal (>20%) and increasing biogas yields by >30% compared to normal digestion. Hydrolysis reduces the HRT in the digester and allows for more compact and efficient designs (Fig. 5.9), with the sludge more viscous and easier to pump. Hydrolysis has also fully replaced thermophilic digestion due to significant cost savings in heating the digester. There are three types of hydrolysis widely employed: (i) enzymatic or biological hydrolysis (42–55°C), (ii) pasteurization (55–70°C), and (iii) thermal hydrolysis (combined high pressure and temperature), although chemical pretreatment is also possible. All these processes ensure complete pathogen destruction as well as making the sludge far more digestible. The presence of extracellular polymeric substances (EPS) that bind activated sludge flocs together makes the penetration of extracellular enzymes produced by hydrolytic bacteria much more difficult, slowing natural hydrolysis and the rate of digestion overall. The EPS is made up of protein, carbohydrates, lipids, and humic acids, which are all potentially degradable (Sun *et al.*, 2016).

5.3.1 *Enzymatic or Biological Hydrolysis*

A wide range of enzymes are able to breakdown complex organic macromolecules, including protease, amylase, lipase, cellulase, urease, and xylanases, while there are many specific enzymes for treating contaminants in wastewaters and sludges as well as specific industrial waste streams. These include oxygenases to treat fossil fuel derivatives, peroxidases for treating aromatic compounds, dyes, and organic peroxides, laccases to treat phenols and dyes, and tyrosinases for treating phenolic compounds. While all these enzymes are commercially available, their use in the pre-treatment or conditioning of biosolids is largely limited to their use in septic tanks or industrial processes due to cost.

Biological enzyme treatment is also being considered as a conditioner for waste-activated sludge to increase dewaterability by acting on the EPS in the sludge flocs. The enzymes destroy the colloidal floc structure, releasing water and breaking down the larger organic molecules. Usually, protease and lipase are used, although α-amylase, alkaline protease, neutral protease, and cellulase are also widely used. Kang *et al.* (2023) used two enzymes to determine the optimal conditions for the use of enzymes for waste activated sludge (WAS) conditioning. They observed that for α-amylase, it was 3 hours at pH 6 and 45°C using a dosage rate of 30 mg g^{-1} TSS (total suspended solids), and 2 hours for neutral protease at pH 6.5 and 40°C, also at a dosage rate of 30 mg g^{-1} TSS. The sludge concentration of proteins and polysaccharides both increased in the supernatant, as did the soluble chemical oxygen demand (sCOD) fraction. This suggests that the problem of EPS degradation during anaerobic digestion can be solved by pre-treating the WAS stream separately from the primary sludge before digestion, ensuring hydrolysis of the EPS. Quintero-García *et al.* (2023) have reviewed the use of enzymes in biosolids treatment and indicate that their future may be in the production of value-added products from biosolids. However, natural enzymes can be released from the actual biosolids themselves by fragmenting the organic material, including the bacteria releasing the enzymes into solution. This can be done mechanically, by microwaves, or most successfully using ultrasonication (Chu *et al.*, 2002; Aldin *et al.*, 2009). The advantage is that not only are the enzymes released immediately, but the solids are reduced in size, allowing better and more rapid degradation overall. The various sludge pretreatment options are compared by Uthirakrishnan *et al.* (2022).

Biological pre-hydrolysis (BPH) separates both hydrolysis and acidogenesis from methane generation by having two or more anaerobic reactors in series, which significantly reduces HRT and also increases solids loading to between 4 and 6 kg VS $m^{-3}d^{-1}$. When digestion is completed in a single reactor, any of the non-methanogenic stages can become the rate-limiting step, with methanogenesis inhibited by a lack of substrate or resulting operational factors such as pH or alkalinity. Hydrolysis, acidogenesis, and acetogenesis operate more efficiently under acidic conditions, while methanogenesis operates better at neutral pH. BPH is achieved by raising the temperature of stored thickened sludge to encourage hydrolytic and acidogenic bacteria to break down large molecules of proteins, lipids, and polysaccharides, including extra cellular polymers (ECPs), into smaller soluble molecules. Ding *et al.* (2017) found that using a temperature of 42°C for 6 days increased the sCOD from 175 to 3,315 mg L^{-1}, with the dominant VFA produced being acetic acid, with the concentration increasing from 41.5 to 786 mg L^{-1}, resulting in a significant increase in methane yield. Wessex Water has installed a successful BPH system at Avonmouth WWTP (wastewater treatment plant) using six completely stirred reactors in series before the sludge finally enters the methane phase digester. Under the trade name of Monsal Enzymic Hydrolysis™, the plug flow nature of the design ensures no short-circuiting and allows hydrolysis, acid production, and acetogenic phases to occur in sequence. The six tanks together give an HRT of 2–3 days and are operated at 33°C, although optimum enzyme production has been recorded at 42°C, with the final methane digesters operated at 35°C. The hydraulic loading is 1,400 m^3d^{-1} at 5.5% DS, which is equivalent to a loading rate of 78 t DS d^{-1}. The system has an average solids reduction of 52%, with a biogas output of between 390 and 430 $m^3 t^{-1}$ DS (Bungay and Abdelwahab, 2008).

5.3.2 *Pasteurization*

Pasteurization requires the sludge to be preheated to 70°C for a minimum of 30 minutes. There are a number of systems available, for example, Veolia's BioPasteur™ system, where biosolids are passed through batch pasteurization tanks and maintained at 75°C for 30 minutes. Heat is recovered using heat exchangers, and the conditioned biosolids are fed continuously into the digester operated either at mesophilic or thermophilic ranges. Anglian Water has used pasteurization in their HpH technology, which is an acronym for heat recovery, pasteurization, and biological hydrolysis, which integrates these three steps. Hot water and steam from existing combined heat and power (CHP) engines, which convert biogas into electricity, are used to heat the thickened biosolids to 40–45°C. The sludge is pumped into the first pasteurization tank and heated to 55°C by recirculating it through a steam injector. Once at temperature, it is stored in a second pasteurization tank for five hours where pasteurization is completed. The final stage is for the pasteurized sludge to be cooled to between 38 and 42°C in a separate tank, where it is held for 36 hours for biological hydrolysis to occur. It is then fed into the digesters as required. The process allows much greater organic loading rates to the digesters with enhanced methane yields 30% greater than conventional digesters, producing on average >1 MWh of renewable electricity per ton of dry solids via CHP, which is equivalent to 2.5 MWh of biogas energy. With 25% more volatile suspended solids (VSS) destruction, fewer biosolids are produced, saving costs on dewatering and producing excellent and pathogen-free Class A biosolids, which can be used for a wide range of uses. Both capital and operational costs are lower than other hydrolysis systems, with all the required heat coming from the CHP engines helping WWTPs to become net zero. HpH technology is marketed by Royal HaskoningDHV under the trade name of Helea™ (Fig. 5.10).

Figure 5.10. The Helea process comprising, from left to right, a heating tank, followed by the two pasteurization tanks, and finally, the larger hydrolysis tank. Hot water and steam are pumped around the system in highly insulated pipework. The heat exchanger can be seen in the foreground in front of the heating tank. The photograph is of Basildon WWTP, where 10,300 tons of dry solids are processed annually. Reproduced with permission of Royal HaskoningDHV. https://www.royalhaskoningdhv.com/en/services/helea

5.3.3 *Thermal Hydrolysis*

Thermal hydrolysis uses a feed sludge with 18% DS, which is diluted with hot water to 12–14% DS. This is then heated to 100°C using process steam and then fed into a batch reactor where it is heated to 165°C at 6 bar pressure for 90–110 minutes (Fig. 5.11). The heat disintegrates and liquefies the sludge, which makes it far more degradable. It is then fed into the digester as required (Kepp *et al.*, 2000; Morgan-Sagastume *et al.*, 2011). The process is energy-intensive, and it is not always possible to supply all the required power via energy generated on the plant.

Figure 5.11. Example of a thermal hydrolysis unit used in Ireland. In this case, the sludge is heated to 180°C at 12 bar of pressure for 40 minutes prior to digestion.

5.4 Digester Operational Management

Successful operational management of anaerobic digesters depends on six major factors: composition of the raw sludge, method of addition of raw sludge to the digester, internal mixing and circulation, temperature, pH and alkalinity, and solids retention time.

5.4.1 *Composition of Raw Sludge*

Anaerobic microorganisms require certain substrates, growth factors, trace elements, and nutrients for successful development. Mosey (1983) compared raw and digested sewage sludge and showed quite clearly that major substrates such as lipids, cellulose, and some proteins are present as solids in suspension and that a substantial fraction of the organic matter is either converted to microbial biomass or not metabolized at all, as is the case with lignin (Table 5.1). Nitrogen and phosphorus are both vital for bacterial growth and are required at

Table 5.1. Typical composition of sewage sludges before and after digestion (g per 100 g total solids).

Constituent or test	Raw sludge	Digested sludge
Suspended solids	95	97
COD	140	100
Organic carbon	40	31
Organic matter	60–80	45–60
Greases and fats	7–35	3.5–17
Cellulose	4	0.6
Hemicelluloses	3	1.6
Lignin	6	8
Protein	22–28	16–21
Anionic detergents	0.5–1.5	0.7–2.2
Zinc	0.09	0.14
Copper	0.035	0.055
Lead	0.016	0.026
Chromium	0.01	0.016
Nickel	0.0092	0.015
Cadmium	0.0022	0.0035

minimum concentrations of 2.5% and 0.5% of the dry organic matter content of the sludge, respectively. This is equivalent to a C:N ratio of between 10 and 16:1 and an N:P ratio of 7:1. Huss (1977) found the optimum BOD:N:P ratio to be 100:0.5:0.1, whereas the COD:N:P ratio was found to range between 42:0.7:0.1 and 150:0.7:0.1 by Henze and Harremoes (1983). In practice, sewage sludges have excess nitrogen and phosphorus, ensuring adequate digestion, although certain trade wastes, including sugar beet and other food-processing wastes, may be deficient in these nutrients, making them less amenable to the digestion process (Lettinga, 1995). Sewage sludges are mixtures of complex organic materials and vary widely both in composition and strength from place to place. The average solids content in the UK for sewage sludge is often quoted as 4.5%, although this will vary even from the same source, depending on the operation of the settlement tank from which the sludge is removed.

Secondary sludges from activated sludge or percolating filter units are less amenable to digestion than primary sludges, as they contain a lower proportion of digestible matter per unit mass of solids as well as generally contain much more water than primary sludge. Only about 30% of the available organic and volatile matter is utilized in activated sludge due to the presence of EPS (Sec. 5.3), compared with 50% in most primary sludges.

A wide variety of compounds normally present in sewage can inhibit the digestion process when present in excessive concentrations, most notably detergents, chlorinated hydrocarbons, heavy metals, and ammonia. The problem is that many toxic substances are separated at the primary settlement stage, with organic compounds such as chloroform being heavier than water and thus collecting at the outlet hopper in the primary settlement tank and many heavy metal salts being precipitated at this stage as hydroxides. All inhibitory substances will affect gas production, as the methanogenic archaea are particularly prone to inhibition, unlike the diverse hydrolytic and acid-forming bacteria. Another important factor is that the hydrolytic and acid-forming bacteria are present in the raw sludge and are constantly replaced; in contrast, the methanogenic bacterial population is self-sustaining, and once the population has been reduced, it will take time for the population density to be restored, and, therefore, re-seeding may be necessary.

Detergents/surfactants

Detergents contain surfactants, which can be classified as non-ionic, cationic, anionic, or amphoteric (Table 5.2). Non-ionic, cationic, and amphoteric detergents have little effect on anaerobic digestion, being largely biodegradable, even at high concentrations (Bruce *et al.* 1966), whereas anionic detergents are much less or non-biodegradable anaerobically, often inhibiting the process (Merrettig-Bruns and Jelen, 2009).

Table 5.2. Common examples of surfactants used in household detergent formulations.

Surfactants			
Anionic	**Non-ionic**	**Cationic**	**Amphoteric**
Soaps	Alcohol ethoxylates (AE)	Mono- or di-alkyl quaternary compounds (DTDMAC)	Alkyl betaine
Alkylbenzene sulfonates (ABS)	Alkylphenol ethoxylates (APE)	Esterified mono- or di-alkyl quaternary compounds (esterquats)	Alkylamidopropyl betaine
Alkyl sulfates (AS)	Fatty acid alkanolamides (FAA)	Imidazoline derivatives	Betaines derived from imidazolines
Alkyl ether sulfates (AES)	Alkylamine oxides (AO)		Alkylamphoacetates
Secondary alkane sulfonates (SAS)	*N*-Methylglucamides (NMG)		
	Alkylpolyglycosides (APG)		

Soaps, which are salts of natural fatty acids, are readily degraded anaerobically up to concentrations of 1,000 mg L^{-1}. Most household detergents comprised a significant proportion of anionic surfactants, primarily alkylbenzene sulphonates (ABS) and, in particular, linear alkylbenzene sulphonates (LAS). LAS are non-degradable anaerobically, and as they are strongly adsorbed onto organic solids, they are invariably present in sewage sludge. Anionic detergents at concentrations >1.5% of the dry raw sludge solids (expressed as Manoxol OT) inhibit anaerobic digestion, even in heated digesters with relatively long retention times of up to 40 days, with gas production being particularly sensitive. Concentrations <1.5% will reduce the tolerance of the process to other operational variables, such as temporary overloading. The effects of ABS on the microbial community are numerous and include direct toxicity. Although degraded aerobically, it is clear that these detergents are not utilized in the anaerobic process because the enzymes responsible for their degradation, such as monooxygenase (Cain, 1981), are not able to function due to the lack of oxygen. These concentrations are quite common in domestic sewage sludges, and although bacterial populations can become acclimatized, serious inhibition will occur if the concentration is allowed to rise to 2.0% of the dry sludge solids. As digestion proceeds, the concentration of anionic detergents will increase from 1.5% in the raw sludge to 2.5% DS in the digested sludge. This is not a problem when the sludge is finally disposed of to agricultural land, as it will rapidly degrade the aerobic soil environment. If inhibition occurs during digestion, then the anionic detergents can be rapidly neutralized by the addition of long-chain fatty amines, such as stearine amine, to the sludge prior to digestion (Swanwick and Shurben, 1969). Some anionic surfactants are biodegradable anaerobically, and these include linear alkyl sulfosuccinates and alkyl sulfates (EU, 2016).

The biodegradability of detergents is summarized in the European Union's Detergents Ingredients Database (DID), which gives both the acute and chronic toxicity levels for all detergents used in the European Union, including details of whether they are degradable under aerobic and anaerobic conditions (EU, 2016). The anaerobic breakdown of surfactants has been reviewed by Merrettig-Bruns and

Jelen (2009), who give more precise biodegradability data than the DID. The current problem with the use of detergents is that they are only required under European Commission (EC) legislation to be biodegradable under aerobic conditions, but with the increased use of anaerobic wastewater treatment, there is a need for them to also be biodegradable anaerobically. However, as can be seen in the EU DID, only a fraction of the permitted surfactants used in detergent formulations have been tested for their biodegradability under anaerobic conditions (EU, 2016).

Chlorinated hydrocarbons

Although high concentrations of these solvents are unusual in sewage sludges, they can cause a problem in certain industrial wastes or where domestic sewage contains industrial discharges. The most frequent source of this group of inhibitors is from dry-cleaning operations that discharge chlorinated hydrocarbons into the sewer. Although their effect on the process depends on several factors, it is the concentration in the sludge that is most critical. It appears that many chlorinated hydrocarbons act selectively on the methanogen organisms, which is why they are so toxic to anaerobic digestion compared to aerobic systems (Blum and Speece, 1992). The concentration at which 20% inhibition occurs is summarized in Table 5.3 for the most widely used chlorinated hydrocarbons, although their inhibitory effect will be increased if other inhibitory substances are also present.

The most common chloro-aliphatic compound, chloroform ($CHCl_3$), is toxic to methanogens, and while most of the compound will be lost by volatilization in the sewers or during aerobic treatment, it can still cause detectable inhibition to methane production at concentrations as low as 10 mg kg^{-1} DS. Hickey *et al.* (1987) reported complete inhibition of methane production at concentrations >1 mg L^{-1} due to the compound binding to intercellular proteins, thereby blocking essential binding sites in the methanogenic pathway (Jha and Schmidt, 2017). Where chloroform is a problem, it should be

Table 5.3. Concentration of chlorinated hydrocarbons in digesting sludge at which 20% inhibition occurs (Institute of Water Pollution Control, 1979). Reproduced with permission of the Chartered Institution of Water and Environmental Management, London.

Chemical	**Concentration ($mg\ kg^{-1}$ dry solids)**
Chloroform	15
Trichlorethane	20
1,1,2-Trichlorotrifluoroethane	200
Carbon tetrachloride	200
Trichloroethylene	1800
Tetrachloroethylene	1800

removed from the effluent by air-stripping prior to discharge into the sewer, although some acclimation to the compound is possible (Lumb *et al.,* 1977). Benzene ring compounds such as benzene, toluene, phenol, and chlorophenols all inhibit methanogens (Patel *et al.*, 1991; Wang *et al.*, 1991). Even when protected by the granulated sludge in floc-based reactors (Chap. 7), organics such as aromatic compounds are still strong inhibitors of methanogens (Donlon *et al.*, 1995; Fang *et al.*, 1997). Chlorophenols are all inhibitory to methanogenesis, the most toxic being pentachlorophenol (PHP), with a 50% inhibition of biogas production at 8 mg L^{-1} reported. This has been related to their hydrophobicity, which allows the compound to adhere to bacteria membranes, affecting their integrity and cellular metabolism (Sikkema *et al.,* 1995). There is some evidence that anaerobic microorganisms can become acclimatized to chlorinated hydrocarbons. For example, Lian *et al.* (2020) demonstrated how hexachlorocyclohexanes can be used as a food substrate during anaerobic digestion, with the genus *Methanosaeta* developing a high tolerance. The effects of aromatic compounds and halogenated aliphatics on anaerobic digestion have been reviewed by Jha and Schmidt (2017) and Nielsen *et al.* (2020).

Heavy metals

The most quoted cause of inhibition of sludge digesters is by heavy metals, in particular, chromium, copper, nickel, cadmium, and zinc, although many metals are essential for the function or activation of critical enzymes and coenzymes during digestion (Takashima and Speece, 1989). Metals are non-degradable and so can accumulate within the digester, and can also be released into solution during fermentation, but in practice, high concentrations of metals are required for a significant inhibitory effect (Table 5.4). It is thought that the heavy metals inactivate enzymes and co-enzymes, interfering with their function and thereby inhibiting microbial activity and growth (Oleszkiewicz and Sharma, 1990).

Toxicity of heavy metals to anaerobic digestion increases as the metal affinity for sludges decreases. Nickel has the lowest affinity for sludge, and so is the most toxic, and *vice versa* for lead (i.e., toxicity to digestion is Ni>Cu>Cd>Cr>Pb). Lin (1992) gives IC_{50} values for Cd^{2+}, Cu^{2+}, Zn^{2+}, and Pb of 7.7, 12.5, 16, and 67 mg L^{-1}, respectively. While using pure cultures of *Methanobacterium thermoautotrophicum* and *Methanobacterium formicicum*, Jarrell *et al.* (1987) found significant differences in the IC_{50} values for the two archaea, respectively, for Cu^{2+} at 4.8 and 3.5 mg L^{-1}, Zn^{2+} 3.0 and 8.2 mg L^{-1}, and for Ni^{2+} 1.2 and >15 mg L^{-1}.

Synergistic effects have been noted with heavy metals and a number of other inhibitory substances, and under certain conditions, even

Table 5.4. Concentrations of heavy metals in digesting sludge that cause a 20% reduction in gas production in laboratory experiments (Institute of Water Pollution Control, 1979). Reproduced with permission of the Chartered Institution of Water and Environmental Management, London.

Metal	Batch digesters: concentration (mg kg^{-1} dry solids)	Typical concentration in digested sludges (mg kg^{-1} dry solids)
Nickel	2,000	30–140
Cadmium	2,200	7–50
Copper	2,700	200–800
Zinc	3,400	500–3,000

low concentrations may cause problems. Apart from the concentration of heavy metals in the raw sludge, other factors such as solubility, pH, and the concentration of sulfide present will all affect their availability in the digester. Anaerobic microorganisms can withstand quite high concentrations of total metals, as a considerable percentage of the metal ions can be precipitated out of solution as either sulfides or carbonates. The concentrations of metals that can be present in wastewater without adversely affecting the anaerobic digestion of the resulting sludge have been summarized in Table 5.5. There is always more than one heavy metal present in domestic wastewater sludges, and the inhibitory effects are additive on an equivalent weight basis (Mosey, 1976). Where the milligram equivalent weight (meq) per kg dry sludge solids (K) exceeds 400 meq kg^{-1}, there is a 50% chance of digester failure, which rises to 90% when K exceeds 800 meq kg^{-1}. To ensure a 90% probability that digestion will not be affected, the value of K should be <170 meq kg^{-1}.

K is measured using the concentrations of the most abundant heavy metals (mg L^{-1}) in sewage sludge, excluding chromium, using the equation:

$$K = \frac{(\text{Zn}/32.7) + (\text{Ni}/29.4) + (\text{Pb}/103.6) + (\text{Cd}/56.2) + (\text{Cu}/47.4)\,\text{meq kg}^{-1}}{\text{Sludge solids concentration}\,(\text{kg L}^{-1})} \tag{5.5}$$

Table 5.5. Highest metal concentrations in sewage that will allow satisfactory digestion of sewage sludge.

	Concentration in influent sewage (mg L^{-1})	
Metal	**Primary sludge digestion**	**Combined sludge digestion**
Chromium (hexavalent)	>50	>50[a]
Copper	10	10
Nickel	>40	>10[a]
Zinc	10	10

[a] Higher dose not studied.

The effects of heavy metals on anaerobic digestion are reviewed by Jha and Schmidt (2017).

Some metals at trace concentrations (e.g., nickel, cobalt, manganese, iron, and molybdenum) stimulate methanogens and increase biogas production (Murray and van der Berg, 1981; Basiliko and Yavitt, 2001). Other metals (e.g., Cu^{2+}, Cd^{2+}, and Cr^{4+}) have also been shown to have stimulatory effects at very low concentrations in experimental studies on the anaerobic degradation of cattail waste using rumen cultures (i.e., 2.4, 1.6, and 4.0 mg L^{-1}, respectively), rapidly accelerating inhibition as the concentration increased, resulting in complete inhibition of biogas production at 30 mg Cu^{2+} L^{-1}, 20 mg Cd^{2+} L^{-1}, and 60 mg Cr^{4+} L^{-1} (Yue *et al.*, 2007). The role of metal in anaerobic wastewater treatment is explored further in Sec. 9.2.2.

Ammonia

Ammonia or ammonium ions are essential nitrogen sources for anaerobic digestion but can be inhibitory when present at high concentrations. The optimum C:N:P:S ratio for methanogenesis is 600:15:5:3, so not all waste streams will have significant excess nitrogen. A minimum concentration of 200 mg L^{-1} is required to prevent nitrogen deficiency. Ammonia also plays an important role in buffering the pH during digestion, requiring a balance between free ammonia nitrogen (NH_3) (FAN) and ammonium ions (NH_4^+), with the total ammonia nitrogen (TAN) the total concentration of both. The relative proportions of the NH_3 and NH_4^+ in the digester are largely dependent on the pH and temperature (Eqs. (3.26) and (3.27)) (Sec. 3.5). However, it is FAN which methanogens are particularly sensitive to, showing the importance of maintaining as near a neutral pH as possible during methanogenesis. Methanogenic activity in anaerobic digestion has been reported as being 50–100% inhibited at FAN concentrations of >600 mg L^{-1} or at TAN concentrations >4,000 mg L^{-1}, although the exact level of inhibition varies depending on digester temperature and pH, the microbial community, and reactor configuration.

Inhibition of methanogenic metabolism by ammonia is probably caused by a number of mechanisms, such as FAN limiting the expression of key enzymes and/or proton imbalance and deficiency in potassium (Duc *et al.*, 2022; Yu *et al.*, 2021; Li *et al.*, 2023). It is generally accepted that acetoclastic methanogens are more sensitive to ammonia than hydrogenotrophic archaea, although there are a number of examples showing the opposite being the case. In a pilot study, Yan *et al.* (2019) studied the acclimatization of methanogens over a period of 127 days, where the ammonia concentration was increased in seven steps up to 8.5 g NH_4^+ L^{-1}. They clearly demonstrated that archaea had different tolerances to ammonia inhibition. There was a discernible shift from the original community structure dominated by both hydrogenotrophic and acetoclastic methanogens to more ammonia-tolerant and metabolically versatile acetoclastic methanogens, with *Methanosaeta concilii* (specialist species only able to use acetate) and *Methanosarcina soligelidi* (general species able to use H_2/CO_2, acetate and methanol), the dominant archaea at low (1.1 g) and high (8.5 g) ammonia concentrations, respectively. What is interesting is the authors were able to show how *M. soligelidi* changed its metabolic pathway from acetoclastic to hydrogenotrophic methanogenesis as acclimatization progressed. Ammonia toxicity also targets the pathways from propionate to acetate (propionate-oxidizing bacteria) and acetate to methane (archaea). Under normal operating conditions, it is expected that hydrogenotrophic methanogens will outcompete acetoclastic methanogens due to higher growth rates and greater ammonia tolerance (Guo *et al.*, 2023). However, this is not always the case. Guo *et al.* (2023) compared two pilot anaerobic digesters operated constantly at 600 mg NH_4^+-N L^{-1} (control, R-C) and at 2,000, 3,000, 4,000, and 5,000 mg NH_4^+-N L^{-1} each for a period of 30 days, starting at the control concentration of 600 mg NH_4^+-N L^{-1} through urea regulation (R-A). They found that lignin- and cellulose-degrading bacteria dominated both reactors R-C and R-A, with increasing ammonia not affecting the species richness of hydrolysis bacteria. The total abundance of methanogens in R-C slowly increased from 1.6% to 3.8% over the test period of 150. In R-C, five species of methanogens were recorded: the acetoclastic methanogens *Methanosaeta harundinacea*, *Methanosaeta concilii*, and *Methanosaeta thermophile*, and the hydrogenotrophic methanogens *Methanolinea tarda* and *Methanolinea* sp. Ammonia reduced the diversity of methanogens as the concentration increased, with *Methanosaeta concilii* and *Methanosarcina barkeri* recorded. Both are acetoclastic archaea, with the former dominant up to 2,000 mg NH_4^+-N L^{-1} and the latter becoming increasingly dominant up to 4,000 mg NH_4^+-N L^{-1}. At 5,000 mg NH_4^+-N L^{-1}, both species were inhibited (Fig. 5.12). Like *Methanosarcina soligelidi*, *Methanosarcina barkeri* is also able to utilize several metabolic pathways, including acetate, H_2/CO_2, methanol, and pyruvate.

Methanogenesis is more sensitive to ammonia inhibition than acidogenesis, so as the former is inhibited, the latter continues to produce VFAs, which build up as the methanogens are unable to utilize them at the same rate. The VFAs, especially propionate acid, cause an increase in acidification of the digestate, which initially inhibits methanogenesis, even further creating a feedback loop that increasingly reduces methane production (Yuan *et al.*, 2016). However, as Mosey (1983) explains, the system is largely self-regulating, with the drop in pH converting FAN to the less toxic ammonium ionic form (NH_4^+), thus alleviating inhibition and allowing some degree of microbial acclimatization. These high TAN and FAN concentrations are mainly found in very thick sewage sludges and undiluted farm slurries. Ammonia N is also generated within the digester by the degradation of proteins, amino acids, urea, and nucleic acids.

Ammonia inhibition remains a common operation problem, with a number of control options available to operators. Where process operation allows, then ammonia concentrations in the feed can be reduced by extraction processes such as air-stripping, blending sludges to reduce overall TAN concentrations, the use of trace element supplements (Cai *et al.*, 2023), or altering operational parameters such as pH, temperature, or C:N ratio (Cai *et al.*, 2021; Chen *et al.*, 2021). Ammonia-stripping (e.g., pre-treatment, in-situ, side stream, and post-treatment) is an effective method for ammonia removal to prevent ammonia inhibition during digestion (Fig. 5.13), producing a range of useful byproducts, and has been reviewed by Yellezuome *et al.* (2022).

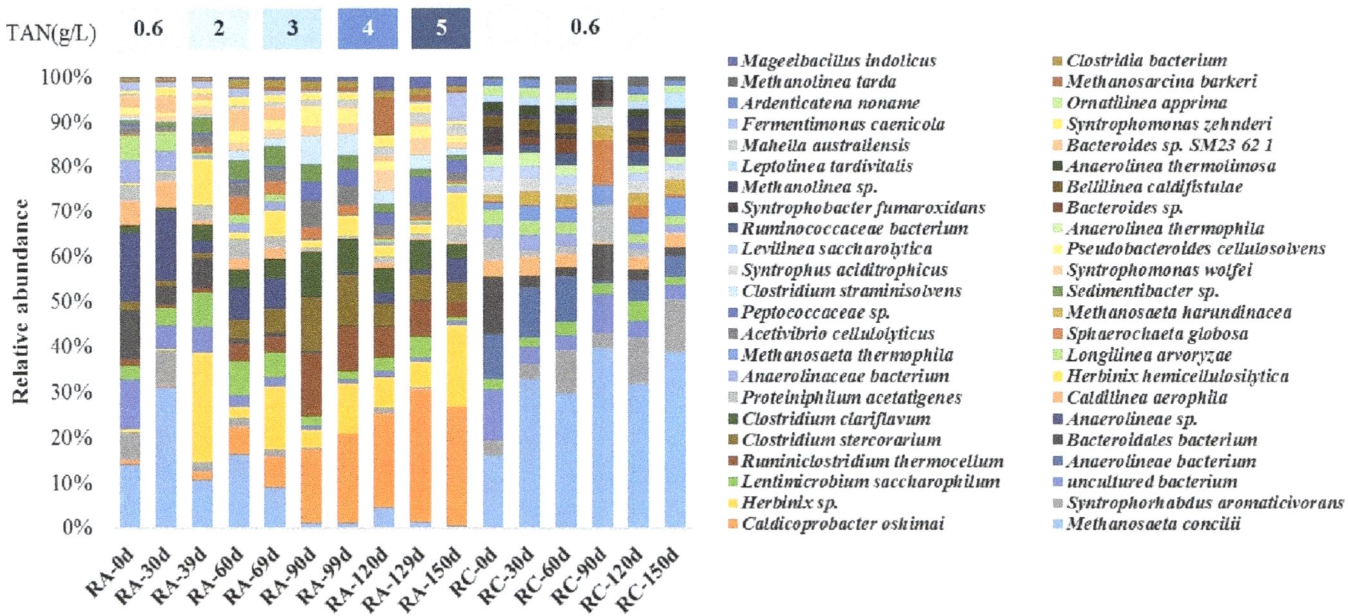

Figure 5.12. Shift in microbial community structure, with ammonia concentration using a pilot-plant anaerobic digester over a period of 150 days. Ammonia concentrations are at 600 mg NH_4^+-N L^{-1} for the control digester (RC) and for the experimental digester (RA) at 600 (-0d, -30d), 2,000 (-39d, -60d), 3,000 (-69d, -90d), 4,000 (-99d, -120d), and 5,000 (-129d, -150d) mg NH_4^+-N L^{-1}, each for a period of 30 days. The values in parentheses are the dates of sampling in days from the start of the experiment (Guo *et al.*, 2023). Reproduced with permission of Elsevier Publishing.

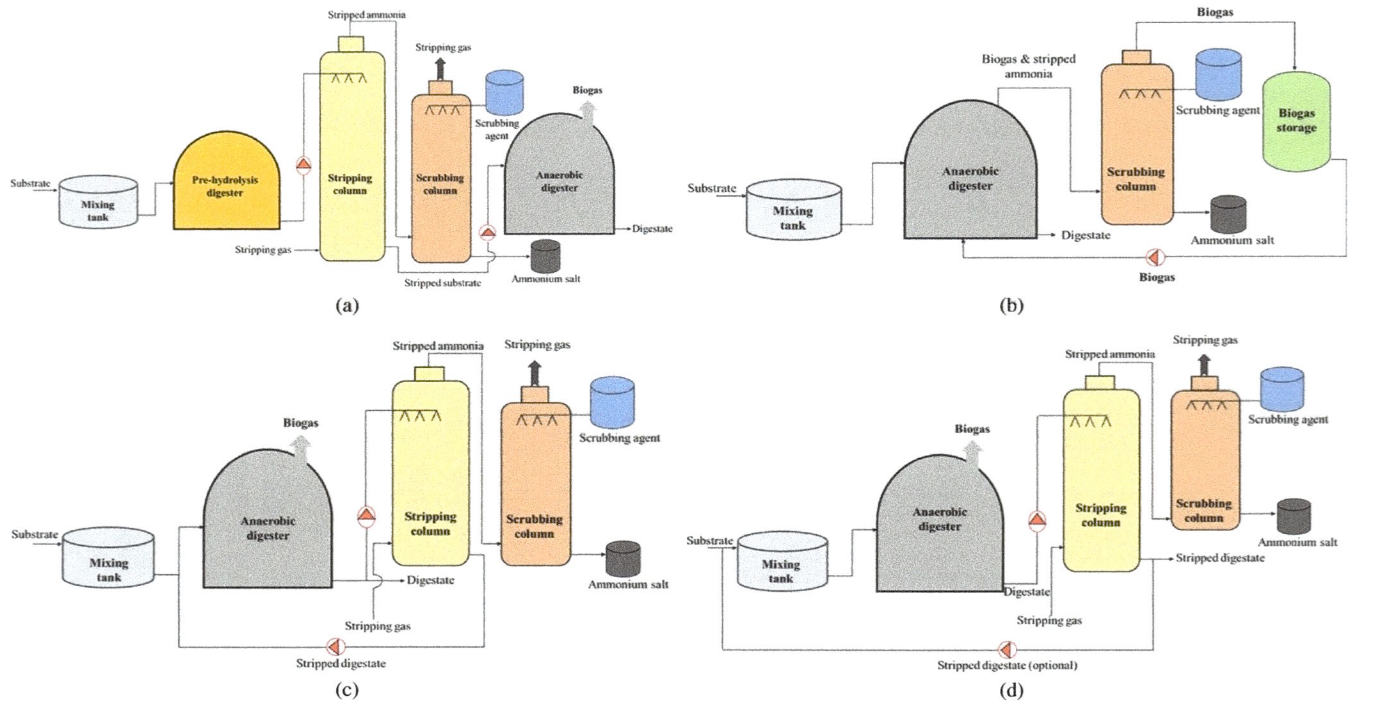

Figure 5.13. Various configurations of ammonia stripping used in anaerobic digestion: (a) Pre-treatment, (b) *in situ*, (c) side-stream, and (d) post-treatment (Yellezuome *et al.*, 2022). Reproduced with permission of Elsevier Publishing.

The addition of carbon- or iron-based materials, including biochar and hydrochar, has also been successful (Shi *et al.*, 2021; Cai *et al.*, 2022; Wang *et al.*, 2024). Usman *et al.* (2021) found that the addition of hydrochar to the feed not only increased methane yields up to an ammonia concentration of 6,000 mg L^{-1} but also promoted methanogenic activity and diversity. They found that under ammonia inhibition conditions, the hydrochar regulated the metabolites in the methane metabolic pathway from hydrogenotrophic methanogens, using hydrogen and carbon dioxide as electron acceptors, to multiple pathways (i.e., acetoclastic, using acetate and/or methylotrophic, using methanol or methylamines) for methanogenesis, thereby maintaining methane yields. Other control methods include microbial acclimatization or bioaugmentation (Shi *et al.*, 2021; Li *et al.*, 2023). Acclimatization to high ammonia concentrations is probably due to changes in the metabolic pathways (Sec. 2.2), although whether this is due to particular species changing pathways or changes in diversity to more tolerant species with different metabolic capabilities is unclear (Niu *et al.*, 2015; Esquivel-Elizondo *et al.*, 2016).

Other ions

Methane formation does not occur readily in the presence of electron acceptors, such as sulfate and nitrate (Jha and Schmidt, 2017). Sulfate can be a particular problem in the digestion process if present in sufficient quantities. The sulfate is reduced to sulfide by bacterial action, with hydrogen sulfide eventually being formed. Sulfate concentrations >500 mg L^{-1} can reduce methane production and generate up to 4% hydrogen sulfide in the biogas. The hydrogen sulfide will form insoluble compounds with heavy metals. Therefore, as long as hydrogen sulfide and heavy metals are present in equivalent proportions, then hydrogen sulfide production will cause no problems and, in fact, be beneficial by removing heavy metals that may cause inhibition. However, two problems can occur: (i) hydrogen is consumed by sulfate reduction and is no longer available for methane formation, which is inhibited, and (ii) hydrogen sulfide itself has a direct toxic effect on the methanogenic bacteria being toxic at concentrations

>200 mg L^{-1} (Mosey, 1976). Remedial action involves precipitating the excess hydrogen sulfide out of the solution as ferric sulfide by the addition of an iron salt, such as ferric chloride or ferric oxide, but not, of course, ferric sulfate (Sec. 8.2).

Nitrates can also cause problems because if denitrification occurs within the digester, there will be a shift in the redox potential, which will suppress methane production (Akunna *et al.*, 1994; Klüber and Conrad, 1998). Methanogens are strict anaerobes and can be completely inhibited by dissolved oxygen concentrations as low as 0.01 mg L^{-1} (Wolfe, 1971). They require a reduced environment with a redox potential within the range of –200 to –420 mv. In a bench scale experiment, Sheng *et al.* (2013) observed that a total ammonia nitrogen (TAN) concentration of <1.0 g L^{-1} had no effect on anaerobic digestion and was, in fact, beneficial. The subsequent addition of nitrate reduced methane production by 12 and 51% at NO_3-N concentrations of 1.0 and 1.5 g L^{-1}, respectively, although at 0.5 g L^{-1}, methane production was enhanced by 12% over the control with no nitrate giving a maximum methane yield of 315 mL g^{-1} VS. This has been a serious problem at many WWTPs in the UK and has been remedied by allowing denitrification to occur before the sludge enters the digester, with the nitrate being converted to nitrogen gas under anoxic conditions. The advent of aerobic-anaerobic-anoxic biological nutrient removal has greatly reduced this problem.

Inhibition testing

Laboratory screening to test the biodegradability of wastewaters and assess the possible inhibitory effects of wastewaters and chemicals should always be carried out prior to anaerobic digestion (HMSO, 1987; 1989).

Screening tests fall into two categories: biodegradability and inhibition/toxicity, with both relying on gas production. Standard tests for anaerobic digestion have been produced by all the main standards organizations, including the International Organization for Standardization (ISO), British Standards Institution (BSI), American Society for Testing Materials (ASTM), Deutsches Institut für Normung (DIN), Comité Européen de Normalisation (CEN), Organization for

Economic Cooperation and Development (OECD), and European Chemical Industry Ecology and Toxicology Centre (ECETOC). Although, in practice, they are all quite similar, individual countries tend to use their own standard methods, with the ISO and OECD procedures the most widely used. There is also an array of research-based tests, some of which are more rapid than the conventional standard test methods (Kawahara *et al.,* 1999; Astals *et al.,* 2015).

Anaerobic biodegradability can be measured using the test method of either the ISO (ISO 11734:1995), the OECD (OECD 311:2006), or ASTM (ASTM E1192-92) (ISO, 1995; ASTM, 1992; OECD, 2006). Two tests have been developed by the ISO for determining the inhibition of gas production by test chemicals, using either a concentrated anaerobic digesting sludge (ISO, 13641-1: 1003) or diluted sludge (ISO 13641-2: 2003) (ISO, 2003a; 2003b). A similar test has been published by the OECD (OECD 224: 2007). In this test, samples of actively digesting anaerobic sludge, with a total solids concentration of between 20 and 40 g L^{-1}, are mixed with a readily degradable soluble substrate, to which a range of different concentrations of the test chemical or wastewater is added and then incubated at 35 ± 2°C in a sealed bottle for 3 days (Fig. 5.14). The gas produced, both methane and carbon dioxide, is measured using a pressure gauge to monitor the increase in pressure in pascals (Pa). The percentage difference between the gas production at different concentrations of the test compound to the control allows the percentage inhibition to be plotted against the Log concentration of the test compound, from which the EC_{50} can be calculated (OECD, 2007).

Test methods have been reviewed by Müller *et al.* (2004) and Rozzi and Remigi (2004), while Gartiser *et al.* (2007) give an excellent example of using ISO tests to assess the anaerobic inhibition and biodegradation of antibiotics.

5.4.2 *Method of Sludge Addition to Digester and Commissioning*

Digesters can be operated either as batch or continuous processes that may incorporate the recycling of solids, gas, or both. For optimum performance, sludge should be added to the digester as frequently as

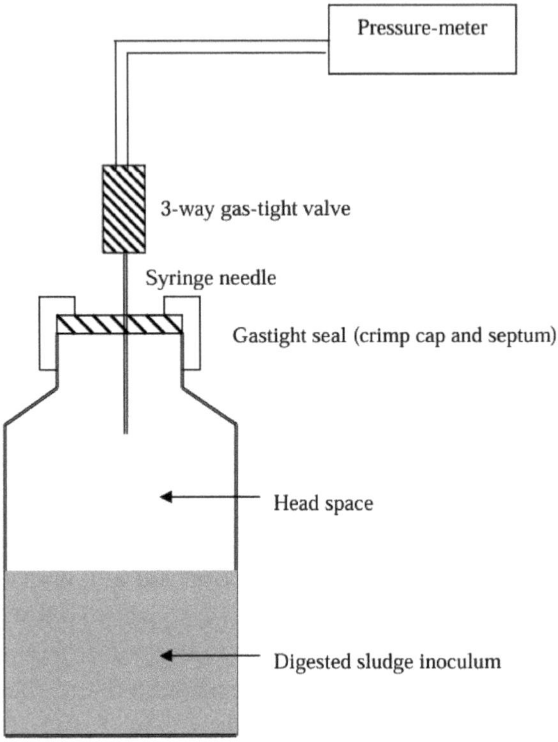

Figure 5.14. Test bottle for anaerobic biodegradability and toxicity testing used in ISO testing.

possible in order to avoid fluctuations in gas production or problems with scouring the active bacteria from the primary digester. It is not generally possible to feed sludge directly from a sedimentation tank on a continuous basis, as it requires consolidation or thickening before it is suitable for use. Although the majority of digesters are fed with raw sludge at least once a day, two or three times a day is preferable.

Commissioning digesters is a topic of some controversy, with two approaches generally used. One is to fill the tank with settled sewage and seed sludge; the other is to fill the tank with primary sludge and add a seed if available. In the first method, the digester is filled with settled sewage to a level where circulation and heating can be operated. While the settled sewage quickly becomes anaerobic, it is advantageous if the headspace and gas piping are purged with nitrogen gas to exclude all oxygen. Once the digester has reached its operating temperature, the seed sludge can be added, equivalent to a minimum of 5% of the total digester volume, with the sludge feed then added slowly over several weeks. The advantage is that the amount of volatile acids is limited, ensuring correct operating pH and alkalinity. However, if the seed sludge becomes too diluted, the correct balance of microorganisms is not maintained, resulting in incomplete digestion and producing unpleasant odors. Although it is possible to use raw sludge, which has been stored for several months as a seed, it is best to use sludge from another digester, preferably from the same plant. The quality of the seed sludge is very important and should only be taken from an operating digester with a good methanogenic population and biogas production rate. The ratio of seed sludge to raw sludge should be between 1:10 and 1:5. Once started, the percentage of seed can be increased, but care must be taken to prevent excessive acid production from inhibiting the development of methanogens. There will be little biogas production for the first 4–6 weeks, so an alternative source of heating the reactor during commissioning will be necessary. Examples of full-scale digester commissioning using this method have been detailed by Hatzigeorgiou *et al.* (2006) and experimentally without using a seed sludge by Dohdoh and Aboullfotoh (2017), who achieved a volatile fatty acid/alkalinity (V/A) ratio of 0.2 after 55 days, and stable operation after 73 days, taking much longer than when a seed was used.

The alternative method is to gradually fill the digester with dilute primary sludge together with the seed sludge. The digester is then heated and mixed (i.e., with a sludge circulation pump and heat exchanger, as there will be no digester gas to operate a gas mixing system), which can take up to 4 days to reach optimal mesophilic temperature. Once at the required temperature, seed sludge can be added if available. The volatile solids concentration, pH, and alkalinity should be monitored throughout. Thickened sludge can be slowly added, ensuring that the <5% new volatile solids are added each day. An increase in the loading rate is governed by the process parameters, and if they fall outside the recommended range, feeding should be temporarily reduced or stopped. Once digester gas becomes available, it can then be used for internal mixing and heating via CHP. Normal loading is achieved after 30 days when a seed has been used, or 60–90 days without the addition of a seed. It is not recommended to gradually fill the digester with sludge as it is not possible to commence circulation or heating until the reactor is full, which can lead to problems in starting up due to the compaction of solids at the base. A range of methods can be found in USEPA (1976).

Key process control parameters should be regularly monitored during commissioning and during normal operation (Bajpai, 2017). The pH should be as close to neutral as possible — if the pH drops, this indicates an increase in volatile acid production. VFA concentration of the digesting sludge should decrease as methanogenesis begins; if it rises, this is an indicator of problems. Biogas composition should show a slow decrease in carbon dioxide composition and a rise in methane; if the opposite starts, again, this indicates a problem. Temperature must stay constant and any instability will impact methanogenesis. Monitoring should include the flow rate (m^3d^{-1}) and solids concentration (%) of the feed sludge. Digester efficiency can be measured by calculating the reduction in volatile solids, which should be 40–60%, although overall gas production per kg VS consumed is also used. The V/A ratio is calculated as the volatile acids concentration in digesting sludge (mg L^{-1}) divided by the alkalinity (mg L^{-1}) and should be <0.1. It is an important parameter that can indicate process instability during commissioning, in particular. If the ratio increases, the pH will drop as the volatile acids build up, more CO_2 is produced in the biogas, and methane formation declines due to the inhibition of methanogenesis (Sec. 5.4.5).

5.4.3 *Internal Mixing and Circulation*

The purpose of mixing and circulation within the digester is: (i) to promote close contact between the raw and digesting sludges; (ii) to maintain a uniform temperature and solids mixture throughout the tank and prevent localized accumulation of inhibitory substances; (iii) to discourage scum formation and settlement of grit and dense solids; and, (iv) most important of all, to encourage the release of gas from the sludge in the lower regions of the digester. Poor mixing will lead to stratification within the digester and will result in partially digested sludge being withdrawn. Efficient mixing turns the conventional plug-flow reactor into a high-rate digestion process by making use of the total reactor volume and ensuring a faster reaction rate due to the removal of mass transfer limitations of food and microorganisms. Mixing can be done either mechanically or by the recirculation of biogas. Mixing is enhanced by the circulation of sludge through the heat exchanger that operates continuously. The evolution of gas, which reaches a maximum of 2–3 hours after the addition of raw sludge, will also supplement the mixing effect. The actual mixing mechanism may not be operated continuously at all plants.

The need for optimal mixing to maximize digestion has led to alternative reactor designs. In small to medium-sized digesters, this has resulted in the development of egg-shaped reactors that can achieve near-perfect mixing, thereby avoiding stagnant areas of settled debris and sludge as well as reducing scum formation (Fig. 5.15). Small to medium-sized conventional tank digesters are very susceptible to particulate settlement and scum formation, seriously reducing mixing

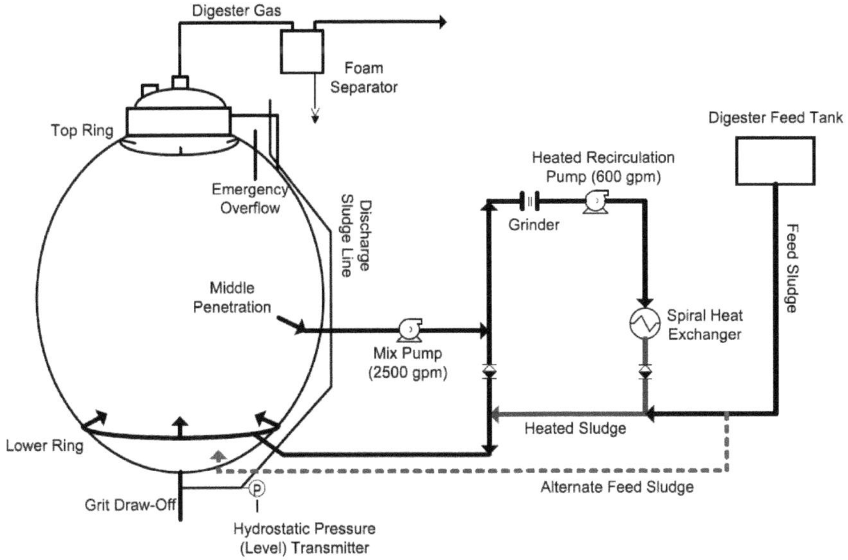

Figure 5.15. Design of an egg-shaped anaerobic digester and heat exchange system (Subramanian *et al.*, 2015). Reproduced with permission of Elsevier Publishing.

and operational efficiency (Figs. 5.16 and 5.17). This problem has become worse as the diameter of shallow digesters is increased, with small ovoid-shaped mixing zones formed within the tanks where the reaction between the liquid sludge and biomass occurs. In contrast, egg-shaped digesters ensure the entire area of the reactor is utilized for treatment (Dichtl, 1997).

Scum, which is composed of grease, oil, and soaps with floating material entrained, will tend to form on the surface of the digestion tank if mixing is not adequate. It causes a number of problems, such as

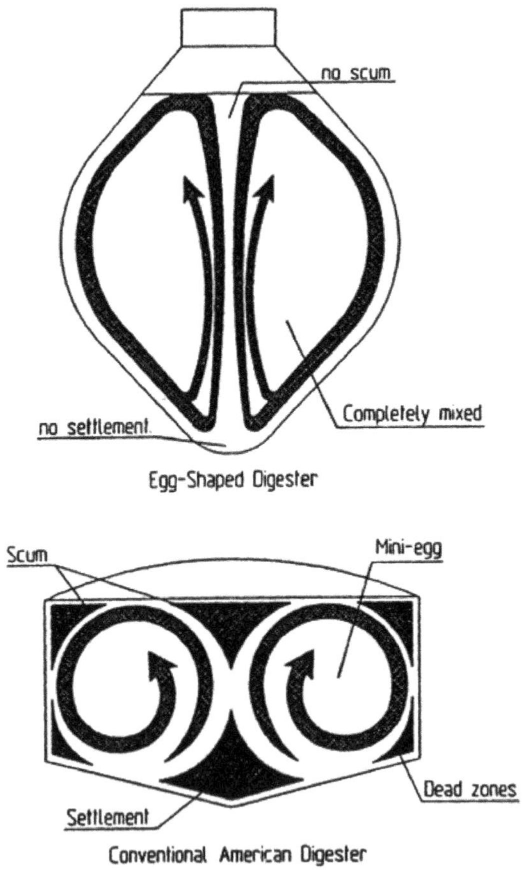

Figure 5.16. Mixing patterns in conventional and egg-shaped digesters showing potential for scum and sediment accumulation (Ditchtl, 1997). Reproduced with permission of the Chartered Institution of Water and Environmental Management, London.

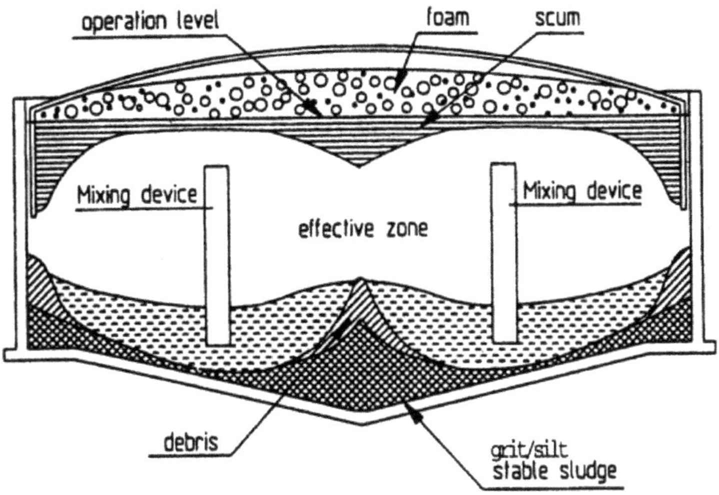

Figure 5.17. Cross-section through a conventional (American) digester (Ditchtl, 1997). Reproduced with permission of the Chartered Institution of Water and Environmental Management, London.

reducing the effective capacity and thus the retention time of the reactor, interfering with mixing and the internal circulation of the sludge, interfering with the movement of the floating gas holder, and restricting the evolution of gas. Most mixing systems withdraw the sludge from the base of the tank and spray it over the surface, preventing scum formation, and screw pumps can be reversed to carry any scum down into the body of the tank. Prevention is best, however, and the sludge should be adequately screened before entry to the digester, and excessive grease and oil should be removed prior to discharge to the sewer. There is also a number of design features aimed at reducing scum formation, including scum removal devices, which disperse sludge by discharging raw or digested sludge into the surface layer via jets or nozzles.

5.4.4 *Solids Retention Time*

The period that solids are retained in the digester is a crucial factor affecting performance. In flow-through systems, the residence time of the waste (HRT) and microbial biomass (MCRT) will be the same. If complete digestion is to be achieved in heated mesophilic digesters, then minimum retention times of more than 15 days are required to ensure that the methanogenic bacteria, which are slow-growing, are able to accumulate to a sufficient population density. The design of digesters normally aims at a minimum retention period of 25–30 days, allowing for some loss in digester capacity due to accumulated non-biodegradable solids within the reactor. Clearly, any increase in the water content of the raw sludge will increase the HRT, and in flow-through systems, this will cause an increased wash-out rate of the microbial biomass and an associated reduction in performance. As mentioned earlier, the minimum MCRT is dependent on the operating temperature of the reactor.

5.4.5 *Temperature*

Anaerobic digestion can occur over a wide temperature range (Table 5.6), which has generally been subdivided into three separate ranges: psychrophilic (5–25°C), mesophilic (25–38°C), and thermophilic (50–70°C) digestion. The rate of anaerobic digestion and gas production is temperature-dependent (Fig. 5.18), with optimum gas production at the higher temperature ranges. Therefore, the warmer the reactor, the shorter the MCRT needs to be for complete digestion (Table 5.7).

Table 5.6. Examples of optimum ranges for pH and temperature for different methanogen genera.

Genus	pH	Temperature (°C)
Methanothermus	6.5	40–85
Methanogenium	6.5–7.8	30–35
Methanolacinia	6.6–7.2	35–40
Methanospirillium	6.6–7.4	37
Methanomicrobium	6.7–7.0	36–38
Methanosphaera	6.8	30–40
Methanolobus	7.0–7.2	37
Methanococcoides	7.0–7.5	30–35
Methanohalobium	7.0–7.5	40-55
Methanothrix	7.1–7.8	37

Table 5.7. Suggested mean cell residence times (MCRT) for the anaerobic digestion of sewage sludge at various temperatures.

Temperature (°C)	MCRT (days)
18	28
20	22
25	18
30	14
35	10
40	8–10

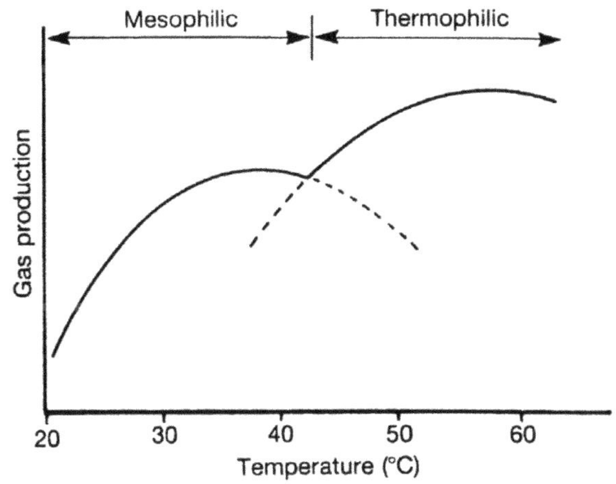

Figure 5.18. Effect of temperature on gas production.

Septic tanks, Imhoff tanks, sludge lagoons, and unheated sludge digesters, which are located mainly at smaller treatment plants, fall into the psychrophilic range. At ambient temperatures in temperate climates, the rate of digestion is so low that the residence time of the bacteria needs to be of the order of 3–12 months. Suitable bacteria for seeding such systems acclimatized to these low temperatures can be obtained from marshlands. The majority of heated sewage sludge digesters operate in the mesophilic range, usually between 30 and 32°C, with maximum residence times of 20–40 days. Sludge residence time (Sec. 5.4.6) is significantly reduced where mixing is enhanced and where the sludge is pre-treated (Sec. 5.3). Thermophilic digestion is technically feasible, and a number of plants have been built and operated successfully. However, the rate of gas production does not increase continuously with temperature but rapidly declines after reaching an optimum at 55°C. A similar situation is found in the mesophilic range at 35°C (Fig. 5.18), indicating that different microbial populations are responsible for thermophilic and mesophilic digestion rather than thermo-tolerant mesophilic species. It appears that between these optimal temperatures, erratic gas production will be encountered. Inhibition of methanogenic archaea occurs at about 63°C (Pfeffer, 1979). The effect of temperature is not great on the first stage of anaerobic digestion, as so many different species are involved that the operating temperature will always fall within the optimum range of some of the microorganisms present. However, the acetogenic and methanogenic bacteria are particularly sensitive to temperature, with even a 2–3°C drop in a mesophilic digester adversely affecting biogas production.

Once operational, a heated digester will be adversely affected if the temperature is allowed to fluctuate by more than just a few degrees. It is general practice to ensure that the operating temperature is near the top of the preferred range before the onset of winter. However, a fall in temperature can be caused by a number of factors, including inadequate heating capacity, scaling of the heat exchanger surfaces, and the raw sludge having a low solids concentration. The temperature can be raised by reducing heat losses from the whole digestion unit by adequate insulation, increasing the heat input, reducing the water content of the raw sludge, and thus reducing the total volume to be heated, and by descaling heat-exchanger surfaces.

Biological hydrolysis can be used as part of a two-stage digestion process, with hydrolysis taking place in the first (thermophilic) digester, followed by mesophilic digestion, a process known as temperature-phased anaerobic digestion (TPAD) (Sec. 5.3.1). The first stage has a short retention time of 2–5 days at 55–65°C, with the subsequent digestion phase operated at 35°C for 10–30 days (Ge *et al.*, 2010;

Wang *et al.*, 2018). Studies on TPAD have provided insight into the effect of temperature on methanogenesis, in particular, with changes in temperature causing significant changes in methanogen abundance, with different mechanisms of methanogenesis also reported at different temperatures. For example, Ho *et al.* (2014) found *Methanosarcina*, an acetoclastic methanogen, dominated at 55–60°C, while at 65°C, the hydrogenotrophic methanogen *Methanothermobacter* dominated. Chen and Chang (2020) also studied the changes in methanogen diversity and abundance from mesophilic to thermophilic conditions. With increasing temperature, there is a discernible shift from acetoclastic methanogenesis to hydrogenotrohic methanogenesis dominance, with changes in diversity discernible. This was confirmed by Steiniger *et al.* (2023) noting a change from acetoclastic (e.g., *Methanosarcina* and *Methanothrix*/*Methanosaeta*) to hydrogenotrohic/methylotrophic methanogenesis (e.g., *Methanothermobacter*/*Methanoculleus*), with temperatures in laboratory digesters operated at 37, 43, 47, and 53°C, with a loss in species diversity for both bacteria and archaea. The bacterial families Caldatribacteriaceae and Coprothermobacteraceae increased with temperature while the Anaerolineaceae decreased while Synergistaceae dominated at the mid-temperature range (Fig. 5.19). The methanogen dominance changed with temperature from *Methanosaeta* (68%) at 37°C to *Methanosarcina* (97%) at 43°C and continued to dominate at 47°C (92%) and 53°C (69%), with *Methanothermobacter* (30%) also common at the highest temperature studied (Fig. 5.20). This study showed that both bacteria and Archaea had a high degree of adaptability over the temperature range between the normal mesophilic and thermophilic operating ranges, intriguingly offering the possibility of successful operation at temperatures between the two. However, acetoclastic methanogenesis can dominate thermophilic digesters under certain conditions, such as high pH or high ammonium concentrations. While numbers of thermophilic methanogens are recorded at mesophilic temperatures, they only dominate under thermophilic conditions. de Vrieze *et al.* (2015) identified three distinct clusters of bacteria with ammonia and temperature. Bacteria in mesophilic digesters are dominated by the phyla Firmicutes,

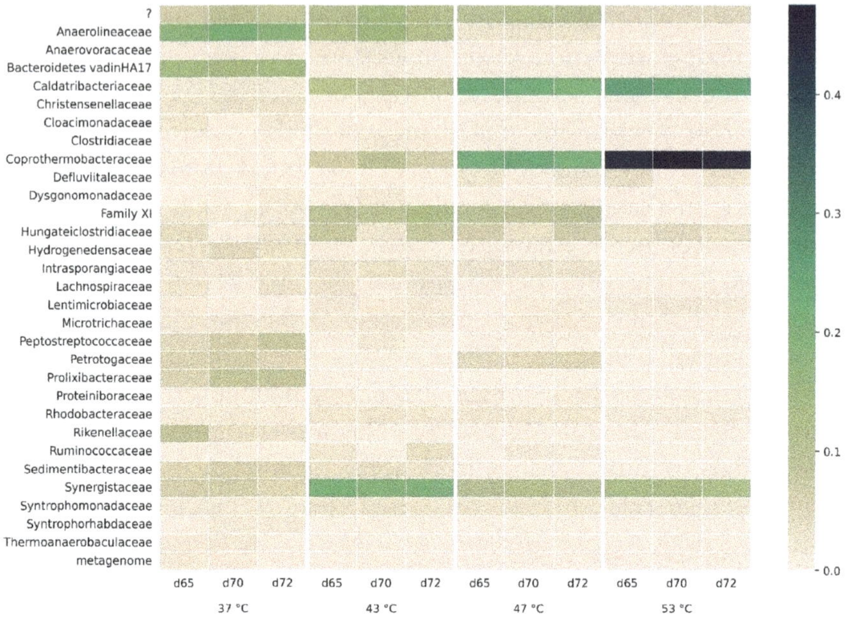

Figure 5.19. Bacterial families detected in biogas reactors operated at four different temperatures after 65, 70, and 72 days. The color code indicates the relative abundance within each sample. Families with an abundance <1% were excluded (Steiniger *et al.*, 2023). Reproduced under the Creative Commons Attribution (CC BY) license (https://creativecommons.org/licenses/by/4.0/).

Proteobacteria, Bacteriodetes, and Chloroflexi, while in thermophilic digesters, Firmicutes, Proteobacteria, Chloroflexi, and Actinobacteria tend to dominate.

5.4.6 *pH and Alkalinity*

Most anaerobic treatment systems have problems with pH control, which arises from differences in the growth rate of the synergistic microbial populations. The activity of the acid-producing bacteria tends to reduce the pH of digesting sludge from the ideal 6.8–7.4 range required by methanogenic archaea, with optimal activity observed at pH 7.0–7.2 (Table 5.6).

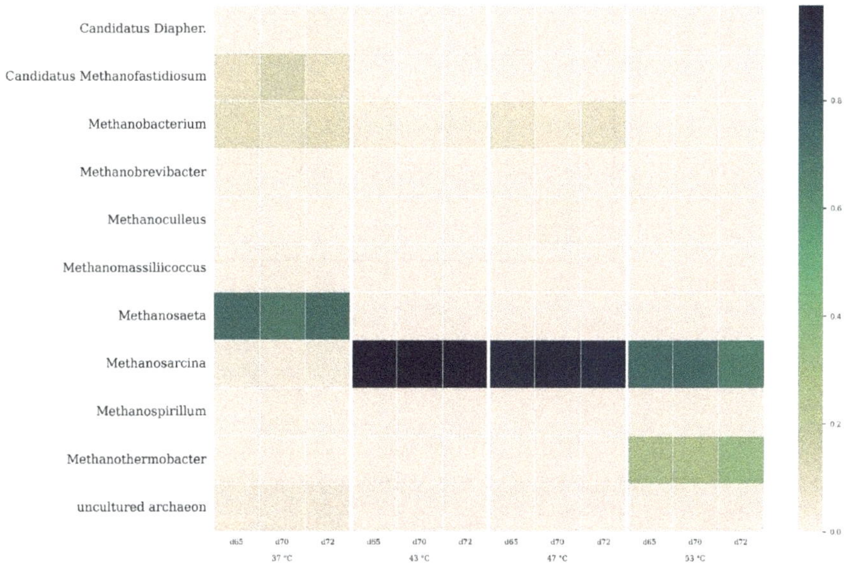

Figure 5.20. Genera of archaea detected in biogas reactors operated at four different temperatures after 65, 70, and 72 days. The color code indicates the relative abundance within each sample. Families with an abundance <1% were excluded (Steiniger *et al.,* 2023). Reproduced under the Creative Commons Attribution (CC BY) license.

Alkalinity is the key driver of pH, so under normal operating conditions, once a stable population of each of the groups has been established, an equilibrium is maintained by the buffering action of ammonium bicarbonate (the bicarbonate alkalinity); hence, no external pH control is required. The bicarbonate ions are derived from carbon dioxide in the digester gas, and the ammonium ions derive from the degradation of proteins in the raw sludge. So, the pH is closely linked to the carbon dioxide concentration in the biogas, which is removed by methanogen metabolism sustaining alkalinity and pH while also improving biogas quality at the same time. Carbon dioxide can form carbonic acid, bicarbonate, or carbonate alkalinity. The three, as well as ammonia and ammonium ions, are in equilibrium with their dominance depending on the pH within the digester, with bicarbonate alkalinity the major source of carbon for methanogens.

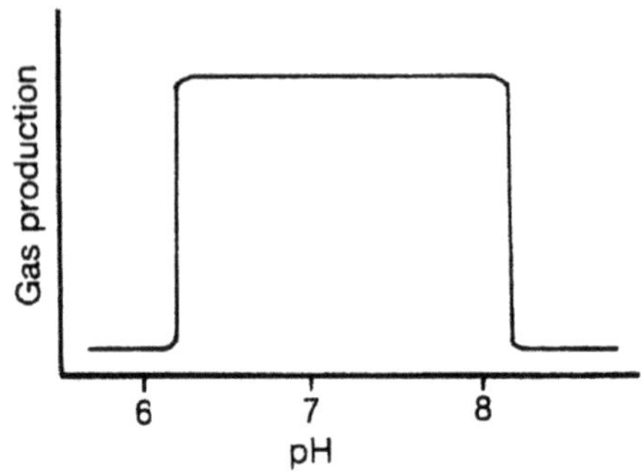

Figure 5.21. The effect of pH on biogas production.

However, the digesting sludge has a tendency to become acidic, especially if the methanogens are inhibited, or the digester is overloaded, which results in an excessive accumulation of VFAs. Under these conditions, the buffering capacity may be exceeded, with the pH rapidly decreasing to 6.0, causing the process to fail, resulting in a sudden decline in gas production (Fig. 5.21). Methanogenic bacteria exhibit a negative response when the pH shifts toward the acid region, as they do when the temperature falls. Growth of methanogens is inhibited below pH 6.2, although the fermentative bacteria will continue to function until the pH has dropped to 4.5–5.0. It should be remembered that the high concentrations of VFAs that are likely to occur in digesters are not toxic to methanogenic bacteria in themselves — it is the pH that is inhibitory, with concentrations of acids up to 11,800 mg L^{-1} observed as non-toxic to methanogens (Velsen and Lettinga, 1979; Newell, 1982). Therefore, any continuous downward trend in pH or alkalinity is an important warning sign requiring immediate attention. The measurement of pH must be done rapidly, as samples of digesting sludge, once exposed to the atmosphere, will rapidly lose carbon dioxide and cause erroneously high pH values. So wherever possible, monitoring systems should be within the reactor itself, giving real-time data and linked to a SCADA system.

As long as the sludge has a fairly high alkalinity, which is especially the case with thick sludges, an increase in acid production will initially produce little effect on the pH. Therefore, in practice, the measurement of volatile acids is a better control factor of the buffering

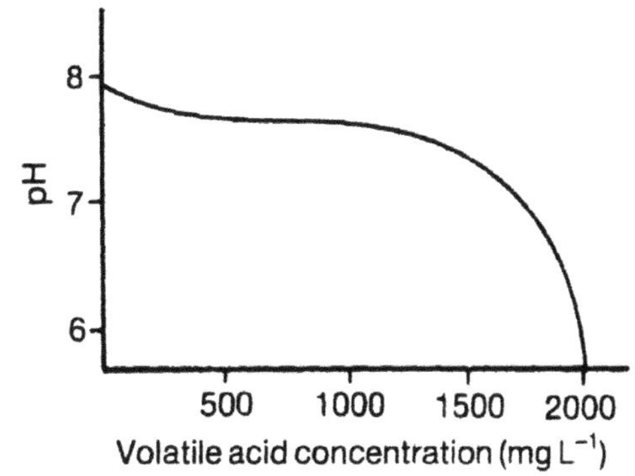

Figure 5.22. Effect of volatile acid production on the pH in a digester.

capacity within an operating digester. Any change in the loading of the digester must be gradual to ensure that the concentration of VFAs does not exceed the normal buffering capacity of the system. Normal VFA concentrations in sewage sludge digesters are between 250 and 1000 mg L^{-1}, with values in excess of 1800–2000 mg L^{-1}, indicating a problem (Fig. 5.22). Determination of individual VFAs is also very useful, as a shift to higher VFAs, such as acetic to butyric, is a sign of instability, and remedial action is required. Provided adequate buffering capacity is available, higher VFA concentrations may be tolerated, although this will lead to incomplete conversion of biodegradable material to gas, with a subsequent increase of the BOD_5 of the digester effluent. Depending on the chemical nature of the sludge, it is possible to have rapid and effective digestion at all pH values between 6.2 and 7.8, although certain trade and sewage wastes restrict rapid digestion to much smaller ranges. Mosey (1983), in his excellent review on anaerobic processes, gives two interesting examples. Ferrous carbonate can become sufficiently soluble below pH 6.4 to release inhibitory concentrations of ferrous ions into solution, whereas, above pH 7.5, an increasing percentage of ammonium ions will be converted to toxic dissolved ammonia gas. The latter example is particularly important in relation to strong sewage and farm wastes.

A loss of alkalinity can also be caused by toxic compounds in the feed sludge, which inhibit the methanogens, or a shock load of organic acids, which can occur in some industrial and food processing scenarios. Proteinaceous wastes tend to produce high alkalinities due to the formation of ammonia and carbon dioxide, while in comparison, non-proteinaceous wastes produce carbon dioxide only. Foaming can also reduce alkalinity, leading to the inhibition of methanogenesis.

The pH value can be neutralized within the reactor by the addition of an alkali. The cheapest and most widely used alkali in full-scale digesters is probably calcium hydroxide (lime). Lime is extremely efficient at increasing the pH to about 6.4–6.5. However, further increases in pH can only be achieved by precipitating most of the dissolved carbonates as calcium carbonate. Therefore, at pH >7.0, the lime reacts with carbon dioxide, which not only results in serious scale formation but reduces the pressure of the gas phase above the sludge by removing the carbon dioxide, which could seriously affect the structural stability of the reactor. Bicarbonates or carbonates of either sodium or potassium are generally used to raise the pH from 6.5 to 7.0, although excessive use of these salts can result in partial or severe inhibition at concentrations of 3,500–5,500 and >8,000 mg Na L^{-1} and 2,500–4,000 and >2,000 mg K L^{-1}, respectively. Other alkalis are less suitable; for example, sodium hydroxide also removes carbon dioxide from the solution, and the use of ammonia or ammonium ions can inhibit the process. For laboratory-scale digestion units, the most useful reagents for adjusting the pH are hydrochloric acid and sodium bicarbonate.

If the sludge contains heavy metals in solution at concentrations likely to cause inhibition, the addition of alkali to raise the pH to 7.5–8.0 will precipitate most of the metal ions out of solution, usually as carbonates, thus reducing the inhibitory effect. Sulfide can also be used to precipitate the heavy metal ions without causing a significant pH change. Apart from chemical buffering, pH control can also be achieved by using other wastes or by increasing the sludge recycle rate where the reactor design allows, such as fluidized beds or downflow biofilters (Chap. 6) (Wheatley, 1985).

5.4.7 *Foaming*

As anaerobic digestion is increasingly employed as an intergral part of a carbon net-zero plan for WWTPs, concerns over foaming have increased globally due to potentially serious impacts on process efficiency and operational costs (Ganidi *et al.*, 2009). The effects of foaming can be extremely costly, with a loss of methane generation leading to an increase in energy costs to operate digesters as well as applying remediation procedures. The main effects of foaming are inefficient gas recovery; an inverse solids profile, resulting with the densest sludge layer at the top of the digester; creation of dead zones, which leads to a reduction of the active volume of digesters, resulting in poor stabilization of sludges; blockages of gas mixing devices and sludge recirculation pipes; possible egress of foam from between the floating covers and the digester wall; and possible damage to the covers of certain designs of digesters. Gas production in full-scale digesters can be reduced by between 30 and 50% during foaming, with 90% reductions in extreme conditions.

Foaming is where the biogas formed during digestion does not escape from the liquid phase to the gas space at the top of the digester; rather, it becomes entrained within the sludge (Yang *et al.*, 2021). Foam is the dispersion of gas bubbles in a liquid or solid. In digesters, the bubbles are dispersed within the sludge, and when they rise to the surface, the water layer (lamella) surrounding the bubble thins so that it bursts, releasing the gas. Surface active agents, commonly referred to as surfactants, stabilize the foam by their molecules becoming attracted to the bubbles, making the lamella stronger so that it must become much thinner than normal before rupturing. So, the presence of surfactants will result in stronger bubbles and more stable foams. Hydrophobic particles also form foams. These particles are poorly wetted by water, producing complex foams made up of dispersed phases of hydrophobic particles and air within the water phase. The particles congregate at the gas–liquid interface, making the foam even stronger. Jenkins *et al.* (1993) described how hydrophobic particles, if they are large enough, bridge the water film between bubbles, creating what is, in essence, a dam that prevents water from draining away, thus preventing the film from thinning and rupturing (Fig. 5.23). As gas bubbles are formed, they become coated with surface active or hydrophobic substances, which increase as they rise and then become attached to solids and microorganisms. The foam slowly expands

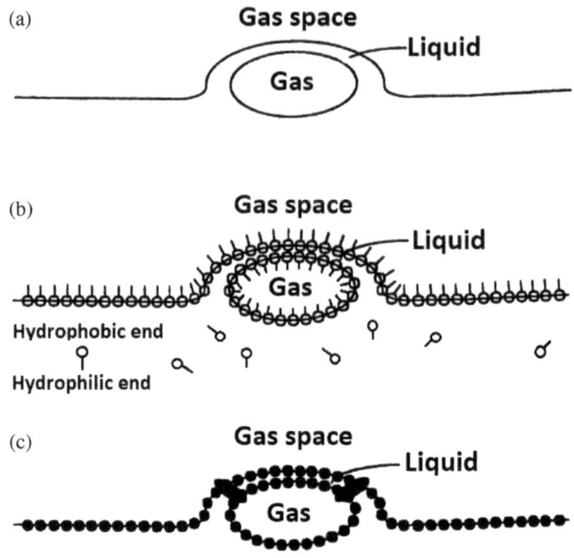

Figure 5.23. Various types of headspace foams in anaerobic digesters. (a) No surface active material is present. (b) Surface active materials present that accumulate at the gas–water interface and stabilize it. (c) Hydrophobic solids present that accumulate at the interface and stabilize it, with larger solid particles forming bridges across the gas bubbles, preventing the drainage of water from it.

upward where it then blocks the gas outlet, which can lead to an increase in pressure, causing the digestate to overflow and even reactor damage, or, in the worst case, an explosion. The foam causes the microbial biomass to become concentrated within the foam, reducing free mixing and contact between the substrate (sludge) and the biomass, leading to a decrease in methanogenic activity and biogas production. When the entrained bubbles cannot rise to the surface due to the viscosity of the sludge, the foam is then found throughout the reactor rather than just at the surface. This phenomenon was first described by Chapman and Krugel (2011), who called it rapid volume expansion, as the foam expands evenly and quickly, similarly to how yeast dough expands (Fig. 5.24). It is generally found in reactors where digestate viscosity is high, as found with food wastes, with the volume of the sludge expanding rapidly, with digestate overflow occurring within a very short time frame, generally less than an hour.

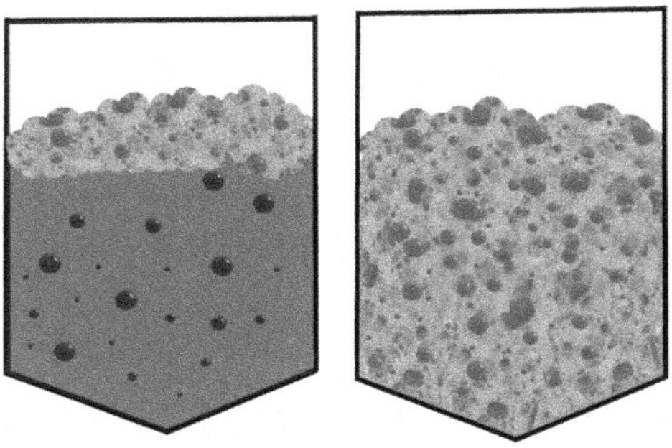

Figure 5.24. Artistist's impression of (a) conventional foaming and (b) rapid volume expansion in anaerobic digesters (Yang *et al.*, 2021). Reproduced with permission of Elsevier Publishing.

Numerous causes of foaming have been suggested, including organic overload, accumulation of acetic acid, inadequate mixing, temperature fluctuations, shock loadings, and the presence of hydrophobic and extracellular polymeric substances (EPS). While all these factors are known to interfere with the operational efficiency of digesters, the presence of microbial hydrophobic filaments appears to be the major cause being directly associated with foam structure, including its formation and stability, although the proliferation of foam-causing microorganisms does require specific abiotic conditions.

Microbial foams produced by surfactant-producing and hydrophobic organisms are more viscous and stable than chemical ones. Foaming is a common phenomenon in the activated sludge process (Gray, 2023), and is associated with the digestion of sludges comprising waste activated sludge (WAS). Two common activated sludge foaming species were identified as causing foaming in digesters in the 1990s: *Giordonia* (formally *Nocardia*) spp. (Hernandez and Jenkins, 1994; Pagilla *et al.*, 1997) and *Microthrix parvicella* filaments (Westlund *et al.*, 1998), and since then, increasing evidence has emerged to support hydrophobic filamentous microorganisms as a key causative factor for anaerobic foaming. *Microthrix* and *Gordonia* are known to only survive under strict anaerobic conditions for a few days (Kirkegaard *et al.*, 2017) with foaming, in this case, linked to the continuous addition of organisms via the WAS. So, dealing with foaming at the activated sludge stage is probably the best approach to the control of foaming in digesters. EPSs can be either hydrophobic or hydrophilic, with the former directly correlated to anaerobic foaming and the latter contributing to foaming by increasing sludge viscosity. Foam stability is linked to hydrophobic proteins, which reduce surface tension, while the role of humic substances, which contain functional groups (e.g., amino hydroxyl), increases the viscosity of the bubble lamella (Yang *et al.*, 2022; 2023). However, not all the filamentous organisms, both bacteria and archaea, isolated from anaerobic foams, have their origin from activated sludge, with a number found only in strictly anaerobic environments. Jiang *et al.* (2021) studied 13 mesophilic digesters at 11 different WWTPs in Denmark and identified the organisms responsible for foaming, using 16S rRNA gene amplicon sequencing with species-level resolution and fluorescence *in situ* hybridization (FISH) for visualization (Fig. 5.25). Foaming was categorized into three categories: non-foaming (Group 1), unstable foams (Group 2), and stable foams (Group 3). They found a significant difference in species composition between non-foaming (Group 1) and foaming (Group 3) sludges. Groups 2 and 3 foams were both dominated by microorganisms producing surfactants (e.g., archaeal species *Methanospirillum*) while Group 3 was also rich in microorganisms with hydrophobic cell surfaces (e.g., *Microthrix*). Non-activated sludge species associated with foaming include *Brevefilum fermentans*, Tetrasphaera, Rikenellaceae, Anaerolineae, Actinomyces, Corynebacterium, and *Lactobacillus* and the archaea genera *Methanospirillum* and *Methanothrix* (formerly *Methanosaeta*). Community structure is also distinctly different between mesophilic and thermophilic digesters (Kirkegaard *et al.*, 2017). Subramanian *et al.* (2015) developed a simple foaming potential test using an air diffuser fitted in the base of a 2 L graduated cylinder into which 200 mL of digested sludge is added. Air is added at 1.5 L min^{-1}, which expands the sludge with bubbles. The maximum height the foam attains is an indicator of the foam potential. Unstable foams collapse after the diffuser is closed, while stable foams remain expanded (Eqs. (5.6) and (5.7)):

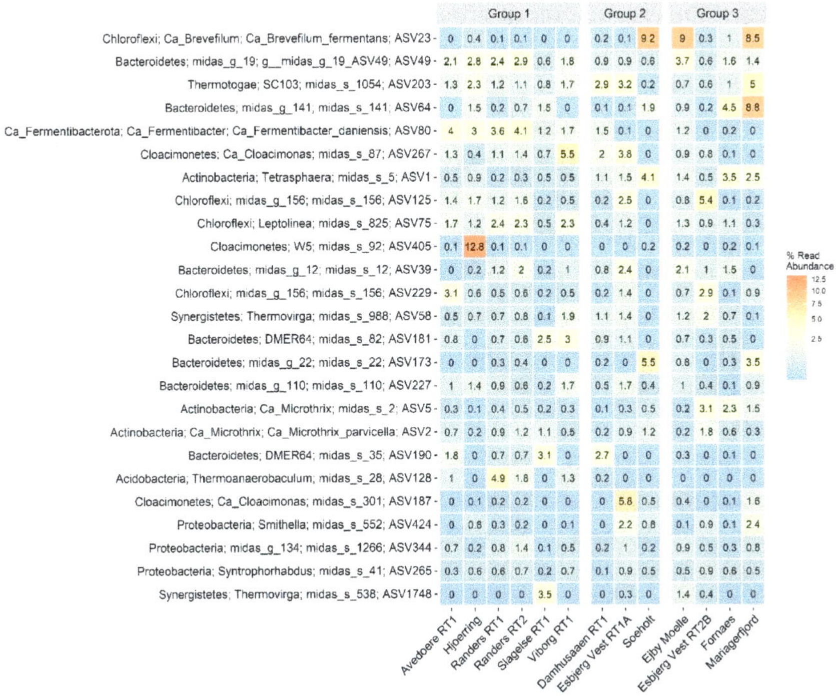

Figure 5.25. Heat map of the 25 most abundant bacterial amplicon sequence variants (ASVs) at 13 mesophilic digesters in Demark, characterized by the foaming propensity test. Non-foaming (Group 1), producing an unstable foam (Group 2), or foaming (Group 3). Taxonomic classification levels presented are phylum, genus, species, and ASV (Jiang *et al.*, 2021). Reproduced with permission of Elsevier Publishing.

$$\text{Unstable foam potential} = \frac{\text{maximum foam height}\,(mL)}{\text{Initial height of sludge}\,(mL)} \tag{5.6}$$

The unstable foam potential gives an indication of the operational or working foam potential, which the authors categorized as: 0–1 (indicating non-foaming), 1–2 (mild foaming), and >2 (average to severe foaming).

$$\text{Stable foam potential} = \frac{\text{Settled foam height}\,(mL)}{\text{Initial height of sludge}\,(mL)} \tag{5.7}$$

The stable foam potential is a measure of permanent foam, in which a value of 0–0.2 is indicative of non-foaming, 0.2–0.5 is indicative of mild foaming, and >0.5 is indicative of average to severe foaming. As with activated sludge foam potential tests, these values will vary between reactor design and operational conditions and so should be used for measuring foam variation at individual plants rather than as a comparative parameter (Gray, 2023).

Foam potential tests are widely used to assess foaming potential and stability in activated sludge, with similar tests used to screen sludge fed into digesters (Jiang *et al.*, 2018). Jiang *et al.* (2021) recorded a strong correlation between foaming potential and the abundance of *Microthrix* spp. in the sludge feed originating from WAS. A similar relationship was seen with the abundance of several true anaerobes that are common in digesters and associated with foaming, including *Brevefilum fermentans*, Tetrasphaera, Rikenellaceae, and the archaeal genus *Methanospirillum* (Fig. 5.26). It is difficult to actually see or monitor foaming within closed digesters, so apart from looking for the presence of indicator microorganisms, the foam potential test is a useful indicator of foaming potential and stability. As in activated sludge, organic overloading may also be a causative factor in foaming due to the accumulation of VFAs, ammonium, and TN (total nitrogen), which can be used as an alternative indicator to the foam potential test as they are all positively and strongly correlated (Jiang *et al.*, 2021). Alfaro *et al.* (2014) observed a direct relationship between foaming and microbial filament density. *Gordonia amarae* was recorded as causing significant foaming in full-scale egg-shaped digesters by Subramanian *et al.* (2015), who found that while excessive induced mixing exacerbated foaming, it could be controlled by reducing mixing or turning it off for short periods, which did not affect overall performance. They were also able to identify a foaming threshold for the actinomycete of 10^6 intersections mg^{-1} VSS in the mixed liquor feed to the digester. The presence of *G. amarae* was found to be seasonal and closely linked to the temperature in the activated sludge reactor, with foaming in the digesters only problematic when the threshold value was exceeded (Fig. 5.27).

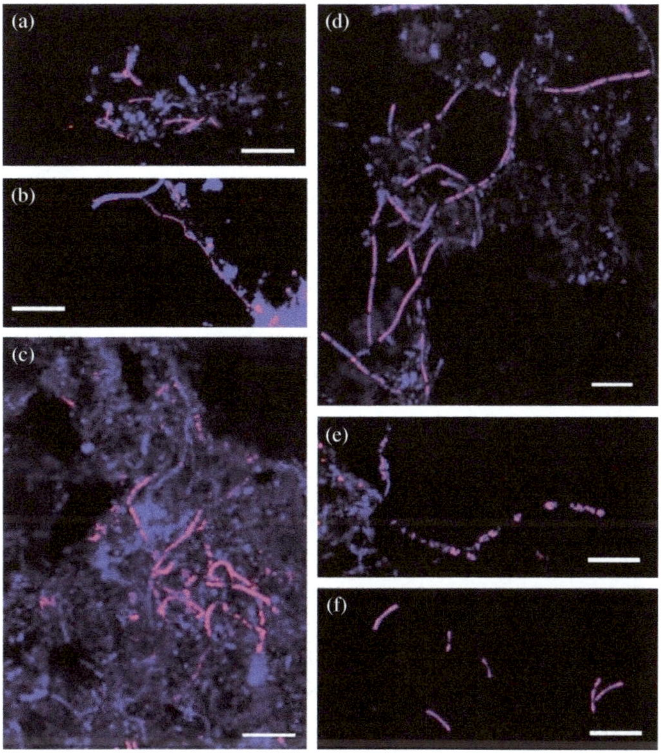

Figure 5.26. Composite FISH micrographs of potential foam-forming microbes isolated in Demark by Jiang *et al.* (2021). (a) *Ca.* Brevefilum spp. (b) *Ca.* Brevefilum spp. (c) *Tetrasphaera* spp. (d) *Ca.* Microthrix spp. (e) midas_g_141 members. (a) to (e) The EUBmix probe set was applied to target most bacteria. Target populations appearing magenta (red + blue), and other bacterial cells appear blue. (f) *Methanospirillum* spp. The *Methanospirillum* cells appear magenta. Scale bar = 10 μm. Reproduced with permission of Elsevier Publishing.

Foaming control is achieved either by preventing foam generation or by controlling foam once it has formed. The best preventative method is to ensure the digester is operated at suitable mixing rates, temperature, and organic loading. Alternatively, the feed can be pre-treated by thermal or ultrasonic hydrolysis, which can destroy those filamentous bacteria associated with activated sludge foaming (e.g., *M. parvicella*) (Alfaro *et al.,* 2014). Emulsification and saponification of lipid-rich substrates have also proven to be effective

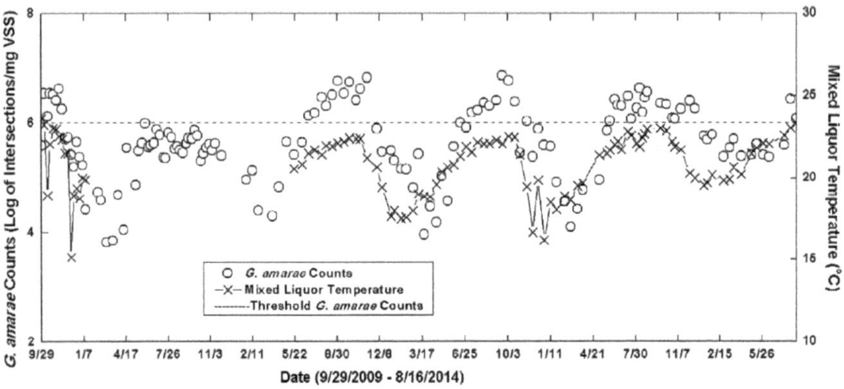

Figure 5.27. Variation of filament density and temperature of *Gordonia amarae* in the activated sludge reactors forming the feed to anaerobic digesters. Foaming in the digesters was triggered when the threshold filament density was reached, which is represented by the dotted line (Subramanian *et al.*, 2016). Reproduced with permission Elsevier Publishing.

(Diamantis*et al.*, 2021). Trace elements have also been successfully added, which are used as co-factors of enzymes that speed up the metabolism of intermediate metabolites. These reduce the viscosity of the digestate, which, in turn, helps to reduce the foam (Moestedt *et al.*, 2016). Foam control methods are not fully effective and are often long-term, making them an expensive option. This involves destabilizing the foam structure, which can be done chemically or physically. Initially, a mechanical approach is taken to disperse the foam, which usually takes the form of either increasing mixing/stirring rates or injecting water under high pressure. The objective is to increase shear stress within the digester, which will weaken the surface tension of the bubble wall, or by removing some of the surfactants at the bubble surface, weakening the bubble and causing them to collapse. However, the addition of large volumes of water lowers the total solids content of the digester, reducing performance. Mechanical solutions are unlikely to solve the problem, which will require either a change in feed or the operating conditions of the digester. The most widely adopted approach is to use anti-foaming or defoaming chemicals, which act directly on the surface of the bubbles, reducing the surface tension or removing the surface surfactants, encouraging the bubbles to rupture and thereby destabilizing the foam structure. There is a wide range of commercial, often synthetic, defoaming chemicals available, although natural oils and buffering agents are also used. Defoaming chemicals are not always successful, and when they are, they can significantly interfere with the performance of the digester in terms of biogas formation. The presence of silicones in synthetic defoamers should be avoided due to the formation of siloxanes in the biogas (Sec. 3.6). Due to the cost of commercial products, organic oils such as vegetable oil or biodiesel are often used in their place. However, over time, the microorganisms present adapt to using them as a substrate, and so their effectiveness quickly declines, requiring increasingly larger dosages for effective control. Lindorfer and Demmig (2016) have suggested that this problem can be overcome by alternating the use of vegetable oil, a commercial defoamer, and biodiesel so that microorganisms do not become too acclimatized to the natural oils as a food source.

Buffering agents include a wide range of compounds and salts, including urea, sodium carbonate, sodium bicarbonate, and ammonium chloride. Generally, large quantities of these compounds are required, with effects taking several days to appear (Kougias *et al.*, 2014).

Biological control is achieved by identifying the main foaming microorganism and targeting it by altering operational conditions, removing it from the feed, or reducing substrate addition, leading to starvation conditions that reduce gas production so that the foaming problem is gradually reduced. In practice, changing the operational conditions of the digester is probably the best and cheapest way that foaming can be controlled; for example, by altering the organic loading by diluting the feed, which also alters the viscosity of the sludge within the reactor (Lindorfer and Demmig, 2016).

The mechanisms of foam formation, assessment, and control have been reviewed by Ganidi *et al.* (2009) and, more recently, by Yang *et al.* (2021).

References

Akunna, J.C., Bizeau, C. and Moletta, R. (1994) Nitrate reduction by anaerobic sludge using glucose at various nitrate concentrations — Ammonification, denitrification and methanogenic activities. *Environmental Technology*, **15**, 41–49. https://doi.org/10.1080/09593339409385402

Aldin, S., Elbeshbishy, E., Nakhla, G. and Ray, M. (2009) Viability of ultrasonication for pre-treatment of biosolids. In *Proceedings of the WEFTEC*, 10–14 October 2009. Water Environment Federation, Orlando, FL, pp. 215–225.

Alfaro, N., Cano, R. and Fdz-Polanco, F. (2014) Effect of thermal hydrolysis and ultrasounds pretreatments on foaming in anaerobic digesters. *Bioresource Technology*, **170**, 477–482. https://doi.org/10.1016/j.biortech.2014.08.013.

Astals, S., Batstone, D.J., Tait, S. and Jensen, P.D. (2015) Development and validation of a rapid test for anaerobic inhibition and toxicity. *Water Research*, **81**, 208–215. https://doi.org/10.1016/j.watres.2015.05.063

ASTM. (1992) *E1192-92. Standard Test Method for Determining the Anaerobic Biodegradation Potential of Organic Chemicals.* American Society for Testing and Materials, Philadelphia.

Bajpai, P. (2017) Process parameters affecting anaerobic digestion. In *Anaerobic Technology in Pulp and Paper Industry.* SpringerBriefs in Applied Sciences and Technology. Springer, Singapore. https://doi.org/10.1007/978-981-10-4130-3_3

Basiliko, N. and Yavitt, J.B. (2001) Influence of Ni, Co, Fe and Na additions on methane production in *Sphagnum*-dominated Northern American peatlands. *Biogeochemistry*, **52**, 133–153. https://doi.org/10.1023/A:1006461803585

Blum, D.J.W. and Speece, R.E. (1992) The toxicity of organic chemicals to treatment processes. *Water Science and Technology*, **25**, 23–31.

Bruce, A.M., Swanwick, J.D. and Ownsworth, R.A. (1966) Synthetic detergents and sludge digestion: Some recent observations. *Water Pollution Control*, **65**, 427–447.

Bungay, S. and Abdelwahab, M. (2008) Monsal enzymic hydrolysis — New developments and lessons learnt. In *13th European Biosolids & Organic Resources Conference & Workshop*. Manchester, UK. https://conferences.aquaenviro.co.uk/wp-content/uploads/sites/7/2018/08/55-Steve-Bungay-1.pdf

Cai, Y., Meng, X., Hu, K., Zhao, X., Usman, M., Esposito, G., Shen, X. and Chen, S. (2023) A novel strategy to reduce trace element supplementation in the semi-solid anaerobic digestion with gradient ammonia concentration: The role of biochar. *Fuel*, **38**, 127332. https://doi.org/10.1016/j.fuel.2022.127332

Cai, Y.F., Zhu, M.M., Meng, X.Y., Zhou, J.L., Zhang, H. and Shen, X. (2022) The role of biochar on alleviating ammonia toxicity in anaerobic digestion of nitrogen-rich wastes: A review. *Bioresource Technology*, **351**, 126924. https://doi.org/10.1016/j.biortech.2022.126924

Cai, Y.F., Gallegos, D., Zheng, Z.H., Stinner, W., Wang, X.F., Pröter, J. and Schäfer, F. (2021) Exploring the combined effect of total ammonia nitrogen, pH and temperature on anaerobic digestion of chicken manure using response surface methodology and two kinetic models. *Bioresource Technology*, **337**, 125328. https://doi.org/10.1016/j.biortech.2021.125328

Cain, R.B. (1981) Microbial degradation of surfactants and 'builder' components. In *Microbial degradation of Xenobiotics and Recalcitrant Compounds* (ed. T. Leisinger *et al.*). Academic Press, London, pp. 325–366.

Casey, T.J. (1981) Developments in anaerobic digestion. *Transactions of the Institute of Engineers in Ireland*, **105**, 25–32.

Chapman, T. and Krugel, S. (2011) Rapid volume expansion — an investigation into digester overflows and safety. *Proceedings of the Water Environment Federation,* **2011**, (4), 1016–1040. https://doi.org/10.2175/193864711802863229.

Chen, H. and Chang, S. (2020) Dissecting methanogenesis for temperature-phased anaerobic digestion: Impact of temperature on community structure, correlation, and fate of methanogens. *Bioresource Technology*, **306**, 123104. https://doi.org/10.1016/j.biortech.2020.123104

Chen, M., Wang, F., Zhang, D.-l., Yi, W.-M. and Liu, Y. (2021) Effects of acid modification on the structure and adsorption NH_4^+-N properties of biochar. *Renewable Energy*, **169**, 1343–1350. https://doi.org/10.1016/j.renene.2021.01.098

Chu, C. P., Lee, D. J., Chang, B. V., You, C. S. and Tay, J.H. (2002) "Weak" ultrasonic pre-treatment on anaerobic digestion of flocculated activated biosolids. *Water Research*, **36**, 2681–2687.

de Vrieze, J., Saunders, A.M., He, Y., Fang, J., Nielsen, P.H., Verstraete, W. and Boon, N. (2015) Ammonia and temperature determine potential clustering in the anaerobic digestion microbiome. *Water Research*, **75**, 312–323. https://doi.org/10.1016/j.watres.2015.02.025

Diamantis, V., Eftaxias, A., Stamatelatou, K., Noutsopoulos, C., Vlachokostas, C. and Aivasidis, A. (2021) Bioenergy in the era of circular economy: Anaerobic digestion technological solutions to produce biogas from lipid-rich wastes. *Renewable Energy*, **168**, 438–447. https://doi.org/10.1016/j.renene.2020.12.034.

Dichtl, N. (1997) Thermophilic and mesophilic (two-stage) anaerobic digestion. *Journal of the Chartered Institution of Water and Environmental Management*, **11**, 98–104.

Ding, H.H., Chang, S. and Liu, Y. (2017) Biological hydrolysis pretreatment on secondary sludge: Enhancement of anaerobic digestion and mechanism study. *Bioresource Technology*, **244**, 989–995. http://dx.doi.org/10.1016/j.biortech.2017.08.064

Dohdoh, A.M. and Aboulfotoh, A.M. (2017) Start-up performance of a mesophilic anaerobic digester without external inoculums. *Environmental Protection Engineering*, **43**, (4), 29–39. https://doi.org/10.37190/epe170403

Donlon, B.A., Razo-Flores, E., Field, J.A. and Lettinga, G. (1995) Toxicity of N-substituted aromatics to acetoclastic methanogenic bacteria in granular sludge. *Applied and Environmental Microbiology*, **61**, 3889–3893.

Duc, L.V., Miyagawa, Y., Inoue, D. and Ike, M. (2022) Identification of key steps and associated microbial populations for efficient anaerobic digestion under high ammonium or salinity conditions. *Bioresource Technology*, **360**, 127571. https://doi.org/10.1016/j.biortech.2022.127571

Esquivel-Elizondo, S., Parameswaran, P., Delgado, A.G., Maldonado, J., Rittmann, B.E. and Krajmalnik-Brown, R. (2016) Archaea and bacteria acclimate to high total ammonia in a methanogenic reactor treating swine waste. *Archaea*, **2016**, 4089684. https://doi.org/10.1155/2016/4089684

EU. (2016) Detergents ingredients database, version 2016. European Union, Brussels. https://api.eublomsten.dk/api/docs/CriteriaDocumentFiles/7883

Fang, H.H.P., Lau, I.W.C. and Chung, D.W.C. (1997) Inhibition of methanogenic activity of starch-degrading granules by aromatic compounds. *Water Science and Technology*, **35**, 247–253.

Ganidi, N., Tyrrel, S. and Cartmell, E. (2009) Anaerobic digestion foaming causes: A review. *Bioresource Technology*, **100**, 5546–5554. https://doi.org/10.1016/j.biortech.2009.06.024

Gartiser, S., Urich, E., Alexy, R. and Kummerer, K. (2007) Anaerobic inhibition and biodegradation of antibiotics in ISO test schemes. *Chemosphere*,

66, (10), 1839–1848. https://doi.org/10.1016/j.chemosphere.2006.08.040

Ge, H., Jensen, P.D. and Batstone, D.J. 2010. Pre-treatment mechanisms during thermophilic–mesophilic temperature phased anaerobic digestion of primary sludge. *Water Research* **44**, 123–130. https://doi.org/10.1016/J.WATRES.2009.09.005.

Gray, N.F. (2023) *Activated Sludge: Developments and Sustainable Solutions.* World Scientific Publishing, Singapore.

Guo, Y., Xiao, F., Yan, M., Tang, S., Duan, Z., Sun, Y. and Li, Y. (2023) Effect of ammonia on anaerobic digestion: Focusing on energy flow and electron transfer. *Chemical Engineering Journal*, **471**, 144638. https://doi.org/10.1016/j.cej.2023.144638

Hatzigeorgiou, M., Owsenek, B., Alkema, T., Sieger, R. and Pallansch, K. (2006) Startup of anaerobic mesophilic digesters. *Proceedings of the Water Environment Federation,* **2006**, (13), 415–412.

Henze, M. and Harremoes, P. (1983) Review paper: Anaerobic treatment of wastewater in fixed film reactors. *Water Science and Technology*, **15**, 1–90.

Hernandez, M. and Jenkins, D. (1994) The fate of *Nocardia* in anaerobic digestion. *Water and Environmental Research,* **66**, (6), 828–835. doi: 10.2175/WER.66.6.10.

Hickey, R.F., Vanderwielen, J. and Switzenbaum, M.S. (1987) The effect of organic toxicants on methane production and hydrogen gas levels during the anaerobic digestion of waste activated sludge. *Water Research*, **21**, 1417–1427. https://doi.org/10.1016/0043-1354(89)90045-6

HMSO. (1987) Determination of the inhibitory effects of chemicals and wastewaters on the anaerobic digestion of sewage sludge. Methods for the Examination of Waters and Associated Materials, Department of the Environment, H.M. Stationery Office, London, UK.

HMSO. (1989) The assessment of biodegradability in anaerobic digesting sludge. Methods for the Examination of Waters and Associated Materials, Department of the Environment, H.M. Stationery Office, London, UK.

Ho, D., Jensen, P. and Batstone, D. (2014) Effects of temperature and hydraulic retention time on acetotrophic pathways and performance in high-rate sludge digestion. *Environmental Science and Technology*, **48**, 6468–6476. https://doi.org/10.1021/es500074j.

Hobson, P.N. (1984) Anaerobic digestion of agricultural wastes. *Water Pollution Control*, **83**, 507–513.

Huss, L. (1977) The Anamet process for food and fermentation industry effluent. *Tribune de CEBEDEAU*, **30**, 390–396.

Institute of Water Pollution Control (1979) *Sewage Sludge. I. Production, Preliminary Treatment and Digestion.* Institute of Water Pollution Control, Maidstone.

ISO. (1995) ISO 11734 Evaluation of the "ultimate" anaerobic biodegradability of organic compounds in digested sludge — method by measurement of the biogas. International Organization for Standardization, Geneva.

ISO. (2003a) ISO 13641-1 Water Quality — Determination of inhibition of gas production of anaerobic bacteria — Part 1: General Test. International Organization for Standardization, Geneva.

ISO. (2003b) ISO 13641-2 Water Quality — Determination of inhibition of gas production of anaerobic bacteria — Part 2: Test for low biomass concentrations. International Organisation for Standardisation, Geneva.

Jarrell, K.F., Saulnier, M. and Ley, A. (1987) Inhibition of methanogenesis in pure cultures by ammonia, fatty acids, and heavy metals, and protection against heavy metal toxicity by sewage sludge. *Canadian Journal of Microbiology*, **33**, 551–554. https://doi.org/10.1139/m87-093

Jenkins, D., Richard, M.G. and Daigger, G.T. (1993) *Manual on the Causes and Control of Activated Sludge Bulking and Foaming*, 2nd edition. Lewis Publishers, Chelsea, MI, USA.

Jha, P. and Schmidt, S. (2017) Reappraisal of chemical interference in anaerobic digestion processes. *Renewable and Sustainable Energy*, **75**, 954–971. http://dx.doi.org/10.1016/j.rser.2016.11.076

Jiang, C., Qi, R., Hao, L., McIlroy, S.J. and Nielsen, P.H. (2018) Monitoring foaming potential in anaerobic digesters. *Waste Management*, **75**, 280–288. https://doi.org/10.1016/j.wasman.2018.02.021

Jiang, C., McIlroy, S.J., Qi, R., Petrigllieri, F., Yashiro, E., Kondrotaite, Z. and Nielsen, P.H. (2021) Identification of microorganisms responsible for foam formation in mesophilic anaerobic digesters treating surplus activated sludge. *Water Research*, **191**, 116779. https://doi.org/10.1016/j.watres.2020.116779

Kang, X, Li, C., Ding, W., ma, Y., Gao, S., Zhou, X., Chen, Y., Liu, W. and Jiang, G. (2023) Optimization of operating conditions in the biological enzymes for efficient waste activated sludge dewatering. *Process Safety and Environmental Protection*, **170**, 545–552. https://doi.org/10.1016/j.psep.2022.12.046

Kawahara, K., Yakabe, Y., Chida, T. and Kida, K. (1999) Evaluation of laboratory-made sludge for an anaerobic biodegradability test and its use for assessment of 13 chemicals. *Chemosphere*, **39**, (12), 2007–2018.

Kepp, U., Machenbach, I., Weisz, N. and Solheim,O.E. (2000) Enhanced stabilisation of sewage sludge through thermal hydrolysis — three years of experience with full scale plant. *Water Science and Technology*, **42**, (9), 89–96.

Kirkegaard, R.H., McIlroy, S.J., Kristensen, J.M., Nierychlo, M., Karst, S.M., Due-holm, M.S., Albertsen, M. and Nielsen, P.H. (2017) The impact of immigration on microbial community composition in full-scale anaerobic digesters. *Scientific Reports*, **7**, (1), 9343. https://doi.org/10.1038/s41598-017-09303-0

Klüber, H.D. and Conrad, R. (1998) Effects of nitrate, nitrite, NO and N_2O on methanogenesis and other redox processes in anoxic rice field soil. *FEMS Microbiology Ecology*, **25**, 301–318. https://doi.org/10.1016/S0168-6496(98)00011-7

Kougias, P.G., Boe, K., O-Thong, S., Kristensen, L.A. and Angelidaki, I. (2014) Anaerobic digestion foaming in full-scale biogas plants: A survey on causes and solutions. *Water Science and Technology*, **69**, (4), 889–897. https://doi.org/10.2166/wst.2013.792.

Lettinga, G. (1995) Anaerobic digestion and wastewater treatment systems. *Antonie van Leeuwenhoek*, **67**, 3–28.

Li, Z.-Y., Inoue, D. and Ike, M. (2023) Mitigating ammonia-inhibition in anaerobic digestion by bioaugmentation: A review. *Journal of Water Process Engineering*, **52**, 103506. https://doi.org/10.1016/j.jwpe.2023.103506

Lian, S., Nikolausz, M., Nijenhuis, I., da Rocha, U.N., Liu, B., Corrêa, F.B., Saraiva, J.P. and Richnow, H.H. (2020) Biotransformation of hexachlorocyclohexanes contaminated biomass for energetic utilization demonstrated in continuous anaerobic digestion system. *Journal of Hazardous Materials*, **384**, 121448. https://doi.org/10.1016/j.jhazmat.2019.121448

Lin, C.Y. (1992) Effect of heavy metals on volatile fatty acid degradation in anaerobic digestion. *Water Research*, **26**, 177–183. https://doi.org/10.1016/0043-1354(92)90217-R

Lindorfer, H. and Demmig, C. (2016) Foam formation in biogas plants — a survey on causes and control strategies. *Chemical Engineering Technology*, **39**, (4), 620–626. https://doi.org/10.1002/ceat.201500297

Lumb, C., Brown, D. and Bottomley, M.K.V. (1977) A chloroform problem at Great Watford and its solution. *Water Pollution Control*, **76**, 459–467.

Merrettig-Bruns, U. and Jelen, E. (2009) Anaerobic biodegradation of detergent surfactants. *Materials*, **2**, (1), 181–206. https://doi.org/10.3390/ma2010181

Moestedt, J., Nordell, E., Shakeri Yekta, S., Lundgren, J., Martí, M., Sundberg, C., Ejlertsson, J., Svensson, B.H. and Björn, A. (2016) Effects of trace element addition on process stability during anaerobic co-digestion of OFMSW and slaughterhouse waste. *Waste Management*, **47**, 11–20. https://doi.org/10.1016/j.wasman.2015.03.007

Morgan-Sagastume, F., Pratt, S., Karlsson, A., Cirne, D., Lant, P. and Werker, A. (2011) Production of volatile fatty acids by fermentation of waste activated sludge pretreated in full-scale thermal hydrolysis plants. *Bioresources Technology*, **102**, (3), 3089–3097.

Mosey, F.E. (1976) Assessment of the maximum concentration of heavy metals in crude sewage which will not inhibit the anaerobic digestion of sludge. *Water Pollution Control*, **75**, 10–20.

Mosey, F.E. (1983) Anaerobic processes. In *Ecological Aspects of Wastewater Treatment*, Vol. 2. *Biological Activities and Treatment Processes* (eds. C.R. Curds and H.A. Hawkes). Academic Press, London, pp. 219–260.

Müller, W.R., Frommert, I. and Jörg, R. (2004) Standardized methods for anaerobic biodegradability testing. *Reviews in Environmental Science and Bio/Technology*, **3**, 141–158. https://doi.org/10.1007/s11157-004-4350-6

Murray, W.D. and Van Den Berg, L. (1981) Effects of nickel, cobalt, and molybdenum on performance of methanogenic fixed-film reactors. *Applied and Environmental Microbiology*, **42**, (3), 502–505.

Newell, P.J. (1982) Anaerobic digestion of organic effluents. In *Biotechnology and the Environment — A Seminar*. National Board for Science and Technology, Dublin.

Nielsen, B.V., Maneein, S., Farid, M.M.A. and Milledge, J.J. (2020) The effects of halogenated compounds on the anaerobic digestion of macroalgae. *Fermentation*, **6**, (3), 85. https://doi.org/10.3390/fermentation6030085

Niu, Q., Kubota, K., Qiao, W., Jing, Z., Zhang, Y. and Yu-You, L. (2015) Effect of ammonia inhibition on microbial community dynamic and process functional resilience in mesophilic methane fermentation of chicken manure. *Journal of Chemical Technology and Biotechnology*, **90**, 2161–2169. https://doi.org/10.1002/jctb.4527

OECD. (2006) OECD Guidelines for testing chemicals 311: Anaerobic biodegradability of organic compounds in digested sludge: method by measurement of gas production. Organization for Economic Co-operation and Development, Paris.

OECD. (2007) Guidelines for testing chemicals 224: Determination of the inhibition of the activity of anaerobic bacteria — reduction of gas production from anaerobically digesting (sewage) sludge. Organization for Economic Co-operation and Development, Paris.

Oleszkiewicz, J.A. and Sharma, V.K. (1990) Stimulation and inhibition of anaerobic processes by heavy metals — a review. *Biological Wastes*, **31**, 45–67. https://doi.org/10.1016/0269-7483(90)90043-R

Pagilla, K.R., Craney, K.C. and Kido, W.H. (1997) Causes and effects of foaming in anaerobic sludge digesters. *Water Science and Technology*, **36**, (6–7), 463–470.

Patel, G.B., Agnew, B.J. and Dicaire, C.J. (1991) Inhibition of pure culture of methanogens by benzene ring compounds. *Applied and Environmental Microbiology*, **57**, 2969–2974.

Pfeffer, J.T. (1979) Anaerobic digestion processes. In *First International Symposium on Anaerobic Digestion*, University College, Cardiff, pp. 15–36.

Quintero-García, O.J., Pérez-Soler, H. and Amezcua-Allieri, M.A. (2023) Enzymatic treatments for biosolids: An outlook and recent trends. *International Journal of Environmental Research and Public Health*, **20**, (6), 4804. https://doi.org/10.3390/ijerph20064804

Rozzi, A. and Remigi, E. (2004) Methods of assessing microbial activity and inhibition under anaerobic conditions: a literature review. *Reviews in Environmental Science and Bio/Technology*, **3**, 93–115. https://doi.org/10.1007/s11157-004-5762-z

Sheng, K., Chen, X., Pan, J., Kloss, R., Wei, Y. and Ying, Y. (2013) Effect of ammonia and nitrate on biogas production from food waste via anaerobic digestion. *Biosystems Engineering*, **116**, 205–212. https://doi.org/10.1016/j.biosystemseng.2013.08.005

Shi, Z., Usman, M., He, J., Chen, H., Zhang, S. and Luo, G. (2021) Combined microbial transcript and metabolic analysis reveals the different roles of hydrochar and biochar in promoting anaerobic digestion of waste activated sludge. *Water Research*, **205**, 117679. https://doi.org/10.1016/j.watres.2021.117679

Sikkema, J., De Bont, J.A. and Poolman, B. (1995) Mechanisms of membrane toxicity of hydrocarbons. *Microbiology and Molecular*

Biology Reviews, **59**, 201–222. https://doi.org/10.1128/mr.59.2.201-222.1995

Steiniger, B., Hupfauf, S., Insam, H. and Schaum, C. (2023) Exploring anaerobic digestion from mesophilic to thermophilic temperatures — operational and microbial aspects. *Fermentation*, **9**, 798. https://doi.org/10.3390/fermentation9090798

Subramanian, B., Miot, A., Jones, B., Klibert, C. and. Pagilla, K.R. (2015) A full-scale study of mixing and foaming in egg-shaped anaerobic digesters. *Bioresources Technology*, **192**, 461–470. https://doi.org/10.1016/j.biortech.2015.06.023

Sun, J., Guo, L., Li, Q., Zhao, Y., Gao, M., She, Z. and Wang, G. (2016) Structural and functional properties of organic matters in extracellular polymeric substances (EPS) and dissolved organic matters (DOM) after heat pretreatment with waste sludge. *Bioresource Technology*, **219**, 614–623.

Swanwick, J.D. and Shurben, D.G. (1969) Effective chemical treatment for inhibition of anaerobic sewage sludge digestion due to anionic detergents. *Water Pollution Control*, **68**, 190–202.

Takashima, M. and Speece, R.E. (1989) Mineral nutrient requirements for high-rate methane fermentation of acetate at low SRT. *Research Journal of the Water Pollution Control Federation*, **61**, (11/12), 645–650. http://www.jstor.org/stable/41480385

USEPA (1976) Anaerobic sludge digestion: Operations manual. EPA 430/9-76-001. Office of Water, US Environmental Protection Agency, Washington. https://nepis.epa.gov/Exe/ZyPURL.cgi?Dockey=00000IKT.TXT

Usman, M., Shi, Z., Ji, M., Ren, S., Luo, G. and Zhang, S. (2021) Microbial insights towards understanding the role of hydrochar in alleviating ammonia inhibition during anaerobic digestion. *Chemical Engineering Journal*, **419**, 129541.

Uthirakrishnan, U., Sharmila, V.G., Merrylin, J., Kumar, S.A., Dharmadhas, J.S., Varjani, S. and Banu, J.R. (2022) Current advances and future outlook on pretreatment techniques to enhance biosolids disintegration and anaerobic digestion: A critical review. *Chemosphere*, **288**, 2–15.

Velsen, A.F.M. van, and Lettinga, G. (1979) Effect of feed composition on digester performance. In *First International Symposium on Anaerobic Digestion*, University College, Cardiff, pp. 113–130.

Wang, G., Dai, X., Zhang, D., He, Q., Dong, B., Li, N. and Ye, N. (2018) Two-phase high solid anaerobic digestion with dewatered sludge:

Improved volatile solid degradation and specific methane generation by temperature and pH regulation. *Bioresource Technology*, **259**, 253–258. https://doi.org/10.1016/J.BIORTECH.2018.03.074.

Wang, L., He, Y., Zhu, Y., Ping, Q. and Li, Y. (2024) Insight into using hydrochar to alleviate ammonia nitrogen inhibition during anaerobic digestion of waste activated sludge: Performance, metagenomic and metabolomic signatures. *Science of the Total Environment*, **916**, 170196. https://doi.org/10.1016/j.scitotenv.2024.170196

Wang, Y.-T., Gabbard, H.D. and Pai, P.-C. (1991) Inhibition of acetate methanogenesis by phenols. *Journal of Environmental Engineering*, **117**, 487–496.

Westlund, A.D., Hagland, E. and Rothman, M. (1998) Operational aspects on foaming in digesters caused by *Microthrix Parvicella. Water Science and Technology*, **38**, (8–9), 29–34.

Wheatley, A.D. (1985) Wastewater treatment and by-product recovery. In *Topics in Wastewater Treatment. Critical Reports on Applied Chemistry*, Vol. 11 (ed. J.M. Sidwick). Blackwell, Oxford, pp. 68–106.

Wolfe, R.S. (1971) Microbial formation of methane. *Advances in Microbial Physiology*, **6**, 107–146.

Yan, M., Fotidis, I.A., Tian, H., Khoshnevisan, B., Treu, L., Tsapekos, P. and Angelidaki, I. (2019) Acclimatization contributes to stable anaerobic digestion of organic fraction of municipal solid waste under extreme ammonia levels: Focusing on microbial community dynamics. *Bioresource Technology*, **286**, 121376. https://doi.org/10.1016/j. biortech.2019. 121376.

Yang, P., Pen, Y., Tan, H., Liu, H., Wu, D., Wang, X., Li, L. and Peng, X. (2021) Foaming mechanisms and control strategies during the anaerobic digestion of organic waste: A critical review. *Science of the Total Environment*, **779**, 145531. https://doi.org/10.1016/j.scitotenv.2021.146531

Yang, P., Peng, Y., Liu, H., Wu, D., Yuan, R., Wang, X., Li, L. and Peng, X. (2022) Multi-scale analysis of the foaming mechanism in anaerobic digestion of food waste: From physicochemical parameter, microbial community to metabolite response. *Water Research*, **218**, 118482. https://doi.org/10.1016/j.watres.2022.118482

Yang, P., Li, L., Zhou, Y., Zhang, S., Zhang, Z., Peng, Y., Liu, H., Ye, W. and Peng, X. (2023) Insights into the effect of extracellular polymeric substances on anaerobic digestion foaming: From perspectives of composition, hydrophobicity, and functional groups. *Chemical Engineering Journal*, **476**, 146812. https://doi.org/10.1016/j.cej.2023.146812

Yellezuome, D., Zhu, X., Wang, Z. and Liu, R. (2022) Mitigation of ammonia inhibition in anaerobic digestion of nitrogen-rich substrates for biogas production by ammonia stripping: a review. *Renewable and Sustainable Energy Reviews*, **157**, 112043. https://doi.org/10.1016/j.rser.2021.112043

Yu, D., Zhang, Q., De Jaegher, D., Liu, J., Sui, Q., Zheng, X. and Wei, Y. (2021) Effect of proton pump inhibitor on microbial community, function, and kinetics in anaerobic digestion with ammonia stress. *Bioresource Technology*, **319**, 124118. https://doi.org/10.1016/j.biortech.2020.124118.

Yuan, H. and Zhu, N. (2016) Progress in inhibition mechanisms and process control of intermediates and by-products in sewage sludge anaerobic digestion. *Renewable and Sustainable Energy Reviews*, **58**, 429–438, https://doi.org/10.1016/j.rser.2015.12.261

Yue, Z.-B., Yu, H.-Q. and Wang, Z.-L. (2007) Anaerobic digestion of cattail with rumen culture in the presence of heavy metals. *Bioresource Technology*, **98**, (4), 781–786. https://doi.org/10.1016/j.biortech.2006.03.017

Section C

Contact (High-rate) Anaerobic Wastewater Treatment Systems

6

Biofilm (Attached Growth) Systems

6.1 Introduction to Contact Systems

Because of the difficulties involved in degrading particulate organic matter and the slow growth of anaerobic bacteria, and of methanogens in particular, digester design has been traditionally based on flow-through stirred tanks with long retention times. Unlike sludge, the organic matter in wastewater is often in solution and is far more amendable to treatment. However, the primary problem with treating these wastes anaerobically has been how to retain sufficient biomass within the reactor and prevent wash-out of the bacteria and archaea. Contact anaerobic systems are specifically designed to treat weaker sludges with low solids concentrations and strong effluents. Unlike the conventional continuously stirred digester, the hydraulic retention time (HRT) and solids retention time (SRT) are independent of each other, with the biomass either retained within the reactor or recycled after separation (Chap. 7). Many contact systems have been developed (Table 6.1) based around anaerobic static media filters, fluidized media, activated sludge, and sludge blanket systems. Data on over 400 anaerobic plants, mainly treating vegetable-processing wastewaters, was collated by Demuynck *et al.* (1983), who gave details on many contact systems, with reactor volumes ranging from 2,000 to 20,000 m^3. Anaerobic contact systems have been a popular choice for industrial wastewater treatment for over 40 years, with an example of the range of applications summarized by Wheatley *et al.* (1997).

Table 6.1. List of anaerobic contact systems used for wastewater treatment.

Reactor group	Example of reactor design	Abbreviation
Anaerobic biofilter	Submerged anaerobic reactor	SMAR
	Up-flow anaerobic filter	ANFIL
	Anaerobic up-flow filter	AUF
	Anaerobic downflow stationary bed reactor	ADSR
	Downflow stationary fixed film reactor	DSFF
	Up-flow stationary fixed film filter	USFF
	Hybrid anaerobic bed reactor	
Anaerobic biological contactors	Anaerobic biological contactors	AnRBC
Anaerobic moving bed biofilm reactors	Anaerobic moving bed biofilm reactors	AnMBBR
Expanded and Fluidized beds	Static or porous medium bed reactor	AnPBR
	Expanded bed reactor	AnEBR
	Fluidized bed reactor	AnFBR
	Anaerobic ammonium oxidation	Anammox

These included food wastewaters such as distilleries, sugar refiners, potato processing, breweries, starch refiners, milk processing, abattoirs, yeast processing, canneries, fruit processing, soft drinks, and non-food wastewaters, such as paper, pharmaceuticals, chemicals, and textiles. At that time, up-flow anaerobic sludge blanket (UASB) reactors (48%) were the most popular, followed by anaerobic filters (23%).

While anaerobic treatment systems, especially UASB reactors and anaerobic filters, are very efficient and cost-effective at treating a wide variety of wastewaters, including sewage, the final effluent still has relatively high concentrations of chemical oxygen demand (COD), nutrients, and suspended solids. This requires further treatment, usually by an aerobic bioreactor, before final discharge to receiving waters. Aerobic treatment is normally used as a post-treatment process for both carbonaceous and nutrient removal to achieve modern discharge standards. All aerobic processes are employed, including activated sludge (Nmñez and Martinez, 2001), biological aerated filters (BAFs) (Lacelle *et al.*, 2001), constructed wetlands (Denny, 1997; Sousa *et al.*, 2001), land treatment (Chernicharo *et al.*, 2001), maturation ponds (Dixo *et al.*, 1995; Catunda and van Haandel,

1996; Cavalcanti *et al.*, 2001), percolating filters (Chernicharo and Nascimento, 2001), submerged aerated filters (SAFs) (Gonçalvez *et al.*, 1999), and sequencing batch reactors (SBRs) (Sousa and Foresti, 1996; Zaloum and Abbott, 1997).

6.2 Anaerobic Biofilters or Static (Fixed) Bed Reactors

Anaerobic reactors using fixed or static media to retain the active biomass are generally known as anaerobic biofilters (Fig. 6.1). There are a number of different anaerobic biofilter designs employing a fixed medium, including the submerged anaerobic reactor (SMAR), the up-flow anaerobic filter (ANFIL) and anaerobic up-flow filter (AUF), the anaerobic downflow stationary bed reactor (ADSR), and the downflow stationary fixed-film reactor (DSFF). The differences in these designs can be quite small, so it is useful to compare two systems: the **up-flow stationary fixed film filter (USFF)**, which is operated with the influent wastewater passing through the filter in an up-flow mode (Fig. 6.1) while (ii) **DSFF systems** operate with the influent passing through the medium in a down-flow direction (Fig. 6.2). This creates two very different reactor environments, even though both have fixed medium on which the microbial community

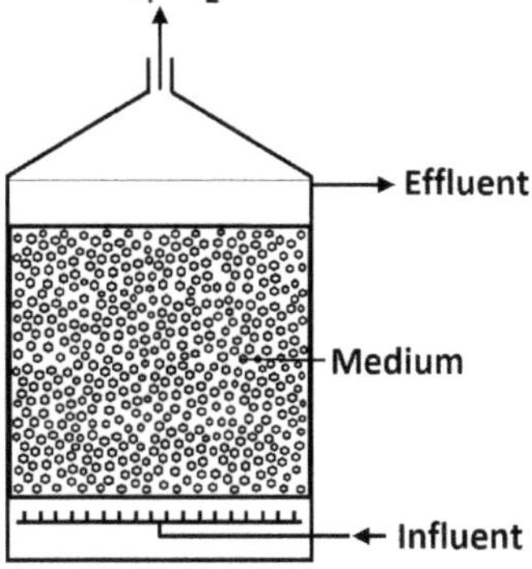

Figure 6.1. Schematic of a typical anaerobic up-flow stationary fixed film filter (USFF) design.

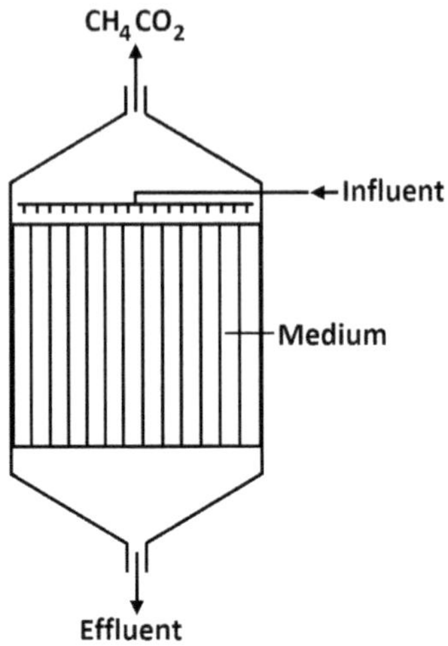

Figure 6.2. Schematic of an anaerobic downflow stationary fixed film reactor (DSFF) design.

develops. In USFF reactors, the majority of the active biomass is found within the interstices of the medium, while in DSFFs, the biomass is predominantly present as attached biomass. While both can treat medium to strong wastewaters, the DSFF system tends to use media with large interstices, often orientated vertically, which allows wastewaters with high suspended solids concentrations to be treated, which would normally block USFF media. In DSFF reactors, there is also a counter-movement of gas and liquid, resulting in intense mixing, which makes them better suited to wastewaters with higher concentrations of suspended solids. Another biofilm system is the anaerobic baffled reactor (AnBR), which is used as an upgrade for septic tanks treating domestic wastewater, especially in Asia (Sec. 7.7). Normally, AnBRs use granular sludge, but a modified version, the packed anaerobic baffled reactors (PABR), includes a fixed medium or carrier to increase biomass concentration to achieve COD removal of up to 90%. Small carriers, such as bamboo sections, are suspended in bunches on rods that sit on the up-flow section of each chamber (normally 4–6 chambers per reactor) (Fig. 7.9).

In design, anaerobic biofilters are submerged trickling filters, full of wastewater so that no oxygen can enter the reactor and are operated using a wide variety of media, from mineral to random and modular plastic filter media. The biomass comprises a thin bacterial film that is firmly attached to this support media. Apart from being operated as an anaerobic process, with some form of biogas collection device, submerged filters can also be operated sequentially or in series aerobically using aeration units or anoxically with a separate carbon input where necessary (Gray, 2021).

The design of anaerobic biofilters varies, although the shape and diameter of both the reactor and the media can have important effects on the stability of the attached film (Berg and Lentz, 1980). Other factors, such as gas production, can also dislodge the film, although its scouring effect may also reduce the chance of clogging within the filter medium. In cylindrical reactors, the most satisfactory diameter-to-height ratio is 1:4. The entire reactor can be filled with the medium, and most anaerobic filters now use plastic media (Anderson *et al.*, 1984), or it may be restricted to the upper part of the reactor only, with the detached biomass settling to the lower chamber, thus providing a dual mixed and fixed system (i.e., hybrid systems) (Fig. 6.3).

Various media have been used. Of the mineral media, porous stones, gravel, glass, and pottery fragments appear to be the favorites, although these have a large bulk density and relatively low surface areas. Mineral media also require strong reactor designs to take the

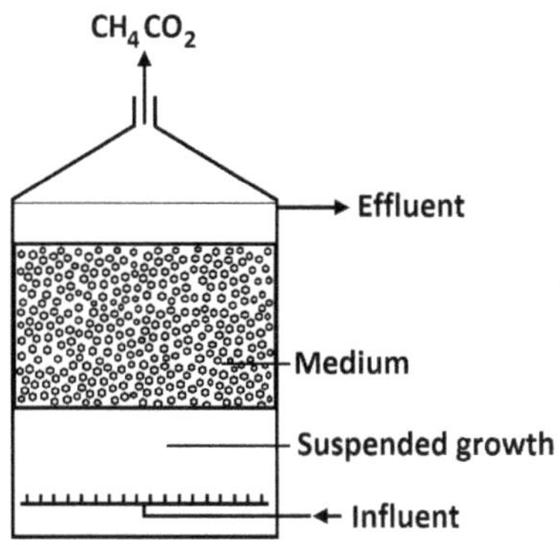

Figure 6.3. Schematic of a hybrid anaerobic biofilter comprising a dual mixed and fixed reactor area.

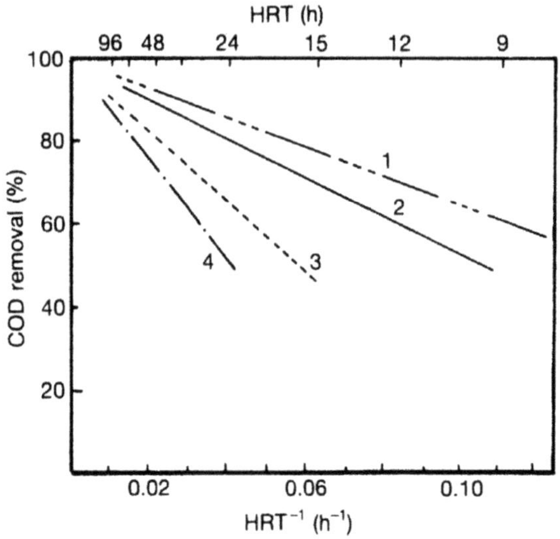

Figure 6.4. COD removal efficiencies of anaerobic filters in relation to different types of media: 1. Large modular; 2. Small modular; 3. Pall rings; 4. Spheres (Young and Dahab, 1982). Reproduced with permission IWA Publishing.

added weight of the medium. In contrast, plastic media have a much greater specific surface area (90–350 m^2m^{-3}) and lower density (50–100 kg m^{-3}), with a voidage in excess of 90% (Characklis and Trlear, 1982). Different media result in very different performance characteristics, and unlike the traditional trickling filter, removal efficiency is not directly related to specific surface areas of the media (Young and Dahab, 1983; Wilkie *et al.*, 1983) (Fig. 6.4). This is because in up-flow filters, a significant portion of the active biomass will be present not as attached film, but as unattached dispersed growths in the interstices (voids) of the medium. The suspended solids flocculate as they travel up through the filter, forming larger particles that finally settle back down the filter column. Therefore, provision must be made for periodic desludging, usually once a year (Young, 1991). The movement of solids and gas bubbles being released introduces an element of mixing into this essentially plug-flow system. When operated in the down-flow mode, all the dispersed solids are washed out of the reactor, leaving only the attached biomass. In these reactors, the media must ensure stable film development as well as prevent excessive solids accumulation. Therefore, modular tube or crossflow media, as opposed to random media, are recommended (Wilkie *et al.*, 1983; Bonastre and Paris, 1989). As all the biomass is attached in DSFFs, the performance in this case is generally directly related to the specific surface area of the medium. In practice, filters employing porous media perform better than those using non-porous material, due not only to higher specific surface areas but to better surfaces for the adhesion and retention of biomass (Fox *et al.*, 1990; Picanço *et al.*, 2001). Needle punched polyester (NPP) medium is widely employed in DSFFs due to its high specific surface area. Biofilter media has been extensively explored in an earlier book in this series on fixed-film reactors (Gray, 2021).

There is increasing interest in the use of biochar as a filter medium in anaerobic biofilters as well as an amendment to digesters and floc-based anaerobic treatment systems. Biochar, which is highly porous with strong adsorptive properties (Gęca *et al.*, 2022), promotes more rapid establishment of the biofilm as well as increasing resistance to inhibitory substances, common in leachate for example, as well as improving overall effluent quality and methane production. It is more commonly used as a secondary stage to another anaerobic process, such as a UASB or a leach bed reactor (LBR) (Collins *et al.*, 2023). Paritosh and Kesharwani (2024) explain that biochar encourages interspecies electron transfer, cell immobilization, and gene level alternations at the microbial level within the biofilm.

Anaerobic biofilters require an initial inoculum of bacteria to speed up the colonization of bacteria and archaea. Although expensive, a start-up inoculum of 10% v/v from a similar unit is recommended, limiting organic loading to <0.1 kg COD kg VSS^{-1} d^{-1}, with a minimum HRT >1 day to limit washout of the biomass. Once established, anaerobic biofilters can achieve 75–85% COD removal when loaded at between 1 and 10 kg COD m^3d^{-1} with HRTs between 18–24 hours. Anaerobic biofilters are ideal for relatively cold and dilute wastes operated in temperate areas, as they have an extremely high MCRT:HRT ratio with an MCRT (mean cell residence time) >100 days. This gives the process a high degree of stability, excellent resistance to inhibitory compounds, and satisfactory performance,

Table 6.2. Comparison of features of various anaerobic reactors. ANFLOW, anaerobic flow-through digester without sludge recycle; ANCONT, contact digester with sludge recycle; ANBIOL, anaerobic biofilters; FANBIOF, fluidized anaerobic biofilter; UASB, upflow anaerobic sludge bed reactor. The more asterisks, the better the process. Adapted from (Oleskiewicz and Olthof, 1982).

Feature	ANFLOW	ANCONT	ANBIOF	FANBIOF	UASB
Ease of start-up	*	*****	****	***	**
Ease of operation after proper acclimatization	**	**	*****	***	***
Good process control possible under transient influent conditions	*	*****	***	***	***
Resistance to shocks due to:					
temperature	*	***	*****	*****	****
toxics	*	***	*****	*****	****
high organic load	*	****	*****	*****	*****
Tolerance for influent quality variations	*	****	**	****	*****
Can tolerate high influent solids fluctuations	****	***	*	**	**
May incorporate sludges from pretreatment and aerobic polishing	*****	****	*	*	*

even at low temperatures (Table 6.2). They can treat similar organic loadings as anaerobic activated sludge systems, even though the MCRT is much higher (Table 6.3). Successful applications include: treating undiluted sewage, starch, whey, cellulose, distillery, pharmaceutical, piggery, and fish-processing wastewaters (Duff and van den Berg, 1982; Genung *et al.*, 1982; Rittman *et al.*, 1982; Sachs *et al.*, 1982; Bedogni *et al.*, 1983; Mathur *et al.*, 1986; Elmitwalli *et al.*, 2001), with an average COD removal rate of between 80 and 90% at a typical organic loading of 3.5 kg $m^{-3}d^{-1}$. DSFF technology is ideal for treating sugar refinery wastewaters as well as producing high biogas yields. In a comparative study using polyvinyl chloride pipe

Table 6.3. Features of various reactors compared with aerobic counterparts. Adapted from (Oleskiewicz and Olthof, 1982). The key is given in Table 6.2.

	Anaerobic processes				**Aerobic processes**		
Feature	**ANFLOW**	**ANCONT**	**ANBIOF**	**FANBIOF**	**UASB**	**Activated sludge**	**Trickling filter**
Loads practised (kg COD m^{-3} d^{-1})	0.5–3	2–8	2–10	0.5–12	1–15	0.5–2	1–3 roughing
Loads used in experimental scale (kg $m^{-3}d^{-1}$)	0.5–10	0.5–100	0.5–25	1–40	1–60	1–10	2–15
HRT used (days)	8+	0.2–8	0.2–4	0.15–3	0.15–8	1–5	0.05–0.2
SRT resulting (days)	8+	15–80	20–300	20–100	30–200	10–30	~30(+)
Temperatures used (°C)	35, 55	35, 55	15–35	35	5–35	15–25(+)	15–25(+)
COD removals attained at practised loads: similar waste assumed (%)	60	90+	90+	90+	90+	90+	60–80 in 1 stage. Multi-stages typical

pieces as a medium, Pradeep *et al.* (2014) found that the BOD and COD removal increased with decreasing HRT (48 to 12 hours), reaching an optimal 79 and 82% removal respectively, at 12 hours, with optimal biogas production but declining at an HRT of 6 hours. A DSFF unit containing fired pottery clay medium with a surface area of 157 m^2m^{-3} was used to treat piggery waste at 35°C (Kennedy and Berg, 1982). It was found to be able to treat higher loadings and have higher rates of methane production than partially or fully mixed reactors and also plug-flow reactors. The support medium was evenly coated with a thin microbial film 2–4 mm thick. Reactor performance, based on the amount of microbial film present, was between 1.1 and 1.4 g COD removed per g of film per day, which is similar to that reported by Berg and Lentz (1980) for other types of media. Canovas-Diaz and Howell (1986) used a modified DSFF reactor to treat cheese whey, which was initially operated flooded as a standard DSFF system, but then the liquid height was reduced by two-thirds. The authors found that partially exposing the biofilm increased gas production while increasing the resistance of the methanogenic archaea to high VFA concentrations (i.e., 3,000 mg L^{-1}). COD removal of up to 90% was achieved using DSFF by Duff and van den Berg (1982) for the treatment of high-protein fish processing wastewater at loading >10 kg COD m^3d^{-1}.

However, USFF systems are limited to treating substrates with relatively low solids, as the interstices of the medium, especially mineral media, can easily become blocked. This is why they are often recommended to be operated in series, with an up-flow anaerobic sludge blanket reactor, as the effluent from this process will be largely free from suspended solids, thus making it ideal for further anaerobic treatment using a USFF. Loading rates are compared to other anaerobic systems in Table 6.4.

6.3 Rotating Biological Contactors

Conventional anaerobic filters require the influent wastewater to be pumped either into or up through the reactor, which requires significant energy for pumping. Also, there is a risk of ponding

Table 6.4. Comparison of loading rates used in anaerobic filters, UASB, and expanded beds (Hobson and Wheatley, 1993). Reproduced with permission of Elsevier Publishing.

Type of unit	Range
Anaerobic filter	
Organic load (kg COD/m^3.d)	2–10
Retention period (h)	10–15
COD removal (%)	70–80
Critical solids concentration in feed (mg/l)	450–1,050
Upflow anaerobic sludge blanket reactor	
Load (kg COD/m^3. d)	2–15
Retention period (h)	10–50
COD removal (%)	70–90
Expanded bed	
Load (kg COD/m^3.d)	2–50
Retention period (h)	0.5–24
COD removal (%)	70–80

(i.e., clogging) of the medium, causing short-circuiting, reduced loading rates, and HRTs. In contrast, anaerobic rotating biological contactors (AnRBCs) have a very small head loss, requiring minimal pumping and low energy requirement, and are simple to operate with no risk of clogging.

The AnRBC is based on standard RBCs, which consist of a series of flat or corrugated discs 2.0 to 3.5 m in diameter and between 10 and 20 mm thick on which the biofilm develops. The discs are mounted on a horizontal shaft, 20 mm apart, which is driven mechanically, usually by a small electric motor, so that the discs rotate at a right angle to the flow of wastewater (Fig. 6.5). The discs are normally fabricated in corrugated polythene, MDPE, PVC, or glass reinforced plastic, although other materials are occasionally used. The surface area of the medium varies between 150 and 200 m^{2}m^{-3}.

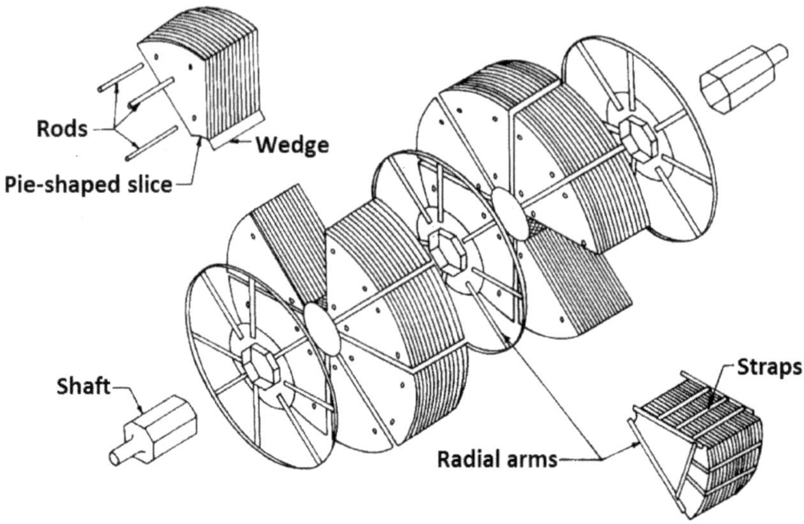

Figure 6.5. Schematic view of disc assembly in a rotating biological contactor showing support rods, used in many designs, that ensure exact spacing between discs (CIWEM, 2000). Reproduced with permission of the Chartered Institution of Water and Environmental Management.

The discs are designed to fit very closely to the outer shell of the tank, in which they are placed to prevent short-circuiting. The majority of RBCs are operated aerobically, with 35–40% of the disc area submerged in the wastewater, which is slowly but continuously rotated, with oxygen transfer taking place while the biofilm passes through the air space above before being re-immersed in the wastewater below. The design, construction, and operation of RBCs have been reviewed by Gray (2021) in the volume on fixed-film reactors in this series. In contrast, the discs of AnRBCs are fully submergered, with the gas space above the water level anaerobic.

While RBCs had been developed in the 1920s, it was not until 1965 that they became commercially available. The AnRBC was developed in the late 1970s and pioneered by the early experimental work of Tait and Friedman (1980). The basic design is the same as the standard aerobic RBC, except that the reactor is airtight and the discs are generally fully submerged, with any biogas collected by a

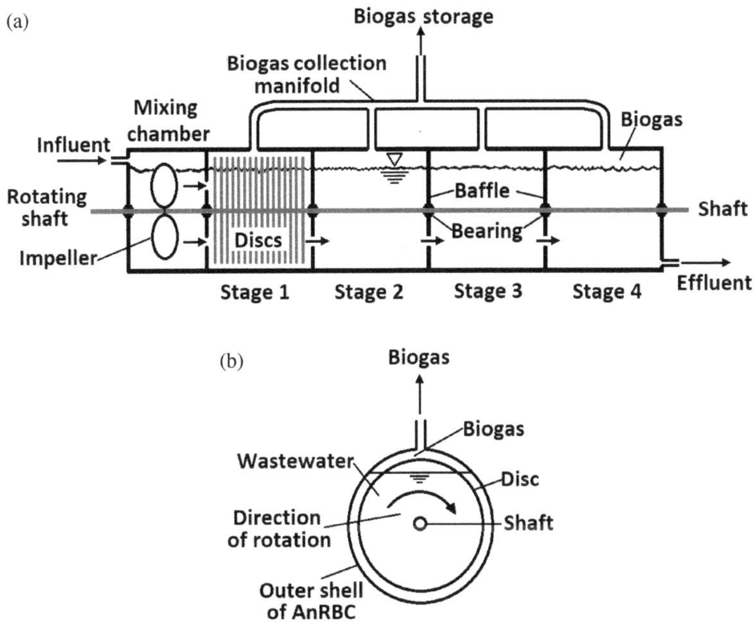

Figure 6.6. Schematic design of an anaerobic rotating biological contactor (AnRBC): (a) side view and (b) cross-section.

series of pipework (Fig. 6.6). The discs are grouped together into sets or stages, which may be housed with separated baffled areas, which minimizes short circuiting while also allowing the development of different microbial consortia to utilize the wastewater as it changes composition as it passes through the reactor. This design particularly lends itself to anaerobic operation, where the different stages of digestion require different consortia of anaerobic bacteria and archaea. This is the design used by Tait and Friedman (1980), who operated their laboratory AnRBCs at 35°C to successfully treat a strong organic wastewater with a COD of 8,500 mg L^{-1} (BOD 5,250 mg L^{-1}). There have been a number of successful AnRBC laboratory and pilot trials. For example, Laquidora *et al.* (1986) achieved COD removal rates of between 10 and 140 g $m^{-2}d^{-1}$ while treating wastewater similar to that from the food processing industry, while Satyanarayan *et al.* (1987) achieved 60–81% COD removal rates treating dairy wastewater. The performance of AnRBCs follows a simple Monod-type expression (Deshpande *et al.*, 1991), with most researchers opting to use a modified Stover–Kincannon model for design and performance estimation. The final effluent quality and reactor volume required can be calculated as:

Final effluent quality (S_e)

$$S_e = S_0 - \frac{U_{max} S_0}{K_B \left(\frac{S_0}{HRT} \right)} \tag{6.1}$$

Reactor liquid volume required (V)

$$V = \frac{QS_0}{\left[\frac{U_{max} S_0}{S_0 - S_e} \right] - K_B} \tag{6.2}$$

where Q is the flow rate (L d^{-1}), V is the reactor liquid volume (L), S_0 is the influent COD (mg L^{-1}), S_e is the effluent COD (mg L^{-1}), U_{max} is the maximum substrate removal rate (mg L^{-1} d^{-1}), and K_B is the saturation value constant (mg L^{-1} d^{-1}). Ebrahimi *et al.* (2018) determined U_{max} and K_B as 7.77 and 8.57 mg L^{-1} d^{-1}, respectively.

The most complete study on AnRBCs was carried out by Lu *et al.* (1995), who recorded an average COD and BOD removal rate of 71 and 76%, respectively, while treating a high-strength organic wastewater at a loading rate of 13.33 kg COD m^{-3}d^{-1} (equivalent to a surface loading rate of 111.4 g COD m^{2}d^{-1}) using a bench-scale reactor (Fig. 6.7). The discs were completely submerged with a rotational speed of 12 rpm and a HRT of 21.6 hours. Biogas was produced from all four baffled sections of the AnRBC, with the highest rate recorded in the first section, which also had the highest biodegradation rate as shown by COD and BOD removal, showing that there were sufficient methanogenic archaea throughout the unit and not just restricted to the final section. Detailed modeling of their system was also carried out by the authors. Disc submergence, rotation speed, and HRT are all significant factors affecting performance.

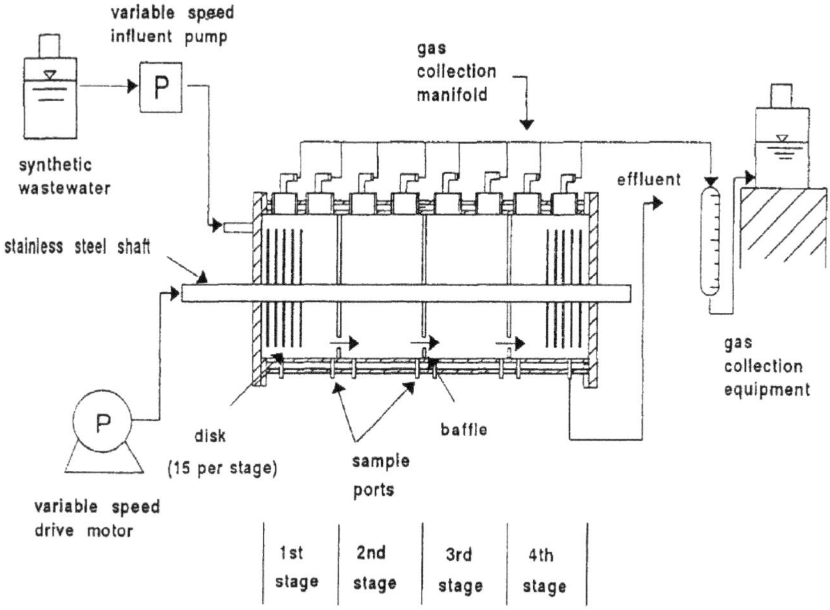

Figure 6.7. Laboratory-scale AnRBC developed by Lu *et al.* (2015). Reproduced with permission of Elsevier Publishing.

There are very few examples of full-scale AnRBCs, with anaerobic biofilters or anaerobic moving bed biofilm reactors being preferred (Sec. 6.4). A full-scale system was used to treat brewery wastewater comprising an AnRBC using random-packed caged medium followed by an aerobic RBC section. The system performed well under fluctuating organic loading and variable pH, but the medium in the AnRBC failed to establish a significant biofilm, thought to be due to variable operating conditions and excessive turbulence exerted on the medium within the cage. Suspended biomass in the anaerobic unit was identified as an important factor in treatment (Ware and Pescod, 1998). An AnRBC followed by an aerobic MBBR to treat a strong organic wastewater (3,500–20,000 mg COD L^{-1}) was tested by Ebrahimi *et al.* (2018). The optimum performance of the AnRBC was achieved using an HRT of 3 days at an organic loading rate of 3.33 kg COD

$m^{-3}d^{-1}$, using a disk rotational speed of 7 rpm to achieve a removal efficiency of 81.5%, with a methane yield of 0.309 L CH_4 g^{-1} COD. The effect of HRT, disc rotational speed, and organic loading rate is illustrated in Fig. 6.8 and shows the importance of optimizing these parameters at full-scale operation. Renewed interest in AnRBC technology has arisen in the potential treatment of high carbohydrate wastewaters for the production of hydrogen using dark fermentation (Mohammadi *et al.*, 2019; 2022) (Sec. 9.1.1). Similarly, the potential of AnRBCs to precipitate metal sulfides from weak mining wastewater by sulfate reducing bacteria (SRB) was successfully demonstrated by Kiran *et al.* (2017). They found that influent metal loading concentrations in excess of 3.64 mg Cu(II) L^{-1} h^{-1} and 1.87 mg L^{-1} h^{-1} for Fe(III), Pb(II), Ni(II), Zn(II), and Cd(II) inhibited SRB activity with optimum removal at an HRT of 48 hours. At metal loading rates of Cu (II) of 100–150 mg L^{-1} and 50–90 mg L^{-1} for the other metals, maximum removal rates of 97% were recorded for Cu(II), 90% for Cd(II), and >77% for the other metals, with removal efficiencies in the order Cu > Cd > Pb > Fe > Zn > Ni (Fig. 6.9). High sulfate and COD removals were also recorded at 53–77 and 68–82%, respectively, at an HRT of 48 hours. Shorter HRTs resulted in reduced removal efficiency. The results offer a potential solution for treating a wide range of metal-contaminated wastewaters (Sec. 5.4.1).

6.4 Anaerobic Moving Bed Biofilm Reactors

Anaerobic moving bed biofilm reactors (AnMBBRs) utilize small plastic media, normally referred to as carriers in these types of reactors. They are normally manufactured from polyethylene and are 10–20 mm in diameter and of a density close to that of water (0.90–0.95 g cm^{-3}), which allows them to move freely within the reactor, on which biofilm readily develops (Fig. 6.10). The media is retained within the reactor by a perforated steel barrier, allowing treated effluent and associated solids to pass freely into a settlement tank. Except for hybrid systems, AnMBBRs do not have sludge return, as all the required biomass is retained on the media, although the suspended biomass also plays a role in treatment. In contrast to other fixed-film systems, the reactor is completely mixed so that the whole reactor

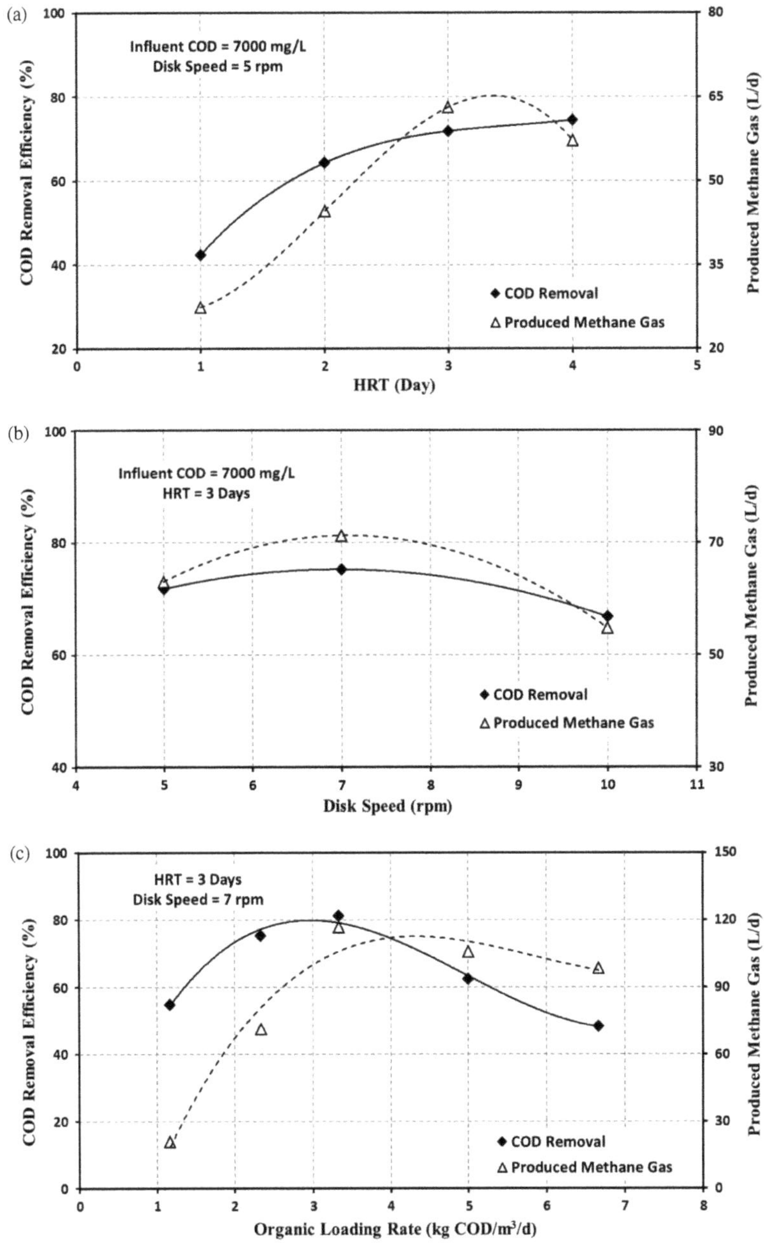

Figure 6.8. The effect of different operating parameters on an anaerobic rotating biological contactor treating a strong organic wastewater: Effect of (a) HRT, (b) disk rotational speed, and (c) organic loading rate on COD removal and methane production (Ebrahimi *et al.*, 2018). Reproduced with permission of the American Society of Civil Engineers, Reston, VA.

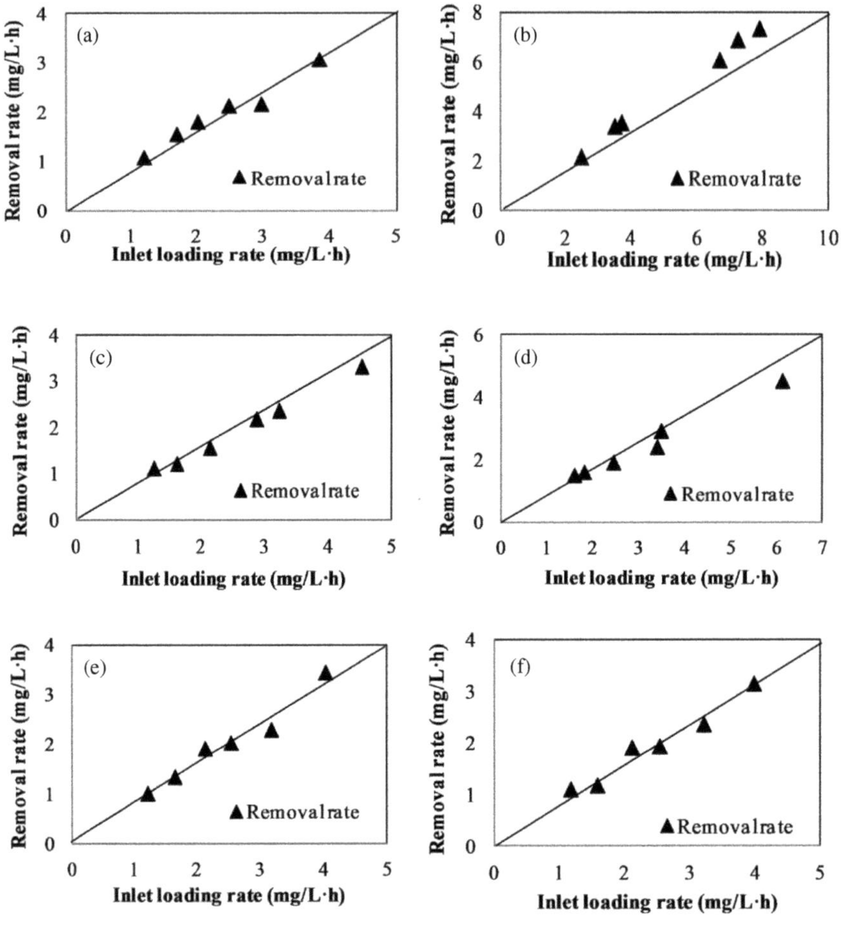

Figure 6.9. Removal rate of heavy metals as a function of inlet loading rate in a laboratory-scale AnRBC operated by Kiran *et al.* (2017): (a) Cd(II), (b) Cu(II), (c) Ni(II), (d) Fe(III), (e) Pb(II), and (f) Zn(II). Reproduced with permission of Elsevier Publishing.

volume is biologically active, resulting in higher biomass activity (Ødegaard *et al.*, 1994), with no clogging problems and a lower head loss than other contact systems.

Originally developed in Norway, with the first reactor installed in Lardnal in 1989, moving bed biofilm reactors are extremely flexible and can be operated as aerobic, anoxic, or anaerobic systems, either as standalone processes, in series, or sequentially within a single tank

Figure 6.10. Example of a moving bed biofilm reactor carrier used in AnMBBRs, although foam pads are also widely employed. Reproduced with permission of Headworks Inc. Houston, Texas.

offering the full range of carbonaceous, nitrification, denitrification, and phosphorus removal options. All MBBR systems have the same advantages over other biological systems, in that they generally have a reduced footprint, low maintenance costs, are robust, can withstand variable loadings, and are easily retrofitted into existing systems. AnMBBRs are generally used as a pretreatment step for stronger wastewaters and biogas generation (Fig. 6.11).

AnMBBRs use the same media as conventional aerobic and anoxic MBBR systems (Table 6.5), with the commercially available MBBR media shown in the review by di Biase *et al.* (2019). Ødegaard *et al.* (2000) analyzed the influence of media size and shape on performance and found that MBBRs should be designed based on surface area loading rates (g COD $m^{-2}d^{-1}$), while the shape and size of the media is not a significant factor as long as an effective surface area is

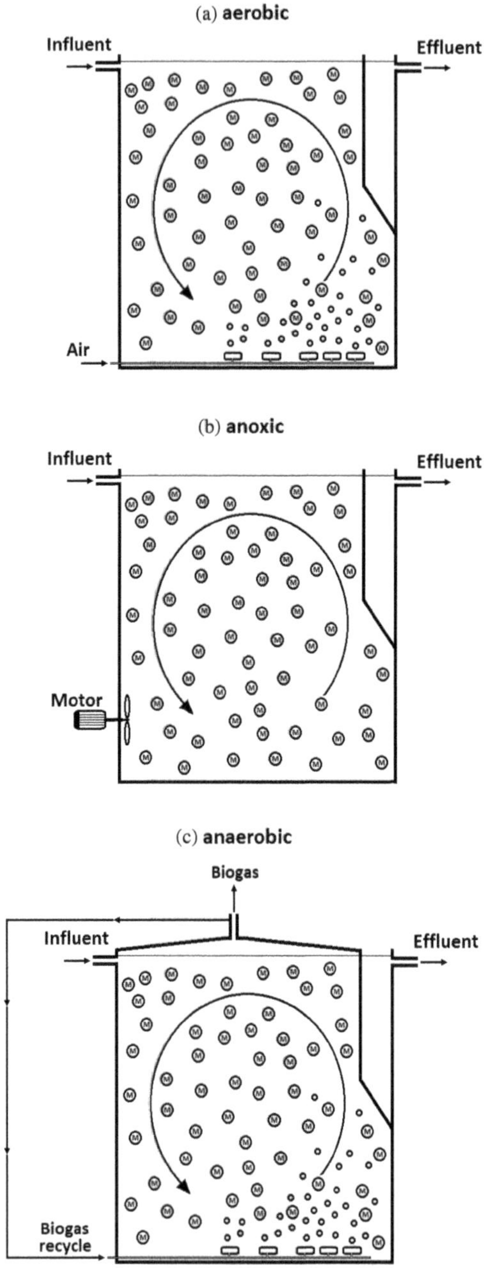

Figure 6.11. Comparison between aerobic (MBBR) and anaerobic moving bed biofilm reactors (AnMBBR). While aerobic systems rely on aeration to keep the carriers in suspension, AnMBBRs rely on base-fed influent, biogas recycling, or a motorized propellor to keep the carriers suspended.

Table 6.5. Characteristics of a typical plastic carrier (Flocor RMP™) used in moving bed biofilm reactors, although there are a wide variety of other types of carriers used in AnMBBRs.

Material	Polypropylene (density = 0.94 g cm³)
Shape	Corrugated cylinder
Dimensions	Length: 20–30 mm Diameter: 15–20 mm
Specific surface	160 m²/m³
Filling rate	70%

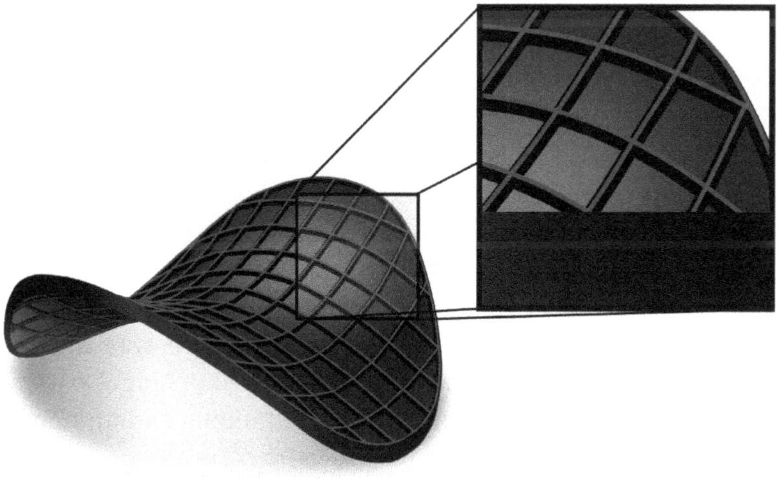

Figure 6.12. The new profiled AnoxKalnes Z-200 carriers developed for AnMBBRS by Veolia (2024). Reproduced with permission of Veolia, Lund, Sweden.

provided. There are several commercial AnMBBRs available, including Veolia's Anoxthane™ Process, which uses the new AnoxK™Z carriers, and Headworks EnergyCell™ Process using ActiveCell ™carriers. AnoxK™Z carriers are unusual in that they do not have holes or gaps in their structure but are rather like potato crisps, flat and oval in shape, and slightly curved, with each side embossed with a raised grid of square cells 225 × 225 μm in size (Fig. 6.12). The biofilm in AnMBBRs develops very slowly, compared to aerobic systems, and so it is important to prevent sloughing to maintain optimum biomass density and diversity. According to the manufacturer, the anaerobic biofilm develops within the surface cells of AnoxK™Z carriers — the depth of the cell wall controlling the overall depth of the biofilm with the excess continuously scoured off. Two cell depths are available at 200 and 400 μm. Flat surface carriers like these that control biofilm development ensure a constant maximum weight of the individual carriers so that they are easily maintained in suspension, also ensuring a fixed reactor biomass density. In contrast, carriers with holes can result in excessive biofilm development and entrapped solids, making them heavy and less easy to maintain in suspension, which requires the carriers to be mechanically cleaned periodically. Where the biofilm development is uncontrolled and irregular, there is a possibility of variable removal efficiencies, although in practice, both types of media work well. This was shown by Arabgol *et al.* (2020; 2022), who compared both AnoxK™Z carriers with a more conventional carrier (AnoxK™K5) under aerobic conditions. They found that the conventional carrier had a significantly higher carbonaceous removal and lower solids production but higher biofilm thickness, which was less dense than both AnoxK™Z carriers. They also noted there was no significant difference between the efficiency of the different cell depths (i.e., 200 and 400 μm), although the thinner grid depth of the AnoxK™Z-200 carrier had maximum density in relation to thickness and was the lightest carrier, making them easier to keep in suspension (Aragol *et al.*, 2022) (Fig. 6.13). However, in AnMBBRs, continuous biofilm depth control, as achieved with the AnoxK™Z carriers, is advantageous where mixing is dependent on mechanical means and shear stress on the carriers is minimal. While the majority of carriers are made out of polyethylene, high-density polyethylene, or polypropylene, Jirasaranporn *et al.* (2022) experimented with PVA gel beads, which are known to provide excellent surfaces for microbial development, in a pilot-scale AnMBBR treating fish canning wastewater. Gel beads are very porous and have a very high specific surface area of 2,500 m^2m^{-3} (i.e., 1,000 m^2m^{-3} surface plus 1,500 m^2m^{-3} internal effective surface area) compared to other carrier materials, which have surface areas between 200 and 1,200 m^2m^{-3}. They found that the beads achieved >70% removal at loading rates between 3.0 and 8.7 kg COD m^3d^{-1}, with the

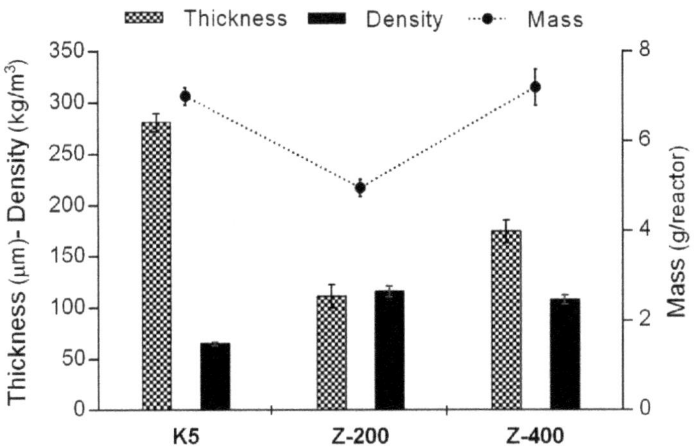

Figure 6.13. Comparison of total biomass, biofilm thickness, and density in identically loaded and operated AnMBBRs using a standard carrier similar to that in Fig. 6.10 (K5) and AnoxKalnes Z-200 and Z-400 (Arabgol *et al.*, 2022). Reproduced with permission of Elsevier Publishing.

acetoclastic methanogen *Methanosaeta* predominating with a relative abundance of 24.0–28.9% and the hydrogenotrophic methanogen *Methanobacterium* (7.7–14.9%), ensuring good biogas production. The latter methanogen is extremely slow-growing, so it appears the very long residence time associated with the gel beads facilitated their establishment. Among the advantages they observed were that the beads had a specific gravity close to water, making them easily suspended in the wastewater, minimizing energy input for mixing. They also reduced biomass loss by sloughing due to the retention of the biomass within the beads, which also protected biological activity from toxic shocks. The authors provide some very interesting 16S rRNA gene amplicon sequencing data on the seed sludge used collected from a UASB using granules (Figs. 6.14 and 6.15) and the AnMBBRs. The dominant phyla in both the seed and the AnMBBRs were Bacteroidota (7.1–10.0% of total sequences), Desulfobacterota (14.1–22.7%), Euryarchaeota (3.5–45.9%), Firmicutes (5.9–10.15%), and Synergistota (4.5–8.9%).

Maximum organic loading rates of 20 kg COD $m^{-3}d^{-1}$ were recorded in laboratory-scale AnMBBRs treating a synthetic protein-based wastewater, achieving 80% COD removal with a maximum

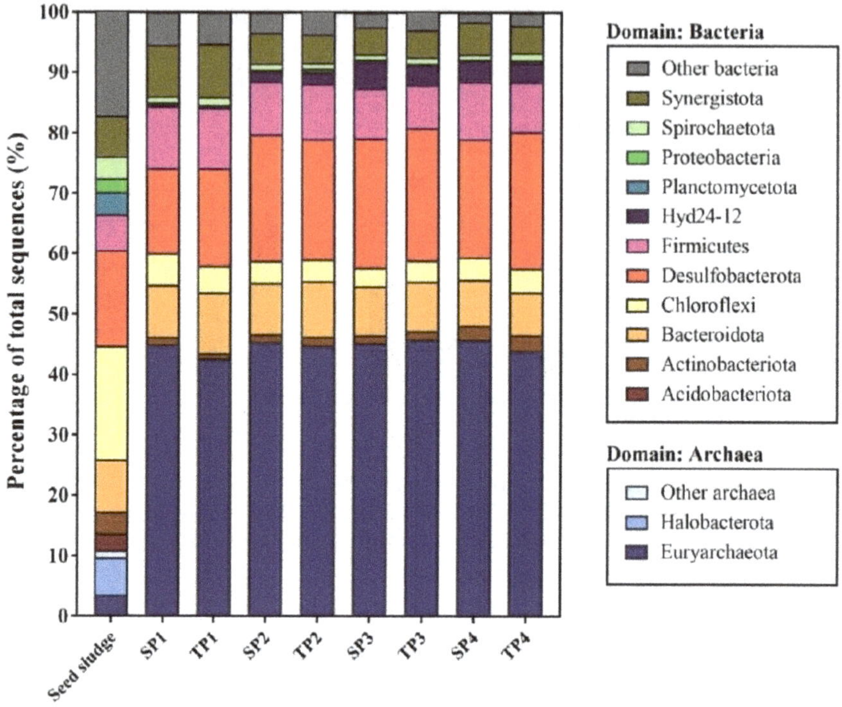

Figure 6.14. Microbial phyla recovered in the seed sludge and on the PVA gel beads used in pilot-scale AnMBBR treating fish canning wastewater. Samples were taken at two different depths of 0.75 m (S) and 1.25 m (T) height within the AnMBBRs, which were operated at organic loading rates of (P1) 4.14 kg COD $m^{-3}d^{-1}$; (P2) 5.38 kg COD $m^{-3}d^{-1}$; (P3) 7.54 kg COD $m^{-3}d^{-1}$; and finally (P4) 4.14 kg COD $m^{-3}d^{-1}$. Reproduced from Jirasaranporn *et al.* (2022) under the Creative Commons Attribution license (CC BY 4.0) (http://creativecommons.org/licenses/by/4.0/).

biogas production of 9.25 m^3 CH_4 m^{-3}, equivalent to an average methane yield of 0.36 m^3 CH_4 kg^{-1} COD removed, with a biogas methane content of 60–70% (di Baise *et al.*, 2016). They used HeadworksBio cubic AC920 plastic carriers and compared two reactors operating at 25 and 35% of the reactor volume, respectively, although no significant difference in performance was recorded between the two reactors. Seed was taken from a sludge digester, with optimum biofilm development occurring after eight weeks. Interestingly, they observed that only 2.5% of COD removal was due to the suspended biomass, with nearly all biodegradation carried out by the carrier biofilm. Shahzad *et al.* (2021) compared the use of an

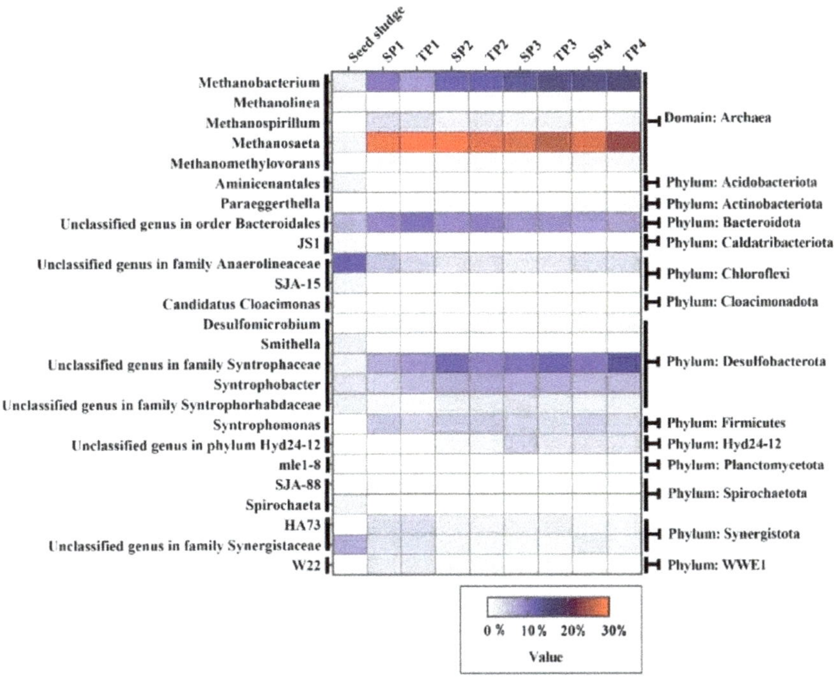

Figure 6.15. The most abundant genera recovered in the seed sludge and on the PVA gel beads used in pilot-scale AnMBBR treating fish canning wastewater. Samples were taken at two different depths of 0.75 m (S) and 1.25 m (T) height within the AnMBBRs, which were operated at organic loading rates of (P1) 4.14 kg COD $m^{-3}d^{-1}$; (P2) 5.38 kg COD $m^{-3}d^{-1}$; (P3) 7.54 kg COD $m^{-3}d^{-1}$; and finally (P4) 4.14 kg COD $m^{-3}d^{-1}$. Reproduced from (Jirasaranporn *et al.*, 2022) under the Creative Commons Attribution (CC BY 4.0) license (http://creativecommons.org/licenses/by/4.0/).

AnMBBR and an up-flow anaerobic hybrid reactor (UAHR) for the treatment of textile desizing wastewater and found optimum loadings of 6 and 8 kg COD m^3d^{-1}, respectively. By increasing the operating temperature from 35 to 55°C, the COD removal, biogas production, and methane content all increased in the AnMBBR by 1.2, 2.6, and 1.1 times, respectively, with similar increases observed for the UAHR. Laboratory scale trials to assess the removal of pharmaceuticals during municipal wastewater treatment using an AnMBBR coupled with a secondary aerobic MBR (Iliopoulou *et al.*, 2023). The plant was operated at two different temperature phases at 35°C (A) and 20°C (B). Together, they produced excellent removal efficiencies (Table 6.6), with the AnMBBR having higher biogas production at 35°C but

Table 6.6. Experimental conditions (pH, temperature, TSS, and biogas production) of each reactor in both phases and removal efficiency of COD, BOD, NH_4–N, and TP (Iliopoulou *et al.*, 2023). Reproduced with permission of Elsevier Publishing.

Methanogenic moving bed biofilm reactor (AnMBBR)

Phase	pH	T (°C)	MLSS (mg/L)	Biogas Production ($mL/L_{reactord}$)	COD_{dis} removal (%)	BOD removal (%)	NH_4–N removal (%)	Tp removal (%)
Phase A	6.9 ± 0.5	35.0 ± 0.5	265 ± 190	176 ± 86	39.6 ± 16.0	51.8 ± 17.0	5.7 ± 9.9	<5%
Phase B	6.7 ± 0.9	20.0 ± 0.5	431 ± 171	68 ± 16	42.3 ± 17.7	54.3 ± 12.2	7.0 ± 10.0	<5%

Aerobic membrane bioreactor (AeMBR)

Phase	pH	T (°C)	MLSS (mg/L)	[a]Total COD_{dis} removal (%)		[a]Total BOD removal (%)	[a]Total NH_4-N removal (%)	[a]Total TP removal (%)
Phase A	7.4 ± 0.6	18.7 ± 1.3	1942 ± 650	86.4 ± 12.5		91.6 ± 5.3	96.4 ± 4.9	<5%
Phase B	7.6 ± 0.2	17.7 ± 1.7	3321 ± 1177	90.9 ± 12.0		95.6 ± 3.9	95.5 ± 6.0	64.4 ± 15.3

[a]Total removal of COD_{dis}, BOD, NH_4–N and TP achieved in the AnMBBR-AeMBR system.

better COD and BOD removal rates at 20°C. Percentage removal ranged from 65 to 100% for metronidazole (MTZ), trimethoprim (TMP), sulfamethoxazole (SMX), and valsartan (VAL). Carbamazepine (CBZ) and diclofenac (DCF) removal efficiencies were much less at <30%. Interestingly, the AnMBBR was solely responsible for the removal of MTZ, TMP, and CBZ, while it also contributed to the removal of SMX and VAL mainly by sorption (Fig. 6.16). Azo dyes are widely used in a wide variety of industries, including paper printing, carpet and textile manufacture, tanneries, polymers, and cosmetics, and are highly polluting. Congo red, for example, is carcinogenic, teratogenic, and mutagenic in animals, including humans, making it a priority-listed pollutant. Maurya *et al.* (2021) used a laboratory scale AnMBBR employing polyurethane and polypropylene foam

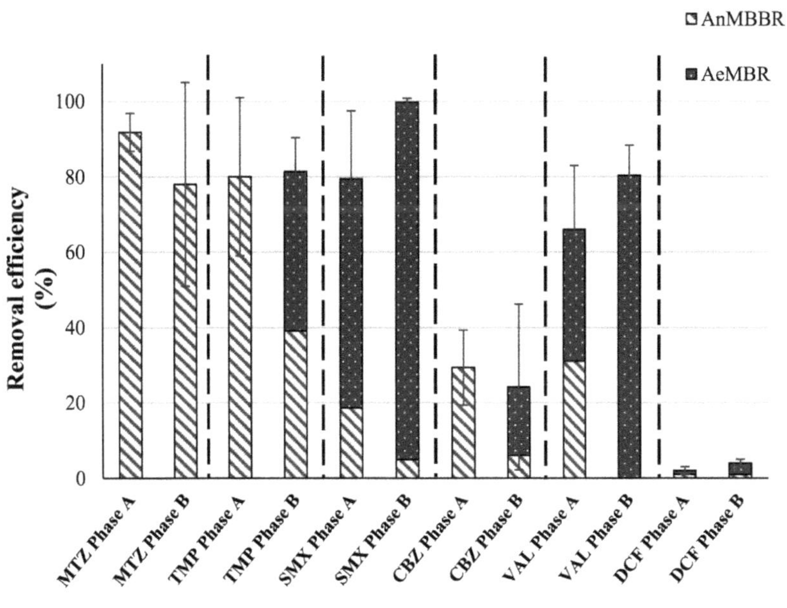

Figure 6.16. Removal (%) of target pharmaceuticals in each experimental phase and reactor of the studied system. The temperature of the Anaerobic Moving Bed Biofilm Reactor (AnMBBR) was set at 35°C at Phase A and 20°C at Phase B (Iliopoulou *et al.*, 2023). MTZ: Metronidazole; TMP: Trimethoprim; SMX: Sulfamethoxazole; CBZ: Carbamazapine; VAL: Valsartan; DCF: Diclofenac. Reproduced with permission of Elsevier Publishing.

carriers into which the bacterium *Lysinibacillus fusiformis*, which is able to biodegrade Congo red dye, had been immobilized. Removal rates of between 58 and 89% were achieved at different loading rates and under different operating conditions. The study opens up the possibility of employing AnMBBRs using immobilized bacteria for the biodegradation of highly toxic waste streams.

AnMBBRs are especially suited for temperatures found in temperate areas, although the optimum temperature range is between 20 and 36°C, making them ideal for food processing wastewaters, which are composed of soluble and highly degradable organic matter. AnMBBRSs have been successfully employed in the dairy industry, sugar beet processing and sugar refining, and vegetable and fruit processing, including brewing, distilling wine production, and general drink and beverage industries. All strong organic wastewaters are amenable to treatment with AnMBBRs, including pulp and paper wastewaters and those from the chemical, textile, oil, and gas industries. The combination of compact, low odor, low maintenance and energy costs, and carbon recovery through biomethane production makes the AnMBBR process an ideal choice for all these industries. Many authors have modeled AnMBBRs and have generally found the Grau second-order and modified Stover-Kincannon models the best fit (Maurya *et al.*, 2021; Shahazad *et al.*, 2021). MBBR technology, in general, is reviewed by Ødegaard (2006) and di Biase *et al.* (2019) and carriers specifically by Zhao *et al.* (2019).

6.5 Expanded and Fluidized Bed Reactors

Anaerobic expanded and fluidized bed reactors (Fig. 6.17) are modifications of the anaerobic biofilter in which fine-grained inert (sand) or reactive (activated carbon or zeolite) random carriers (i.e., media) are used (Yoda *et al.*, 1989). The carrier, which is coated with bacteria and archaea, is expanded or fluidized by applying a high-mixing velocity. However, it is difficult to maintain the optimum mixing velocity in order to ensure the carrier remains in suspension without causing shear forces strong enough to strip off the majority of the accumulated biomass. Therefore, mixing is achieved within the

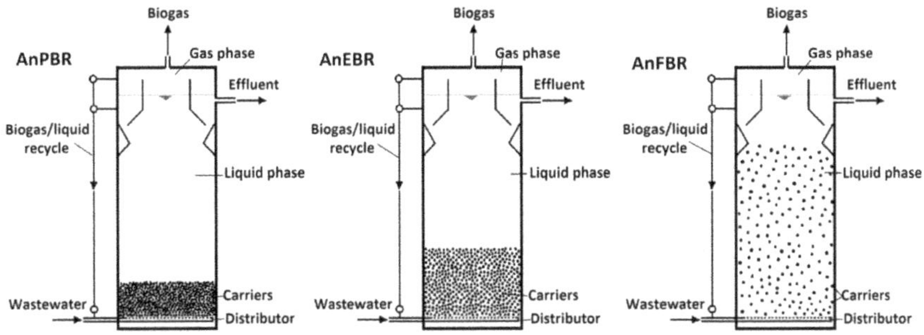

Figure 6.17. Schematic comparison of a porous medium bed reactor (AnPBR), expanded bed reactor (AnEBR), and an anaerobic fluidized bed reactor (AnFBR).

reactor by applying high rates of gas or effluent recycle, along with the influent wastewater, from the base of the reactor. The carrier particles can be fully mixed (fluidized), which increases the volume of the reactor occupied by the medium by 20–25%, or just expanded, which increases the reactor volume occupied by the medium by <10% (Switzenbaun, 1983; Iza, 1991).

When influent wastewater or recycled biogas passes upward through the voids (i.e., interstices) in a layer of an appropriate carrier settled at the base of a reactor without disturbing the carriers, this is known as a static, fixed, or porous medium bed reactor (AnPBR). When the rate of flow through the settled carriers is increased, the carriers begin to be suspended within the reactor, the buoyancy balancing the gravitational and drag forces exerted on the induvial carrier particles, increasing the size of the interstices (i.e., expanded bed reactor — AnEBR). As the flow rate increases even further, the minimum fluidization velocity is reached, and the bed of the medium expands and becomes fluidized (i.e., the solids-liquid mixture behaves as a liquid), the density decreases, and the bed voidage rapidly increases (i.e., fluidized bed reactor — AnFBR) (Fig 6.17). The required minimum fluidization velocity decreases as the size of the medium decreases, so less energy is required where finer carrier particles are used. The hydrodynamics and mass transfer characteristics of fluidized beds are reviewed by Sharma *et al.* (2021).

The particle size of carriers ranges from 0.3 to 1.0 mm, depending on the type of carrier used. The decrease in carrier particle size has resulted in a significant increase in the specific surface area available for biological growth, making such systems highly efficient. However, smaller sizes are readily lost from the reactor, while larger-sized carriers are difficult to fluidize. Due to the difficulty of maintaining an optimum mixing velocity to ensure the medium remains in suspension without causing biomass stripping due to excessive shear, it may be that expanded bed reactors are the more effective system. In expanded beds, the biofilm-coated carrier remains on the floor of the reactor and is expanded by the movement of the liquid forced up through the medium. Fluidized beds are normally much taller, with a height-to-diameter ratio of 5–6:1. Although difficult to start up, fluidized beds are highly resistant to temperature, toxic compounds, and high organic loadings. They are also tolerant to wide fluctuations in influent quality including high solids concentrations (Table 6.2). Expanded and fluidized beds have large solids-liquid contact areas, no clogging, a compact design with a small footprint, and no moving parts or internal structures, making them an attractive option for more complex and strong wastewaters, especially at ambient temperatures (10–20°C), although expanded beds are also highly efficient in treating settled sewage. Details on the design and operation of fluidized bed reactors are given by (Heijnen *et al.*, 1989; Iza, 1991; Marín *et al.*, 1999). Typical loading rates are given in Table 6.4.

6.5.1 *Anaerobic Expanded Bed Reactors*

Anaerobic expanded beds reactors (AnEBRs) were developed in the early 1970s, and Jewell *et al.* (1981) successfully used a laboratory scale unit operated at 20°C to treat settled municipal wastewater. They found that even at HRTs of <30 minutes, the majority of the biodegradable organic matter was removed, producing a final effluent of 40 mg COD L^{-1} and 5 mg SS L^{-1} (Fig. 6.18). This is a much shorter HRT compared to similar anaerobic systems. They recorded optimum removal efficiencies up to a maximum loading of 4 kg BOD $m^{-3}d^{-1}$ and that the AnEBR was able to withstand significant temperature differences as well as rapid changes in organic loading and flow rate, making it a very versatile and robust system.

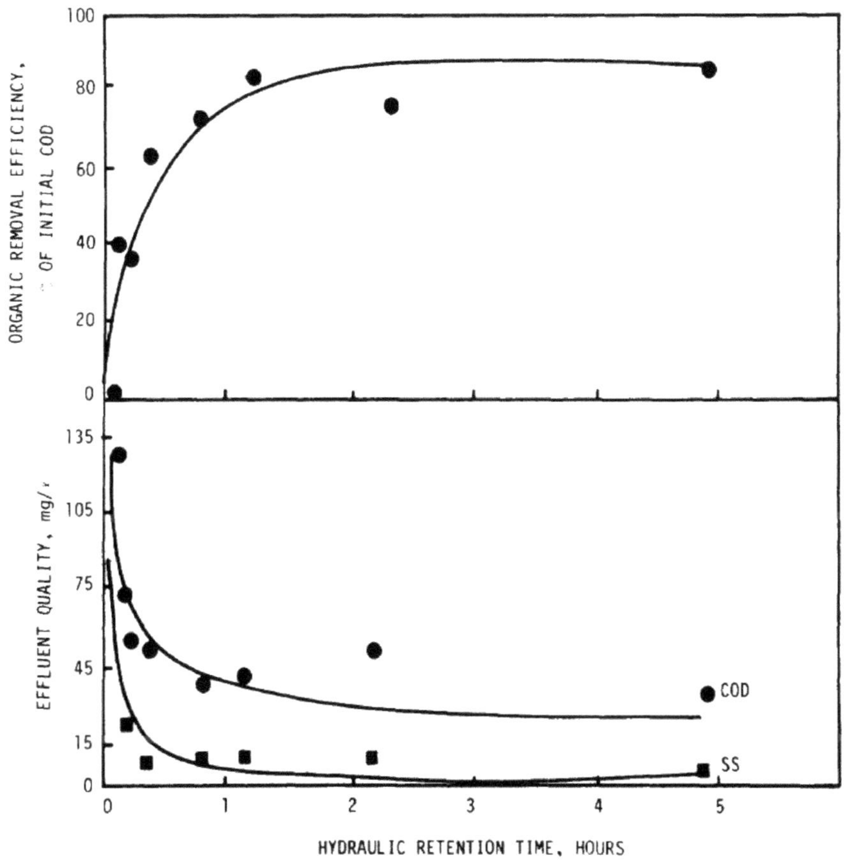

Figure 6.18. Influence of HRT on COD and suspended solids (SS) removal of an anaerobic expanded bed reactor treating settled domestic wastewater (Jewell *et al.*, 1981). Reproduced with permissions of the Water Environment Federation, Alexandria, VA.

AnEBRs have been successfully used to treat a wide range of industrial and agricultural wastewaters (Sharma *et al.*, 1999), offering compact and efficient COD reduction prior to aerobic treatment. They have also been successfully employed to remove trace micropollutants from drinking water using a carbon source such as methanol, although it may be possible to use the natural humic acids in the water as the electron acceptor, with very short HRTs of 1–2 hours, a process known as reductive dechlorination (Carter and Jewell, 1993).

The long sludge retention times mean that the biomass often becomes naturally granular in nature, offering better settleability and treatment potential. Much research was carried out in the 1990s to understand how granulation could be stimulated and maintained (Ye *et al.*, 2005), and today, expanded beds have now been largely superseded by expanded granular sludge bed (EGSB) technology, which is a hybrid between a UASB and AnEBR. They are available commercially, for example, the BioBed™ EGSB reactor from Veolia (Sec. 7.2).

6.5.2 *Anaerobic Fluidized Bed Reactors*

Primary and thickened waste-activated sludges were successfully treated in mesophilic fluidized bed reactors (AnFBRs) using zeolite as media, achieving almost double the treatment rates achieved by conventional digestion (Mustafa *et al.*, 2014). At an HRT of 9 days, COD and VSS removal was 85 and 88%, respectively, for primary sludge at an organic loading rate of 4.2 COD $m^{-3}d^{-1}$ and a 68 and 69% reduction in COD and VSS, respectively, for the waste-activated sludge at the same HRT and loading rate. Methane yields were 0.47 and 0.51 L CH_4 g^{-1} VSS removed for the primary and waste-activated sludges, respectively. Andalib *et al.* (2012) successfully used a zeolite medium (diameter of 425–610 μm and specific surface area of 26.5 m^2 g^{-1}) to treat the byproduct of corn bioethanol production (i.e., stillage) with a COD of 130,000 mg L^{-1} and suspended solids concentration of 47,000 mg L^{-1} in an AnFBR at a loading rate of 29 kg COD $m^{-3}d^{-1}$ with an HRT of 3.5 days. The system achieved an 88 and 78% reduction in COD and TSS (total suspended solids), respectively, with a methane yield of 0.31 L CH_4 g^{-1} COD (equivalent to 40 L CH_4 L^{-1} stillage d^{-1}). They subsequently compared this with primary sludge and found similar results with an 82% reduction in both COD and TSS at a loading rate of 9.5 kg COD $m^{-3}d^{-1}$ and a maximum methane yield of 0.25 L CH_4 g^{-1} COD at an HRT of 4 days (Andalib *et al.*, 2014).

The design and operation of AnFBRs are explained in detail by Deng *et al.* (2016), who give details of how the key design and operating parameters are selected and calculated, including HRT, organic loading rate, sludge loading rate, inclined settler dimensions, sludge discharge frequency, and liquid and gas recycle ratios, and then translated to a full-scale design. They give an example of the design and operating parameters for a full-scale AnFBR for treating 300 tonnes of synthetic dyeing wastewater per day. The influent to the AnFBR contains 100 mg L^{-1} (ppm) of the dye Reactive Blue 250 and a COD of 1,000 mg L^{-1}, with a COD/BOD ratio of 0.42. The wastewater has a pH of 5.5–8.0, a temperature of 20–23°C, and a total nitrogen and phosphorus content of 50 mg N L^{-1} and 20 mg P L^{-1}. The design is required to achieve a 90% color reduction and 80% COD removal. The reactor design is shown in Fig. 6.19 and the full-scale design and operating conditions are given in Table 6.7. Carriers for AnFBRs should be porous, have a large surface area, be non-reactive but able

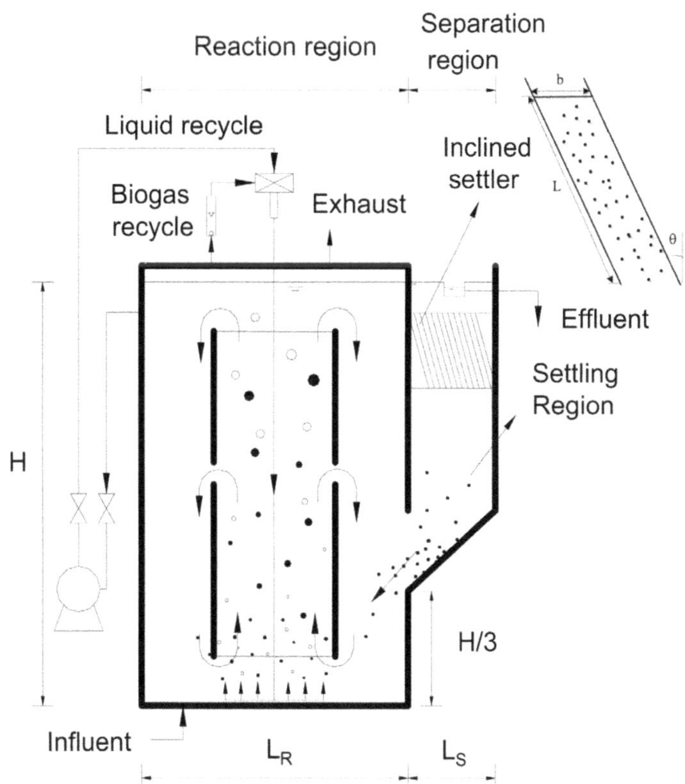

Figure 6.19. Schematic design for an AnFBR treating a synthetic dyeing wastewater (Deng *et al.*, 2016). Details of design values are given in Table 6.7. Reproduced with permission of Elsevier Publishing.

Table 6.7. Design and operational parameters for a full-scale AnFBR treating a synthetic dyeing wastewater (Deng *et al.*, 2016). Reproduced with permission of Elsevier Publishing.

Influent flowrate, Q_0	300 m^3/d
Hydraulic retention time, HRT	14h
Organic loading rate, OLR	1.46 kgCOD/m^3 d
Sludge loading rate, SLR	0.29 kgCOD/kgVSS d
Percentage of bioparticles volume	15%
Safety factor	1.5
Volume of the reaction region, V	309 m^3
Number of reactors, N	1
Reactor height, H	12.1 m
Length of reaction region, L_R	5.5 m
Reactor width, W	4.6 m
Time between bioparticles discharge, t_d	67 days
Biogas production rate	112.7 m^3/d
Liquid recycle ratio	5.8
Biogas recycle ratio	4.8
Length of the inclined channel filled with suspension, L	1.2 m
Width of the inclined channel, w	0.05 m
Spacing between the channels, b	0.05 m
Angle of inclination of the channel from the vertical, θ	30°
Number of inclined channels, N_C	2376
Arrangement of inclined channels in the inclined settler	33 × 72
Length of the separation region, L_S	1.8 m

to adsorb pollutants, be ideally between 0.1 and 0.7 mm, and have a density of 1,500 to 2,500 kg m^3 (Heijnen *et al.*,1989). Granular activated carbon (GAC) is an ideal carrier and can reduce the start up-time from 100 to 15 hours when used instead of sand (Wright and Raper, 1996). Typical loading values and HRT based on maximum influent COD and BOD/COD ratio have been summarized by Deng *et al.* (2016) (Table 6.8).

As the COD is consumed, a portion is used for biomass production, approximately 1.42 kg COD per kg biomass, and the remainder

Table 6.8. Typical operating values for the solids loading rate (SLR), organic loading rate (OLR), hydraulic retention time (HRT), and volume of bioparticles for AnFBRs based on influent BOD/COD ratio and maximum COD (Deng *et al.*, 2016). Reproduced with permission of Elsevier Publishing.

Initial BOD/COD	Maximum influent COD (ppm)	SLR (kgCOD/kgVSS d)	OLR (kgCOD/m³ d)	HRT (h)	Percentage of bioparticles volume in reactor
≤0.3	≤1500	0.05–0.3	<3.0	>12	5–20
0.3–0.5	≤2000	0.1–0.55	1–8	8–48	5–20
>0.5	≤15,000	0.6–2	5–40	3–24	5–20

for biogas production. The amount of methane in biogas can be approximated as 65%, with a production rate of 0.35 m^3 methane per kg COD consumed. Therefore, the amount of methane generated (G_{methane}) can be calculated as the amount of COD removed by subtracting the amount of COD used for biomass production:

$$G_{\text{methene}} = 0.35(Q_0(S_0 - S) - 1.42m_t) \tag{6.3}$$

where $\boldsymbol{Q_0}$ is the influent flow rate (m^3d^{-1}), $\boldsymbol{S_0}$ is the influent COD concentration (kg m^{-3}), $\boldsymbol{S}$ is the effluent COD concentration (kg m^{-3}), and $\boldsymbol{m_t}$ is the amount of biomass produced per day (kg VSS d^{-1}). The parameter m_t is calculated by measuring the amount of biomass at the start of degradation to final discharge:

$$m_t = \frac{V(Mt_d - M_i)}{t_d} \tag{6.4}$$

where $\boldsymbol{V}$ is the total volume of the reaction region of the reactor (m^3), $\boldsymbol{Mt_d}$ is the biomass concentration at the point of discharge (kg VSS m^{-3}), $\boldsymbol{M_i}$ is the initial biomass concentration in the reactor (kg VSS m^{-3}), and $\boldsymbol{t_d}$ is the time between each bioparticles discharge (d). Carriers will need to be recovered and biomass removed by washing before the carriers are returned to maintain the optimum sludge loading rate (Table 6.8). The biogas (G_{biogas}) production rate is calculated as:

$$G_{\text{biogas}} = \frac{G_{\text{methane}}}{0.65} \tag{6.5}$$

This is further explained by van Lier *et al.* (2008) and Deng *et al.* (2016).

In Cuba, fluidized beds are used to treat vinasse, the final waste from the alcohol distillation process from sugar cane molasses. With organic loading rates of up to 10 kg COD $m^{-3}d^{-1}$ and COD removals >70%, the waste produces methane-rich biogas but with sulfide concentrations >1% (Conde *et al.*, 1993). However, sulfide concentrations can be controlled by using either granular activated carbon (GAC) or zeolites as carriers. Other researchers have also observed the simultaneous removal of sulfate and ammonia in anaerobic GAC fluidized beds treating vinasse (Fdz-Polanco *et al.*, 2000; 2001). This resulted in high concentrations of molecular nitrogen in the biogas but low concentrations of hydrogen sulfide. They also identified an unusual anaerobic nitrogen removal mechanism, resulting in the loss of nitrogen in the liquid phase and the presence of nitrogen in the biogas, with up to 55% of the total Kjeldahl nitrogen removed by this process, which was identified as the anammox process reaction. The removal mechanism of sulfur is less clear, with elemental sulfur found within the reactor and associated with the effluent suspended solids. The function of the GAC appears to be to increase the localized concentration of nitrogenous and sulfur species by adsorption close to the biofilm, encouraging the biochemical reactions to occur.

The overall reaction has been postulated as:

$$SO_4^{=} + 2NH_4^{+} \rightarrow S + N_2 + 4H_2O \ (\Delta G° = -47.8 \text{ kJ mol}^{-1}) \quad (6.6)$$

Fdz-Polanco *et al.* (2001) have suggested that this reaction comprises three separate reactions:

$$3SO_4^{=} + 4NH_4^{+} \rightarrow 3S^{=} + 4NO_2^{-} + 4H_2O + 8H^{+} \quad (6.7)$$

$$3S^{=} + 2NO_2^{-} + 8H^{+} \rightarrow 3S + N_2 + 4H_2O \quad (6.8)$$

$$2NO_2^{-} + 2NH_4^{+} \rightarrow 2N_2 + 4H_2O \quad (6.9)$$

The last equation being the Anammox reaction.

6.5.3 *Anammox Process*

Although both nitrification and denitrification were observed in the 19th century, anammox was not discovered until the 1990s. The reaction is carried out by anaerobic ammonium oxidation (anammox) bacteria, where nitrite (the electron acceptor) is converted to nitrogen gas by autotrophic bacteria using ammonium as the electron donor:

$$NH_4^+ + NO_2 \rightarrow N_2 + 2H_2O \quad (6.10)$$

As this is an autotrophic reaction, no external carbon source (e.g., methanol) is required (Jetten *et al.*, 1997), with activity inhibited at oxygen concentrations >0.5% (Strous *et al.*, 1997a; 1997b). To facilitate the anammox reaction, nitrite must be present and can be produced by a separate nitritation process known as a partial nitritation (PN). Most commercial anammox systems employ PN as an integrated part of the main process (Lackner *et al.*, 2014), including when used as a tertiary step for nitrate removal from domestic wastewater treatment plants (Cao *et al.*, 2019). The process is normally used in conjunction with the PN SHARON® process (Gray, 2021), which produces an ammonium-nitrite feed effluent ideal for the anammox process (Sec. 4.4.7).

The autotrophic bacteria responsible for the oxidation of ammonia and the simultaneous reduction of nitrite belong to the widely distributed phylum *Planctomycetes*. Not all the bacteria in this phylum are able to oxidize ammonium, this being restricted to the anammox-*Planctomycetes*, which have a characteristic membrane-bound cell compartment called the anammoxosome (Fuerst and Sagulenko, 2011). Ammonium is used as an energy source by the anammox bacteria and carbon dioxide as a carbon source, with NO_2^-, and NO_3^-, if present, used as electron acceptors. Kartel *et al.* (2011) have described the process in detail where the enzyme nitrite oxidoreductase reduces nitrite to NO, which then reacts with ammonium to form hydrazine (N_2H_2), enabled by the enzyme hydrazine synthase, which is oxidized by hydrazine dehydrogenase to nitrogen. The growth rate (doubling time 11 days) and growth yield (0.11 g VSS g^{-1} NH_4-N) for these microorganisms are very low, resulting in low sludge production.

Therefore, efficient biomass retention is required to maintain suitable anammox biomass concentrations within reactors, favoring up-flow fixed bed reactors but especially expanded and fluidized systems, and also AnMMBRs (Strous *et al.*, 1997b). Anammox reactors are very compact due to a high maximum specific nitrogen consumption rate (0.82 g N g^{-1} VSS d^{-1}), a high affinity for ammonium and nitrite, and granular growth, allowing efficient biomass retention. They use far less energy (60%) than alternative nitrification-denitrification processes and do not require a separate organic carbon source, making them a more cost-effective treatment option for ammonia-rich effluents (Lackner *et al.*, 2014). Also, as nitrous oxide and carbon dioxide are not an intermediate reaction product in the anammox reaction, greenhouse gas emissions are minimized.

A key disadvantage to the system is the sensitivity of the anammox bacteria to changes in environmental conditions, in particular, the nature of the wastewater feed. While organic matter, nutrients, sulfide, and salinity concentrations are all possible sources of inhibition, it is the toxic nature of ammonium and nitrite that can act synergistically in terms of toxicity, which is most often the cause of process inhibition and failure. Therefore, the feed ratio must be carefully controlled, as well as the pH. It has been established by Zhang *et al.* (2016) that influent nitrogen concentrations of below 320 mg N L^{-1} guaranteed stable conditions are maintained in expanded bed reactors used for anammox. The development of the anammox granule can take a very long time and have a settling velocity in the region of 100–150 m h^{-1} (Fig. 6.20).

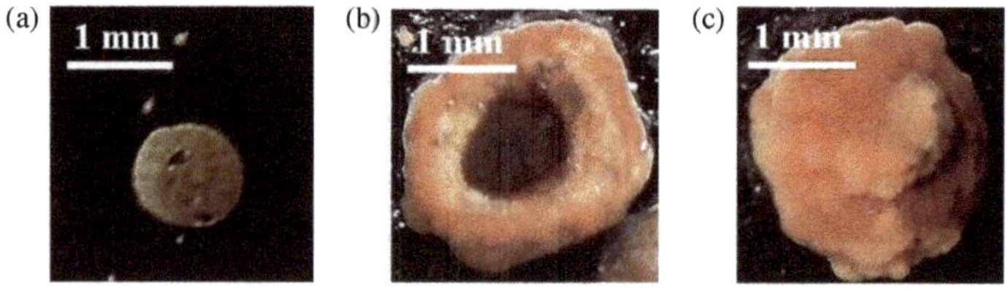

Figure 6.20. Structure of the carrier used by Zhang *et al.* (2016) in their anaerobic expanded bed anammox reactor. (a) The carrier, (b) the carrier with anammox biofilm development (1.0–1.2 mm depth), and (c) the actual anammox granule (average size 2.4 mm) with a high settling velocity of >200 m h^{-1}. Reproduced with permission of Elsevier Publishing.

References

Andalib, M., Hafez, H., Elbeshbishy, E., Nakhla, G. and Zhu, J. (2012) Treatment of thin stillage in a high-rate anaerobic fluidized bed bioreactor (AFBR). *Bioresource Technology*, **121**, 411–418. http://dx.doi.org/10.1016/j.biortech.2012.07.008

Andalib, M., Elbeshbishy, E., Mustafa, N., Hafez, H., Nakhla, G. and Zhu, J. (2014) Performance of an anaerobic fluidized bed bioreactor (AFBR) for digestion of primary municipal wastewater biosolids and bioethanol thin stillage. *Renewable Energy*, **71**, 276–285. http://dx.doi.org/10.1016/j.renene.2014.05.039

Anderson, G.K., Donnelly, T. and McKeown, K. (1984) The application of anaerobic packed bed reactors to industrial wastewater treatment. *Water Pollution Control*, **83**, 491–498.

Arabgol, R., Vanrollenghem, P.A. and Delatolla, R. (2022) Influence of MBBR carrier geometrical properties and biofilm thickness restraint on biofilm properties, effluent particle size distribution, settling velocity distribution, and settling behaviour. *Journal of Environmental Sciences*, **122**, 138–149. https://doi.org/10.1016/j.jes.2021.09.029

Arabgol, R., Vanrollenghem, P.A., Piculell, M. and Delatolla, R. (2020) The impact of biofilm thickness-restraint and carrier type on attached growth system performance, solids characteristics and settleablity. *Environmental Science: Water Research and Technology*, **6**, 2843–2855. https://doi.org/10.1039/D0EW00314J

Bedogni, S., Bregoli, M. and Biglia, A. (1983) Full-scale anaerobic filter treating sugar-mill anionic effluents. In *Proceedings of the Anaerobic Wastewater Treatment Symposium*, pp. 313–314.

Berg, L. van den and Lentz, C.P. (1980) Effects of film area-to-volume ratio, film support, height and direction of flow on performance of methanogenic fixed film reactors. *Proceeding of the US Department of the Environment Workshop Seminar on Anaerobic Filters.*

Bonastre, N. and Paris, J.M. (1989) Survey of laboratory pilot and industrial anaerobic filter installations. *Process Biochemistry*, **24**, (1), 15–20.

Canovas-Diaz, M. and Howell, J.A. (1986) Stability of a percolating anaerobic downflow fixed film reactor under overloading conditions. *Biotechnology Letters*, **8**, 379–384. https://doi.org/10.1007/BF01040871

Cao, S., Du, R., Peng, Y., Li, B. and Wang, S. (2019) Novel two stage partial denitrification (PD)-Anammox process for tertiary nitrogen removal from low carbon/nitrogen (C/N) municipal sewage. *Chemical*

Engineering Journal, **362**, 107–115. https://doi.org/10.1016/j.cej.2018.12.160

Carter, S.R. and Jewell, W.J. (1993) Biotransformation of tetrachloroethylene by anaerobic attached-films at low temperatures. *Water Research*, **27**, (4), 607–615. https://doi.org/10.1016/0043-1354(93)90170-M

Catunda, P.F.C. and von Haandel, A.C. (1996) Improved performance and increased applicability of waste stabilization ponds by pre-treatment in a UASB reactor. *Water Science and Technology*, **33**, (7), 147–156.

Cavalcanti, P.F.F., van Haandel, A. and Lettinga, G. (2001) Polishing ponds for post-treatment of digested sewage part 1: Flow through ponds. *Water Science and Technology*, **44**, (4), 237–245.

Characklis, W.G. and Trulear, M.G. (1982) Dynamics of biofilm processes. *Journal of the Water Pollution Control Federation*, **54**, 1288–1301.

Chernicharo, C.A.L. and Nascimento, M.C.P. (2001) Feasibility of a pilot-scale UASB/trickling filter system for domestic sewage treatment. *Water Science and Technology*, **44**, (4), 1–314.

Chernicharo, C.A.L., Silveira Cota, R. da, Zerbini, A.M., von Sperling, M. and Novy de Castro Brito, L.H. (2001) Post treatment of anaerobic effluents in an overland flow system. *Water Science and Technology*, **44**, (4), 229–236.

CIWEM (2000) *Biological Filtration and Other Fixed Film Processes*. The Chartered Institution of Water and Environmental Management, London, UK.

Collins, B.A., Birzer, C.H., Harris, P.W., Kidd, S.P., McCabe, B.K. and Medwell, P.R. (2023) Two-phase anaerobic digestion in leach bed reactors coupled to anaerobic filters: A review and potential of biochar filters. *Renewable and Sustainable Energy Reviews*, **175**, 113187. https://doi.org/10.1016/j.rser.2023.113187

Conde, J.L., Travieso, L. and Sánchez, E. (1993) Biogas purification process using intensive microalgal cultures. *Biotechnology Letters*, **15**, 317–326. https://doi.org/10.1007/BF00128326

Demuynck, M.G., Naveau, H.P., and Nyns, E.I. (1983) Anaerobic fermentation technology in Europe. In *Environmental Biotechnology: Future Prospects* (ed. J.M. Sidwick). FAST Report FOB51, CEC, Brussels.

Deng, Z., Fung, K.Y., Ng, K.M. and Wei, C. (2016) Design of anaerobic fluidized bed bioreactor — Dyeing effluents. *Chemical Engineering Science*, **139**, 273–284. http://dx.doi.org/10.1016/j.ces.2015.09.029

Denny, P. (1997) Implementation of constructed wetlands in developing countries. *Water Science and Technology*, **35**, (5), 27–34.

Deshpande, V.P., Kaul, S.N. and Deshpande, C.V. (1991) An evaluation of a simple kinetic model for the treatment of domestic sewage by means of an anaerobic rotating biological contactor. *Bioresource Technology*, **38**, 31–38. https://doi.org/10.1016/0960-8524(91)90218-9

di Baise, A., Devlin, T.R. and Oleszkiewicz, J.A. (2016) Start-up of an anaerobic moving bed-biofilm reactor and transition to brewer wastewater treatment. *Journal of Environmental Engineering,* **142**, (12), 06016008. https://doi.org/10.1061/(ASCE)EE.1943-7870.000114

di Biase, A., Kowalski, M.S., Devlin, T.R. and Oleszkiewiez, J.A. (2019) Moving bed biofilm reactor technology in municipal wastewater treatment: A review. *Journal of Environmental Management*, **247**, 849–866. https://doi.org/10.1016/j.jenvman.2019.06.053

Dixo, N.G.H., Gambrill, M.P., Catunda, P.F.C. and van Haandel, A.C. (1995) Removal of pathogenic organisms from the effluents of an upflow anaerobic digester using waste stabilization ponds. *Water Science and Technology*, **31**, (12), 275–284.

Duff, S.J.B. and van den Berg, L. (1982) Treatment of fish processing waste using the downflow stationary fixed film (DSFF) reactor. *Biotechnology Letters,* **4**, 821–826. https://doi.org/10.1007/BF00131160

Ebrahimi, M., Kazemi, H., Mirbagheri, S. and Rockaway, T.D. (2018) Integrated approach to treatment of high-strength organic wastewater by using anaerobic rotating biological contactor. *Journal of Environmental Engineering*, **144**, 04017102. https://doi.org/10.1061/(ASCE)EE.1943-7870.0001312

Elmitwalli, T., Zeeman, Gr. and Lettinga, G. (2001) Anaerobic treatment of domestic sewage at low temperature. *Water Science and Technology*, **44**, (4), 33–40.

Fdz-Polanco, F., Fdz-Polanco, M., Fernández, N., Ureña, M. and Villaverde, S. (2000) Simultaneous organic nitrogen and sulfate removal in anaerobic GAC fluidised bed reactor. In *Proceedings of the VI Latin-American Workshop-Seminar on Wastewater Anaerobic Treatment*, Recife, Brazil, pp. 36–43.

Fdz-Polanco, F., Fdz-Polanco, M., Fernández, N., Ureña, M.A., García, P.A. and Villaverde, S. (2001) Simultaneous organic nitrogen and sulfate removal in anaerobic GAC fluidised bed reactor. *Water Science and Technology*, **44**, (4), 15–22.

Fox, P., Suidan, M.T. and Bandy, J.T. (1990) A comparison of media types in acetate fed expanded-bed anaerobic reactors. *Water Research*, **24**, 827–834.

Fuerst, J.A. and Sagulenko, E. (2011) Beyond the bacterium: Planctomycetes challenge our concepts of microbial structure and function. *Nature Reviews Microbiology*, **9**, 403–413.

Gęca, M., Wiśniewska, M. and Nowicki, P. (2022) Biochars and activated carbons as adsorbents of inorganic and organic compounds from multi-component systems — a review. *Advances in Colloid and Interface Science*, **305**, 102687. https://doi.org/10.1016/j.cis.2022.102687

Genung, R.K., Hancher, C.W., Rivers, A.L. and Harris, M.T. (1982) Energy conservation and methane production in munipal wastewater treatment using fixed-film anaerobic bioreactors. *Biotechnology Bioengineering Symposium*, **12**, 365–380.

Gonçalves, R.F., de Araújo, V.L. and Bof, V.S. (1999) Combining upflow anaerobic sludge blanket (UASB) reactors and submerged aerated bio-filters for secondary domestic wastewater treatment. *Water Science and Technology*, **40**, (8), 71–7

Gray, N.F. (2021) *Fixed-Film Reactors in Wastewater Treatment.* World Scientific Publishing, Singapore.

Heijnen, J. J., Mulder, A., Enger, W. and Hoeks, F. (1989) Review on the application of anaerobic fluidized bed reactors in waste-water treatment. *Chemical Engineering Journal*, **41**, (3), B37–B50. https://doi.org/10.1016/0300-9467(89)80029-2

Hobson, P.N. and Wheatley, A.D. (1993) *Anaerobic Digestion: A Modern Theory and Practice.* Elsevier Applied Science, London, UK.

Iliopoulou, A., Arvaniti, O.S., Deligiannis, M., Gatidou, G., Vyrides, I., Fountoulakis, M.S. and Stainakis, A.S. (2023) Combined use of strictly anaerobic MBBR and aerobic MBR for municipal wastewater treatment and removal of pharmaceuticals. *Journal of Environmental Management*, **343**, 118211. https://doi.org/10.1016/j.jenvman.2023.118211

Iza, J. (1991) Fluidized bed reactors for anaerobic treatment. *Water Science and Technology*, **24**, (8), 109–132.

Jetten, M.S.M., Horn, S.J. and van Loosdrecht, M.C.M. (1997) Towards a more sustainable municipal wastewater treatment system. *Water Science and Technology*, **35**, (9), 171–80.

Jewell, W.J., Swizenbaum, M.S. and Morris, J.W. (1981) Municipal waste-water treatment with the anaerobic attached microbial film expanded bed process. *Journal of the Water Pollution Control Federation*, **53**, (4), 482–490.

Jirasaranporn, A., Jantharadej, K., Wongwailikhit, K., Chawengkijwanich, C., Lohwatcharin, J. and Suwannasilp, B.B. (2022) A pilot-scale anaerobic moving-bed biofilm reactor with PVA gel beads as media for the treatment of fish canning industry wastewater. *Water Reuse*, **12**, (2), 206–222. https://doi.org/10.2166/wrd.2022.090

Kartal, B., Maalcke, W.J., de Almeida, N.M., Cirpus, I., Gloerich, J., Geerts, W., den Camp, H.J.M.O., Harhangi, H.R., Janssen-Megens, E.M., Francoijs, K.-J., Stunnenberg, H.G., Keltjens J.T., Jetten, M.S.M. and Strousm, M. (2011). Molecular mechanism of anaerobic ammonium oxidation. *Nature*, **479**, 127–130

Kennedy, K. J. and Berg, L. van den. (1982). Anaerobic digestion of piggery waste using a stationary film reactor. *Agricultural Wastes*, **4**, 151–158.

Kiran, M.G., Pakshirajan, K. and Das, G. (2017) A new application of anaerobic rotating biological contactor reactor for heavy metal removal under sulfate reducing condition. *Chemical Engineering Journal*, **321**, 67–75. http://dx.doi.org/10.1016/j.cej.2017.03.080

Lacalle, M.L., Villaverde, S., Fdz-Polanco, F. and García-Encina, P.A. (2001) Combined anaerobic/aerobic (UASB+UBAF) system for organic matter and nitrogen removal from a high strength industrial wastewater. *Water Science and Technology*, **44**, (4), 255–262.

Lackner, S., Gilbert, E.M., Vlaeminck, S.E., Joss, A., Horn, H. and van Loosdrecht, M.C.M. (2014) Full-scale partial nitritation/anammox experiences — an application survey. *Water Research*, **55**, 292–303.

Laquidara, M.J., Blanc, F.C. and O'Shaughnessy, J.C. (1986) Development of biofilm, operating characteristics and operational control in the anaerobic rotating biological contactor process. *Journal of the Water Pollution Control Federation*, **58**, 107–114.

Lu, C., Yeh, A. C. and Lin, M.-R. (1995) Treatment of high-strength organic wastewaters using an anaerobic rotating biological contactor. *Environment International,* **21**, 313–323. https://doi.org/10.1016/0160-4120(95)00027-I

Marín, P., Alkalay, D., Guerrero, L., Chamy, R. and Schiappacasse, M.C. (1999) Design and startup of an anaerobic fluidised bed reactor. *Water Science and Technology*, **40**, (8), 63–70.

Maurya, K.L., Swain, G., Sonwani, R.K., Verma, A. and Singh, R.S. (2021) Bioremediation of Congo red in an anaerobic moving bed bioreactor: Process optimization and kinetic modelling. *Bioresource Technology Reports,* **16**, 100843. https://doi.org/10.1016/j.biteb.2021.100843

Mohammadi, M. and Mohammadi, P. (2022) Developing single-substrate steady-state model for biohydrogen production in continuous anaerobic activated sludge — rotating biological contactor system: Novel insights on the process. *International Journal of Energy Research,* **46**, 2041–2050.

Mohammadi, M., Mohammadi, P., Karami, N., Barzegar, A. and Annuar, M.S.M. (2019) Efficient hydrogen gas production from molasses in hybrid anaerobic-activated sludge-rotating biological contactor. *International Journal of Hydrogen Energy*, **44**, 2592–2602. https://doi.org/10.1016/j.ijhydene.2018.12.033

Mustafa, N., Elbeshbishy, E., Nakhla, G. and Zhu, J. (2014) Anaerobic digestion of municipal wastewater sludges using anaerobic fluidized bed bioreactor. *Bioresource Technology*, **172**, 461–466. http://dx.doi.org/10.1016 /j.biortech.2014.09.081

Núñez, L.A. and Martinez, B. (2001) Evaluation of an anaerobic aerobic system for carbon and nitrogen removal in slaughterhouse wastewater. *Water Science and Technology*, **44**, (12), 271–277.

Ødegaard, H. (2006) Innovations in wastewater treatment: The moving bed biofilm reactor process. *Water Science and Technology*, **53**, (9), 17–33.

Ødegaard, H., Gisvold, B. and Stickland, J. (2000) The influence of carrier size and shape in the moving bed biofilm process. *Water Science and Technology*, **41**, (4–5), 383–392.

Ødegaard, H., Rusten, B. and Westrum, T. (1994) A new moving bed biofilm reactor — Applications and results. *Water Science and Technology*, **29**, (10–11), 157–165.

Paritosh, K. and Kesharwani, N. (2024) Biochar mediated high-rate anaerobic bioreactors: A critical review on high strength wastewater treatment and management. *Journal of Environmental Management*, **355**, 120348. https://doi.org/10.1016/j.jenvman.2024.120348

Picanço, A.P., Vallero, M.V.G., Gianotti, E.P., Zaiat, M. and Blundi, C.E. (2001) Influence of porosity and composition of supports on the methanogenic biofilm characteristics developed in a fixed bed anaerobic reactor. *Water Science and Technology*, **44**, (4), 197–204.

Pradeep, N.V., Anupama, S., Arun Kumar, J.M., Vidyashree, K.G., Lakshmi, P., Ankitha, K. and Pooja, J. (2014) Treatment of sugar industry wastewater in anaerobic downflow stationary fixed film (DSFF) Reactor. *Sugar Tech*, **16**, 9–14. https://doi.org/10.1007/s12355-013-0227-8

Rittmann, B.E., Strubler, C.E. and Tuzicka, T. (1982) Anaerobic filter pretreatment kinetics. *Journal of the Sanitary Engineering Division, American Society of Civil Engineers*, **108**, 900–911.

Sachs, E.F., Jennett, J.C. and Red, M.C. (1982). Pharmaceutical waste treatment by anaerobic filter. *Journal of the Sanitary Engineering Division, American Society of Civil Engineers*, **108**, 297–313.

Satyanarayan, S., Thakar, K., Kaul, S.N., Badrinath, S.D. and Swarnkar, N.G. (1987) Anaerobic rotating biological drum contactor for the treatment of dairy wastes. *Indian Chemical Engineer*, **29**, 3–6.

Shahzad, H.M., Khan, S.J., Zeshan, Jamal, Y. and Habib, Z. (2021) Evaluating the performance of anaerobic moving bed bioreactor and up-flow anaerobic hybrid reactor for treating textile desizing wastewater. *Biochemical Engineering Journal*, **174**, 108123. https://doi.org/10.1016/j.bej.2021.108123

Sharma, M., Mohapatra, T. and Ghosh, P. (2021) Hydrodynamics, mass and heat transfer study for emerging heterogeneous Fenton process in multiple fluidized-bed reactor system for wastewater treatment — A review. *Chemical Engineering Research and Design*, **171**, 48–62. https://doi.org/10.1016/j.cherd.2021.04.019

Sharma, V.K., Teta, C. and Castelluccio, G. (1999) Anaerobic treatment of semi-solid organic waste. *Energy Conversion and Management*, **40**, 369–384. https://doi.org/10.1016/S0196-8904(98)00128-9

Sousa, J.T. and Foresti, E. (1996) Domestic sewage treatment in a UASB-SBR reactor system. *Water Science and Technology*, **33**, (3), 199–206.

Sousa, J.T. de, van Haandel, A.C. van and Guimaráes, A.A.V. (2001) Post-treatment of anaerobic effluents in constructed wetland systems. *Water Science and Technology*, **44**, (4), 213–219.

Strous, M., van Gerven, E., Kuenen, J.G. and Jetten, M.S.M. (1997a) Effects of aerobic and microaerobic conditions on anaerobic ammonium-oxidizing (anammox) sludge. *Applied Environmental Microbiology*, **63**,(6), 2446–3448.

Strous, M., van Gerven, E., Zheng, P., Kuenen, J.G. and Jetten, M.S.M. (1997b) Ammonium removal from concentrated waste streams with the anaerobic ammonium oxidation (ANAMMOX) process in different reactor configurations. *Water Research*, **31**, (8), 1955–1962.

Switzenbaum, M.A. (1983) A comparison of the anaerobic filter and anaerobic expanded/fluidised bed process. *Water Science and Technology*, **15**, 399–413.

Tait, S.J. and Friedman, A.A. (1980) Anaerobic rotating biological contactor for carbonaceous wastewaters. *Journal of the Water Pollution Control Federation*, **52**, (8), 2257–2269.

van Lier, J.B., Mahmoud, N. and Zeeman, G. (2008) Anaerobic wastewater treatment. In *Biological Wastewater Treatment: Principles, Modeling and Design* (eds. M. Henze, M.C.M. van Loosdrecht, G.A. Ekama and D. Brdjanovic). IWA Publishing, London.

Veolia. (2024) *Z-MBBR:* Adding a new dimension to biofilm technology. Veolia Water Technologies, Lund. https://www.anoxkaldnes.com/sites/g/files/dvc4056/files/document/2020/09/AnoxKaldnes_Z-MBBR_Folder.pdf

Ware, A.J. and Pescod, M.B. (1998) Full-scale studies with an anaerobic aerobic RBC unit treating brewery wastewater. *Proceedings of the Fourteenth Biennial Conference of the International Association on Water Pollution Research and Control*, Brighton, UK, 18–21 July 1988, pp 197–208. https://doi.org/10.1016/B978-1-4832-8439-2.50023-7

Wheatley, A.D., Fisher, M.B. and Grobicki, A.M.W. (1997) Applications of anaerobic digestion for the treatment of industrial wastewaters in Europe. *Journal of the Chartered Institution of Water and Environmental Management*, **11**, 39–46.

Wilkie, A., Reynolds, P.J. and Colleran, E. (1983) Media effects in anaerobic filters. In *Proceedings of Anaerobic Treatment Symposium*, pp. 242–258.

Wright, P.C. and Raper, J.A. (1996) A review of some parameters involved in fluidized bed bioreactors. *Chemical Engineering Technology*, **19**, 50–64. https://doi.org/10.1002/ceat.270190109

Ye, F.-X., Chen, Y.-X. and Feng, X.-S. (2005) Advanced start-up of anaerobic attached film expanded bed reactor by pre-aeration of biofilm carrier. *Bioresource Technology*, **96**, 115–119. https://doi.org/10.1016/j.biortech.2003.03.002

Yoda, M., Kitagawa, M. and Miyaji, Y. (1989) Granular sludge formation in the anaerobic expanded micro-carrier bed process. *Water Science and Technology*, **21**, 109–120.

Young, J.C. and Dahab, M.F. (1983) Effects of media design on the performance of fixed-bed anaerobic filters. *Water Science and Technology*, **15**, 321–335.

Zaloum, R. and Abbott, M. (1997) Anaerobic pre-treatment improves single sequencing batch reactor treatment of landfill leachates. *Water Science and Technology*, **35**, (1), 207–214.

Zhang, Y., Niu, Q., Ma, H., He, S., Kubota, K. and Li, Y.-Y. (2016) Long-term operation performance and variation of substrate tolerance ability in an anammox attached film expanded bed (AAFEB) reactor. *Bioresource Technology*, **211**, 21–40. http://dx.doi.org/10.1016/j.biortech.2016.03.055

Zhao, Y., Liu, D., Huang, W., Yang, Y., Ji, M., Nghiem, L.D., Trinh, Q.T. and Tran, N.H. (2019) Insight into biofilm carriers for biological wastewater treatment processes: Current state-of-the art, challenges, and opportunities. *Bioresource Technology*, **288**, 121619. https://doi.org/10.1016/j.biortech.2019.121619

7
Floc and Granular-based Systems (Dispersed Growth)

The drive toward more sustainable and low-carbon wastewater treatment has focused attention on the exploitation of processes utilizing anaerobic granulation. Show and Lee (2016) compared the major advantages that such systems have over their more common aerobic treatment options. These included: (a) higher organic loadings enabling the treatment of high-strength wastewaters, (b) lower operational costs due to no aeration, (c) tall column or rectangular-shape reactor designs requiring much lower bioreactor footprint and land space, (d) no noise, aerosol, and odor problems, (e) higher flexibility in the range of wastewaters that can be treated in terms of strength, toxicity, etc., (f) significantly lower sludge volume generation leading to significant cost sludge production, resulting in reduction in sludge treatment and disposal costs, (g) high carbon capturing potential through biogas production coupled with high chemical oxygen demand (COD) removal capabilities, and (h) anaerobic granules can be stored unfed and without need of aeration for long periods of time without loss of viability.

Granular anaerobic sludge has opened up new opportunities for sustainable, low-cost, zero-carbon wastewater treatment, from the early up-flow anaerobic sludge bed (UASB) reactors (Sec. 7.1) through the transition to expended granular bed reactors (EGBRs) (Sec. 7.2) and now the third-generation anaerobic systems such as Internal Circulation (IC) (Sec. 7.3). There are many other designs using granular sludge, including anaerobic continuously stirred tank reactors (AnCSTRs) (Sec. 7.4), anaerobic sequencing batch reactors (AnSBRs) (Sec. 7.5), anaerobic baffled reactors (AnBRs) (Sec. 7.6), anaerobic migrating blanket reactors (Sec. 7.7), and hybrid reactors (AnHRs) (Sec. 7.8). All these systems are commercially available with many different often-patented modifications.

7.1 Up-flow Anaerobic Sludge Blanket Reactors

In the UASB system, the biomass is retained in the reactor by flocculation, using a similar process technology to that used in sludge blanket clarification in potable water production. Originally developed in the Netherlands (Lettinga *et al.*, 1980), it is now the most popular type of anaerobic digester design for industrial wastewater treatment. A heavily flocculated sludge develops within the special reactor (Fig. 7.1), which acts as a separate fluidized bed, able to withstand high mixing forces. It should be stressed that no support medium (carrier) is added to the reactor and that the biomass that is produced flocculates to produce dense flocs or granules up to several

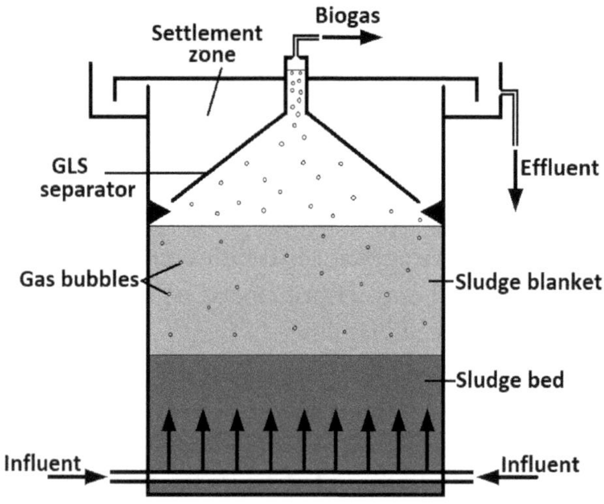

Figure 7.1. Schematic design of an up-flow sludge blanket reactor (UASB) with a simplified gas-liquid-solids (GLS) separator.

millimeters in diameter and a settling velocity of 10 m h^{-1} (Laguna *et al.*, 1999). Granules are composed of a wide range of proteins and polysaccharides (extracellular polymeric substances (EPS)), both bacteria and methanogenic archaea (e.g., *Methanothrix*, *Methanobacterium*, *Methanobrevibacter*, *Methanosarcina*) as well as calcium precipitates (Wu *et al.*, 1987; Vissier *et al.*, 1991; Brummeler *et al.*, 1985). Granules have three distinct layers, with the exact microbial composition dependent on the growth substrate (Hulshoff-Pol *et al.*, 1982; 1983; 1986; Grotenhuis *et al.*, 1991: Harmsen *et al.*, 1996). The external layer is a heterogenous layer of hydrogen-producing microorganisms consisting of rods, cocci, and filaments. The middle layer contains predominantly bacterial rods, including hydrogen-producing acetogens and hydrogen-consuming organisms. In the center is a homogenous bacterial core with a large number of cavities. It consists of *Methanothrix*-like cells that are thought to act as the nucleation centers for granule development (MacLeod *et al.*, 1990) (Sec. 9.2.1).

The reactor comprises the sludge bed, the sludge blanket, and a series of gas-liquid-solids (GLS) separators at the top of the reactor (Fig. 7.1). The biomass takes up 40–60% of the reactor volume, with a solids concentration range of 40–100 g L^{-1}. While the sludge is expanded by the passage of influent wastewater from a distribution system at the base of the reactor, forming the sludge bed at an upflow velocity of 0.5 to 1.0 m h^{-1}, mixing is also caused by gas production bubbles, which can attach to granules carrying them upward along with smaller sludge particles forming the sludge blanket, which occupies between 15 and 25% of the reactor volume.

The GLS separator provides a series of inclined plates or baffles on which the particles rising to the surface come into contact, causing the gas bubbles to be released, thereby allowing the granules to fall back to the sludge bed. The gas that is constantly being released from the sludge bed, as well as the bubbles detached by the GLS separator, are subsequently collected by an inverse hopper at the top of the reactor and passed to the biogas storage facility. The effluent will contain some fine particulate matter, mainly flocs, as well as small granules, which are removed in a separate settlement zone, allowing them to fall back into the UASB reactor. COD removal efficiencies of

70–85% are possible at 20–25°C with organic loadings of up to 10 kg COD $m^{-3}d^{-1}$ and an hydraulic retention time (HRT) of 4–20 hours, depending on wastewater characteristics. All wastewaters must first pass through screens and grit separation, and also, if necessary, primary sedimentation. Design and loading rates are given by Lettinga and Holshoff-Pol (1991).

Mixing is achieved by pumping the feed wastewater through the base of the reactor up through the sludge blanket. Above the blanket, finer particles flocculate, and in the upper settlement zones, they settle as sludge back to the blanket, thus preventing the washout of biomass. The settlement zones occur below the gas collection bowls that slope at an angle of 50°, which promotes the return of the sludge from the settlement areas. The process is characterized by very high mean cell residence times (MCRT), which are even higher than anaerobic filters, and a low HRT (<1 day). It takes up to six months for a suitable granular sludge to develop from a soluble wastewater, so a suitable inoculum is required for rapid start-up. The process is dependent entirely on the ability of the biomass to form granules using particular wastewaters and is widely used for strong industrial wastewaters with low suspended solids content, such as creamery wastes and the waste from the manufacture of soft drinks. Reactors are typically 3–5 m in height. A 200 m^3 UASB successfully treated sugar beet waste at a rate of 16 kg COD $m^{-3}d^{-1}$, achieving a 90% removal at an HRT of only 4 hours. UASB reactors have been used to treat a variety of complex wastewaters (Imai *et al.*, 2000), including wastewaters from fish canning (Puñal and Lema, 1999), fiberboard manufacture (Fernández *et al.*, 1995; Puñal *et al.*, 1999), landfill leachate (Chang, 1989; Borzacconi *et al.*, 1999), agro-industries (Driessen and Yspeert, 1999), chemical industry wastewaters such as polyethylene terephthalate (PET) (Fdz-Polanco *et al.*, 1999), malting and brewing (Fang *et al.*, 1990; Yan and Tay, 1996; Martínez *et al.*, 2001), slaughterhouses (Manjunath *et al.*, 2000; Del Nery *et al.*, 2001), and papermills (van Lier *et al.*, 2001). Compared to contact processes, it is difficult to operate and maintain the structure of the sludge blanket (Lettinga and Hulshoff-Pol, 1991) (Table 6.2). Details of loading rates are summarized in Table 6.4. Sludge blanket reactors have been used as a separate methanogenic phase, with the first two stages of the digestion process being carried out in a separate reactor with increased biogas production (Morris and Burgess, 1984). The most well-known commercial system is the Biothane® UASB process. A development of the UASB reactor is the expanded granular sludge bed (EGSB) reactor, which is designed to treat low-strength wastewaters at high velocities at temperatures of 10°C and less (Kato *et al.*, 1994; Rebac *et al.*, 1995; Jeison and Chamy, 1999; van Lier *et al.*, 2001) (Sec. 7.2).

The structure of anaerobic granules was captured by Wiegant and de Man (1986) in a series of remarkable images (Fig. 7.2), and

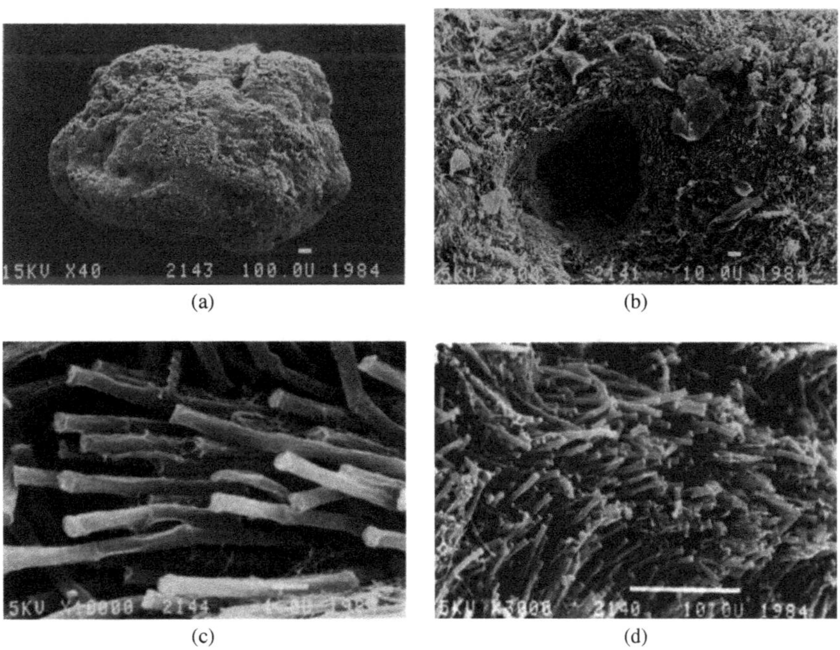

Figure 7.2. Scanning electron micrographs of thermophilic granular sludge grown on acetate from digested sewage sludge as inoculum. (a) Large granule showing several cavities through which the biogas is excreted. (b) Detail of (a) showing one such cavity. (c) Detail showing blunt-ended filaments typical of the acetoclastic methanogen *Merhanorhrix*. (d) Large magnification showing the dense structure of the granule (Wiegant and de Man, 1986). Bars indicate 100, 10, 1, and 10 pm, respectively. Reproduced with permission of John Wiley and Sons Inc.

while the process of granulation remains only partially understood, a number of different hypotheses have been proposed (Show *et al.*, 2020) (Sec. 9.2.1). An amazing in-depth study of the granulation — regeneration cycle in the anammox process, including the formation and structure of granules and the role of the different mechanisms involved including quorum sensing, has been carried out by Xue *et al.* (2023).

7.2 Expanded Granular Sludge Bed Reactors

The development of EGSB technology arose from the need to treat chemical industry wastewaters that were inhibitory or toxic to UASBs (Zoutberg and de Been, 1997). Like the UASB, EGSB technology relies on granular sludge development, although when compared to the UASB reactor, they are much taller: 12–18 m compared to 5–7 m, with a height:width ratio of 4–5:1 and increased up-flow velocities at 6–10 m h^{-1}, with high recycle ratios (van Lier, 2008). This ensures a truly expanded bed of biomass with a high degree of mixing, maximizing wastewater–granule contact to achieve high COD removal rates, and biogas production (Fig. 7.3). The increased recycling rates ensure a high degree of dilution so that toxic compounds can be safely treated. For example, an EGSB successfully treated a chemical wastewater containing formaldehyde and methanol, from which the former was being produced from the latter, at concentrations of up to 10 g L^{-1} and 20 g L^{-1}, respectively, achieving removal efficiencies of >98% for both compounds (Zoutberg and de Been, 1997). Organic loading rates of 30 kg $m^{-3}d^{-1}$ are achievable with both low and high-strength wastewaters able to be treated, including normal municipal wastewaters at low temperatures (Lettinga *et al.*, 1997). High biogas production can be achieved when treating food processing wastewaters, such as brewery, yeast, sugar, fruit and vegetable processing, and corn ethanol production wastewaters.

The higher loading rates, simpler operation, and lower operational costs have seen a shift from the popular UASB to EGSB systems for industrial treatment (van Lier, 2008). The relative loading rates of different reactor types are compared in Fig. 7.4. Biothane systems

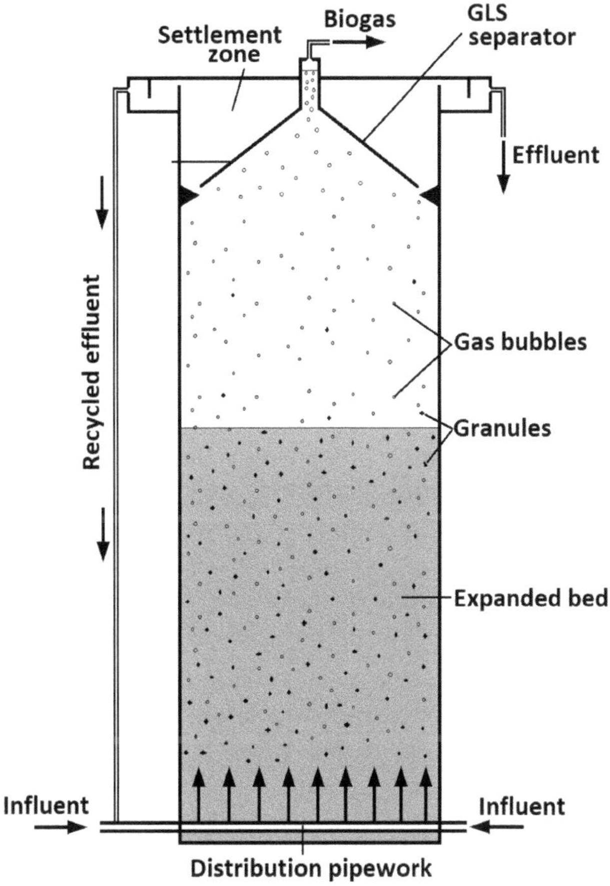

Figure 7.3. Schematic design of an expanded granular bed reactor (EGBR) with a simplified GLS separator.

have developed a commercial EGSB process called Biobed®, operating at high liquid (10 m h^{-1}) and gas velocities (7 m h^{-1}) (Zoutberg and de Been, 1997).

A down-flow version of the EGBB reactor has been developed by Professor Eyyup Ellis in the United States (Mach, 2000) and has been shown to be effective in treating a wide range of wastewaters, including municipal wastewater, which achieved a final effluent with a BOD of <30 mg L^{-1} at an HRT of 24 hours (Evans, 2004). The static granular bed reactor (SGBR) is simpler in operation to a standard

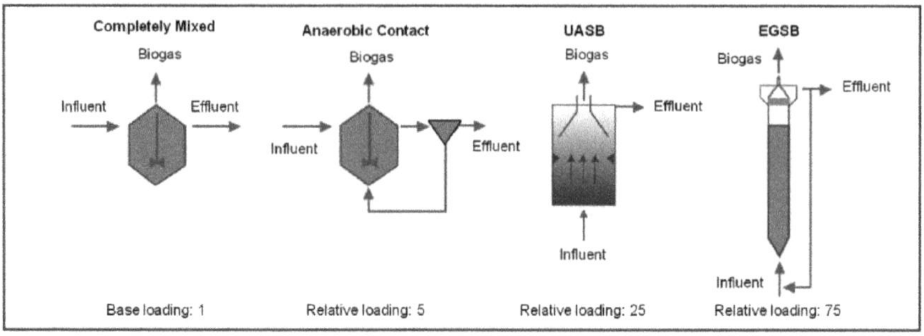

Figure 7.4. Comparison of major floc-based anaerobic wastewater treatment systems in terms of loading capacity (Show *et al.*, 2020). Reproduced with permission of Elsevier Publishing.

EGBR or UASB, requiring significantly less energy input, except for occasional backwashing to remove solids accumulation, with no granule washout in the effluent. Debrik and Coskun (2009) successfully used a SGBR to treat poultry wastewater, achieving a 95% COD reduction using laboratory-scale reactors producing 0.25 L CH_4 g^{-1} COD removed at an operating temperature of 22°C (Fig. 7.5). The process followed modified Stover–Kincannon and Grau second-order models. Debrik also tested the system on leachate at a loading rate of 15kg COD m^{-3} d^{-1}, recording >90% COD removal (Debrik *et al.*, 2005). However, the long-term operation of the process, especially in maintaining the structure of the granular sludge, remains less clear.

7.3 Internal Circulation Systems

Internal Circulation (IC) has been pioneered in the Netherlands by the company Paques and comprises two reactor chambers, one on top of the other, each with its own GLS separator (Fig. 7.8). Essentially, the IC process is two UASB reactors: the lower unit treating a high organic loading and the upper acting as a polishing unit treating a low organic loading. Enhanced GLS separation results in better biomass retention and enhanced treatment, resulting in high-quality effluents (Mao *et al.*, 2015). The reactors are more complex than either standard UASB or EGSB reactors but with a smaller footprint and the

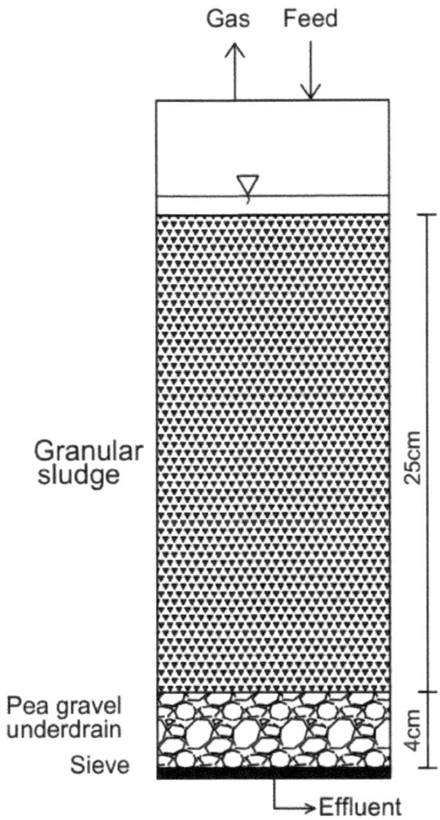

Figure 7.5. The laboratory scale static granular bed reactor (SGBR) used by Debrik and Coskun (2009). Reproduced with permission of Elsevier Publishing.

ability to treat much higher organic loadings of between 20 and 50 kg COD $m^{-3}d^{-1}$. They are very tall compared to other anaerobic contact systems, being 2–3 times higher than normal UASBs and cheaper to construct in terms of treatment capacity (Wang *et al.*, 2018). The influent enters the base of the reactor and is mixed with the granular sludge via the distribution system. Biogas produced from the organic matter in the lower chamber is collected by the first GLS separator, resulting in a gas lift effect, which pulls the biomass and partially treated wastewater upward via the riser tube into the gas-liquid separator at the very top of the reactor (Fig. 7.6). Here, the

Figure 7.6. Paques BIOPAQ®IC internal circulation (IC) reactor comprising two reactor chambers on top of each other, each with its own GLS separator: 1. Wastewater enters the reactor and mixes with granular biomass via the distribution system, 2. Organic fraction converted to biogas, 3. Biogas collected in a lower separator, creating a gas lift, 4. Liquid rises upward in the riser with biogas, 5. Biogas leaves the reactor in a liquid–gas separator, 6. Water returns to the distribution system via the downer, 7. Effluent undergoes further treatment (polishing) in the upper reactor chamber, 8. Biogas collected from the upper chamber by the upper phase separator, 9. Effluent leaves the reactor, 10. Biogas leaves the reactor. https://cdn.opptylab.com/hg/assets/brochurebiopaqic.pdf. Reproduced with permission of Paques Global, https://www.paquesglobal.com/.

biogas is removed and the liquid returned to the lower chamber via the downer tube into the distribution system, causing effective mixing and ensuring good expansion of the granules in the lower chamber, hence the concept of internal circulation. The second chamber acts as a polishing area for the effluent, with any biogas removed by the second GLS separator, after which the effluent is discharged (Fig. 7.6). IC is described as a third-generation anaerobic treatment process, being a development from the UASB and EGSB processes. It is slowly replacing the UASB as the first choice for the treatment of strong and complex wastewaters, including leachate (Habets and Driessen, 2007). It is marketed by Paques under the trade name BIOPAQ® IC (https://www.paquesglobal.com/products/biopaq-ic), with numerous successful plants in operation globally treating a wide range of industrial wastewaters. A full-scale IC reactor was successfully employed to treat leachate from a municipal solid waste incineration plant. At an organic loading rate (OLR) of 21–25 kg COD $m^{-3}d^{-1}$, high COD removals were recorded at between 89 and 93% and biogas yields of 0.42–0.50 m^3 kg^{-1} COD (Wang *et al.*, 2018). The zeta potential and hydrophobicity of the granular sludge were correlated to the protein/polysaccharide ratio of the EPS, with the protein content increasing over time, showing that protein was an important part of EPS development and the process granulation (Figs. 7.7 and 7.8).

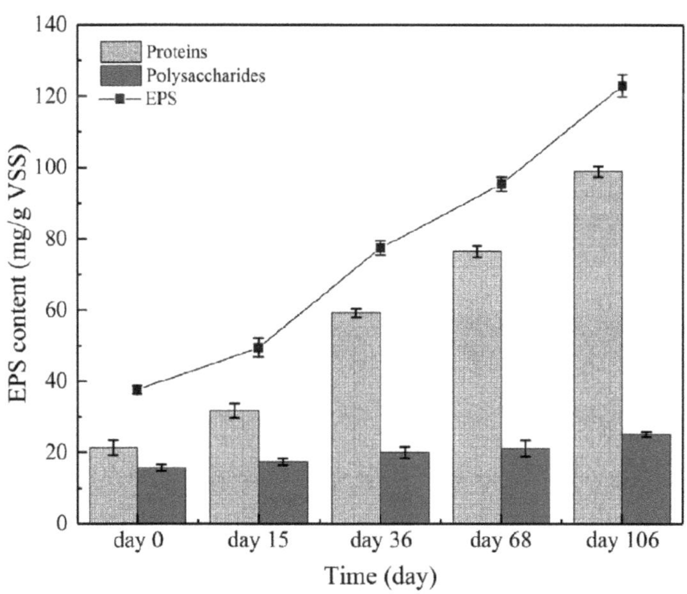

Figure 7.7. Development of protein and polysaccharide content in EPS as granules mature in an IC reactor (Wang *et al.*, 2018). Reproduced with permission of Elsevier Publishing.

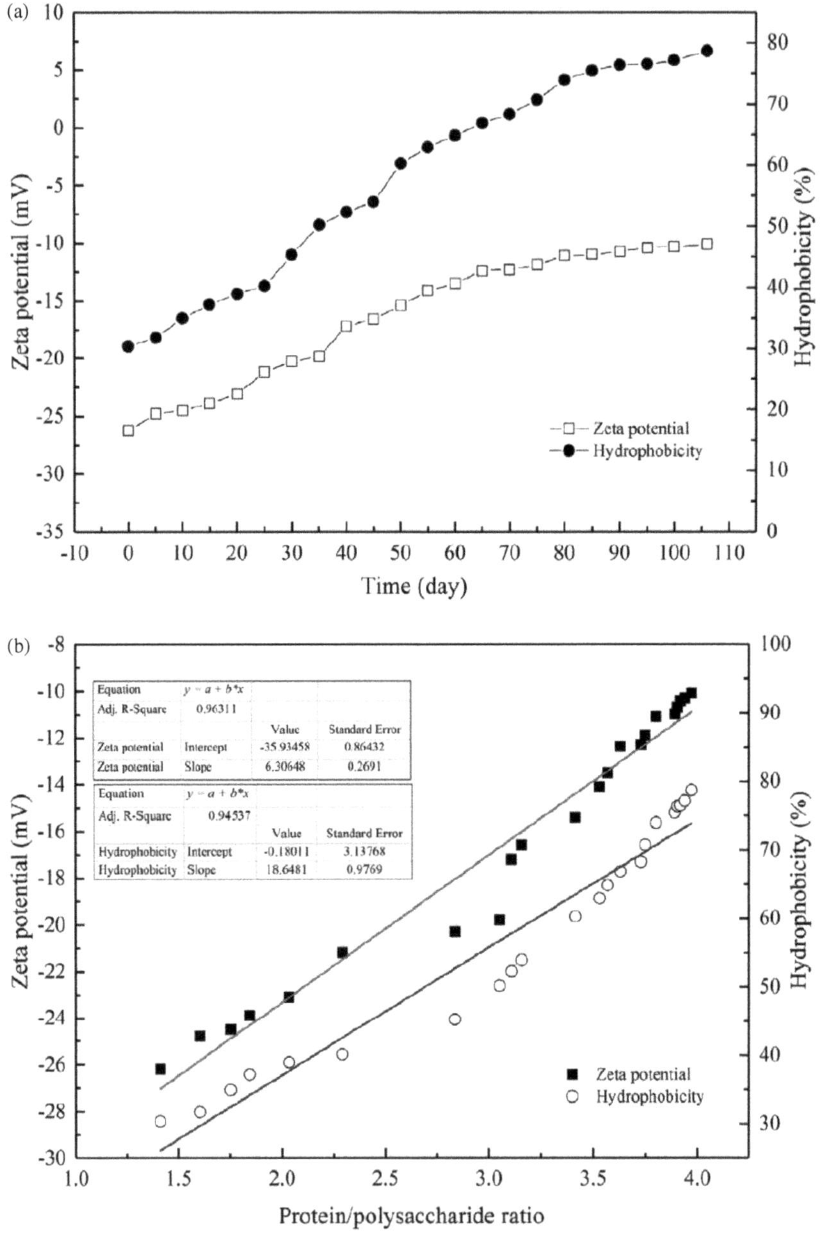

Figure 7.8. Relationship between zeta potential, cell surface hydrophobicity, and EPS during sludge granulation in an IC reactor treating leachate (Wang *et al.*, 2018). Reproduced with permission of Elsevier Publishing.

All anaerobic processes are susceptible to inhibition due to inconsistency in loading or organic overloading or the presence of toxic and inhibitory compounds. It is important to ensure real-time monitoring to ensure consistency of operation; this is especially so with IC systems (Jimenez *et al.*, 2015; Irizar *et al.*, 2018).

7.4 Anaerobic Continuously Stirred Tank Reactors

The simplest of all contact systems is the anaerobic continuously stirred tank reactor (AnCSTR), which is used for traditional sludge digestion. The design ensures excellent mixing of the biomass and the wastewater, ensuring complete breakdown and good biogas production. It is especially well-adapted to treating wastes with high FOG (fats, oil, and grease) content, as foaming is minimized by stirring rather than by the expansion of the sludge blanket. Biomass retention is low due the completely mixed nature of the liquid/biomass, so a solids separation step ensures biomass concentration is maintained, ensuring operational stability even under variable loading conditions (Fig. 7.9). Internal baffles can also be installed to enhance retention of the granular sludge or a membrane separator (Sec. 9.2.3). Loading rates depend on the nature of the substrate but are normally between 2 and 5 kg COD $m^{-3}d^{-1}$. The AnCSTR is often operated in series with another bioreactor, such as an AnBR, and can be used for all types of strong organic wastewaters and unthickened sludges, such as waste-activated sludge and flotation sludges with high lipid contents. For example, Lutze and Engelhart (2020) explored the use of an AnCSTR to treat waste-activated sludge and lipid-rich flotation sludge from milk processing, which achieved a maximum biogas yield of 280 L kg^{-1} COD at a SRT of 20 days as well as good COD removals, although biogas yields were affected by lipid concentration requiring longer sludge ages as lipid content increased. Lipid contents >0.8 kg COD_{lipid} kg^{-1} COD required a minimum SRT of 15 days, while those at <0.8 kg COD_{lipid} kg^{-1} COD were effectively treated at an SRT of 10 days. Among the commercially available AnCSTRs are the ADI® CSTR from Evoqua and Veolia's Biobulk-CSTR.

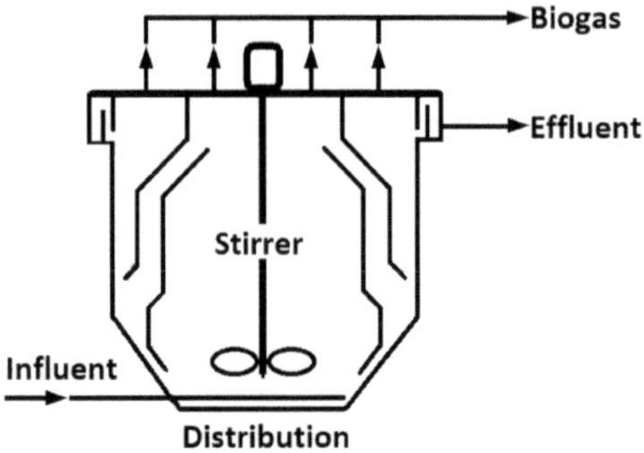

Figure 7.9. Schematic diagram of an anaerobic continuously stirred tank reactor (AnCSTR).

7.5 Anaerobic Sequencing Batch Reactor

The original activated sludge process was developed as a single-unit process batch reactor where aeration and settlement occurred within the same tank (Gray, 2023). Today, a number of sequencing batch reactors are used in parallel so while one tank is filling, another is in the aeration stage and another is in the settlement/decanting phase, thus allowing for continuous loading. Coordinating the phases to ensure that influent wastewater can always enter a reactor requires computer-aided technology.

In anaerobic sequencing batch reactors (AnSBRs), the aeration phase is omitted and replaced with a reaction time during which sludge granules adsorb and biodegrade the organic fraction of the wastewater under anaerobic conditions. Mixing is usually achieved by a paddle, although gas and/or liquid recycling can be used. While the mixing stage is longer than the aeration phase seen in activated sludge, settlement is much faster due to the granular nature of the sludge. In practice, it is far more difficult to synchronize cycle time between AnSBR reactors, permitting continuous loading, especially where hydraulic or organic rates vary, so they are generally operated as single batch reactors treating waste streams with consistent characteristics and predictable flow patterns or which have been stored in an equalization basin (Akil and Jayanthi, 2012). This makes them ideal for the pharmaceutical industry, where manufacturers rely on small batch processes based on predictable work-shift patterns and more complex wastewaters, such as those from the dye and textile industries and leachate, which generally require longer reaction times (Timur and Özturk, 1999; Kumar *et al.*, 2022). It has been suggested that AnSBRs are not as efficient as UASBs at high loading rates, although while COD removal is comparable, more biomass is produced (Angenent and Sung, 2001). The time for a suitable granular sludge to develop depends on the seed used and the time required for the acclimatization of the imported seed sludge to the wastewater being treated, which can be as long as five months (Wirtz and Dague, 1996). The design is very simple and does not include a feed distribution or GLS separation system as in UASB, with no up-flow hydraulic pattern (Fig. 7.10).

Co-digestion of wastewaters has many benefits, such as optimizing buffering capacity and C:N ratio, diluting inhibitory and toxic compounds, improving COD removal and methane production, increasing operational stability, and reducing costs. For example, the treatability of tannery wastewaters was improved by mixing 50: 50 with dairy wastewater (Berhe and Leta, 2023). Tannery wastewater is difficult to treat anaerobically due to its high alkalinity and low COD:TN ratio, leading to ammonification and an increase in pH, both of which inhibit methanogenesis. In contrast, dairy wastewater has a low alkalinity content, which quickly becomes acidified,

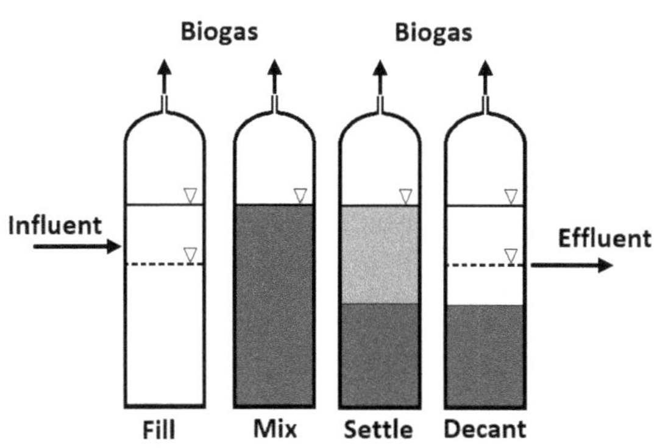

Figure 7.10. Four phases of an anaerobic sequencing batch reactor.

increasing volatile fatty acid (VFA) concentration that lowers the pH but with a high COD:TN (total nitrogen) ratio.

Berhe and Leta (2023) also used two-stage AnSBRs to treat their wastewaters in order to split acidogenesis and methanogenesis, with the working volume split at a ratio of 2:3, respectively, which further increased methane yields and also improved operational stability. Separating phases in this way increases the plug flow nature of the system, which is already a feature of the AnSBR that promotes a more diverse microbial community structure, allowing optimum control of each reactor for the key consortia of bacteria and archaea present, such as preventing intermediate products such as VFAs or inhibitory compounds to accumulate that may affect methanogenesis. A final effluent COD removal rate of 85.6% was achieved at an HRT of 15 days, with a methane yield of 892 mL CH_4 g^{-1} COD at a biogas methane content of 72%. A similar problem was identified by Deng *et al.* (2024) in treating high-strength slaughterhouse wastewater, which is rich in protein and FOG, which accumulate in the reactor, causing foaming or sludge flotation and other problems. The typical strength of the wastewater is COD (4,700–6,500 mg L^{-1}), proteins (500–1,100 mg L^{-1}), FOG (100–600 mg L^{-1}), TSS (1,100–3,700 mg L^{-1}), and NH_4^+-N (175–420 mg L^{-1}). Proteins are hydrolyzed to peptides and amino acids, then deaminated to ammonium and VFAs (i.e., ammonification), while FOG is hydrolyzed into glycerol and long-chain fatty acids (LCFAs). Interestingly, the degradation of LCFAs involves a syntrophic relationship between acetogenic β-oxidizing bacteria and acetate and hydrogen-utilizing methanogenic archaea. Lipid hydrolysis occurs far more quickly than LCFA degradation and so can result in elevated concentrations of LCFAs, which can inhibit both hydrogenotrophic and acetolactic methanogens as well as the syntrophic β-oxidizing bacteria (Sousa *et al.*, 2009). So, where protein-rich wastewaters are treated, high ammonium concentrations can occur, with ammonium inhibiting methanogens at concentrations >1,500 mg L^{-1}, with VFAs also increasing in concentration, causing acidification. Similar problems are seen when treating wastewaters rich in FOG, which has high methane-producing potential. In practice, carbohydrates are degraded much more quickly than either proteins or FOG, so where they are constantly being added to a completely mixed reactor, protein degradation is reduced, leading to protein accumulation similarly with FOG. To a large extent, this is overcome by the feast-famine regime associated with batch reactors, such as AnSBRs, which ensures all are fully degraded each cycle. The AnSBR system used by Deng *et al.* (2024) employed a buffer tank before the AnSBR, which increased the rate of ammonification, achieved a 90 and 80% removal of COD and protein, respectively, and a COD to methane conversion rate of >70% at an optimum OLR of 6.2 kg COD m^{-3} d^{-1}. While all operating parameters play a part in determining the microbial community structure within the reactor, the feeding frequency and OLR were identified by Zeils *et al.* (2017) as the most important. Fly ash has been used as an enhancer in AnSBRs, treating winery wastewater, and was found to increase biogas yield by 12% and COD removal by 42% (Lauzurique *et al.*, 2023).

The problem of maintaining a granular sludge is overcome by using carriers in AnSBRs, which also reduces or eliminates the time required for settlement, reducing overall cycle time, and reduces solids loss as well as increasing both COD removal and biogas generation per unit reactor volume. Anaerobic sequencing batch biofilm reactors (AnSBBRs) were compared to a standard AnSBR in laboratory tests, using synthetic wastewater as feed and granular activated carbon (GAC) and zeolites as carriers (Fig. 7.11). Operating a 24-hour cycle of 12 minutes of feeding, 23 hours of anaerobic biodegradation, 30 minutes of settling, and 15 minutes of decanting at 35°C, gas production was greater in the AnSBBRs, with an increase over the AnSBR of 7% with GAC only as a carrier, 4% using only zeolite as a carrier, and 21% using combined GAC and Zeolite carriers; while mean COD removals were 89.5, 92.3, 96.5, and 96.5%, respectively (Dutta *et al.*, 2014). They have been used successfully to treat a wide range of wastewaters. Sarti *et al.* (2007) used a pilot-scale AnSBBR to treat domestic wastewater with polyurethane foam cubes with 50 mm sides as a carrier in an 8-hour cycle (feeding (2 hours), reaction (5 hours), discharge (0.5 hours), and idle (0.5 hours)), although the mean COD removal was only 66%. No settlement

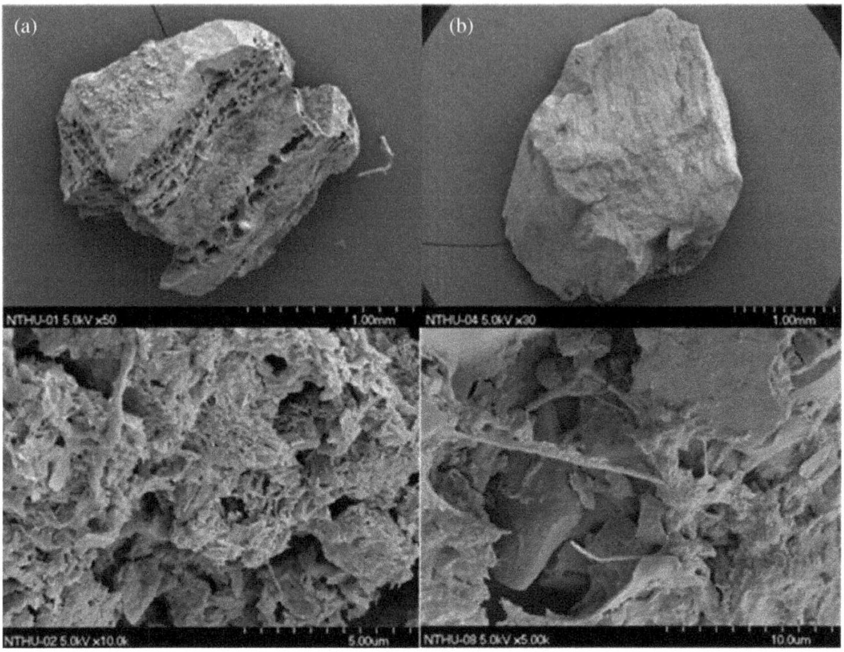

Figure 7.11. Scanning electron micrograph of granular activated carbon (GAC) and zeolite carriers used in an anaerobic sequencing batch reactor treating synthetic municipal wastewater: (a) GAC outer surface (top) and inside (bottom) and (b) zeolite outer surface (top) and inside (bottom) (Dutta *et al.*, 2014). Reproduced with permission of Elsevier Publishing.

period was used during the cycle, with the carriers held within a perforated wall against the reactor wall (Fig. 7.12). An unusual example is a full-scale pilot AnSBBR used to successfully treat sulfate-rich wastewaters by a consortium of acidogenic bacteria, acetoclastic methanogenic archaea, and sulfate reducing bacteria, dominating the biofilm that developed on a bed of mineral coal (40–80 mm diameter), which acted as a medium (Sarti *et al.*, 2011). The wastewater was batch-fed into the AnSBBR from a storage tank and then continuously recycled through the bed during the entire reaction period. Using a cycle time of 48 hours (feeding (1 hour), reaction (46 hours), and discharge (1 hour)), sulfate concentrations of 0.58 to 5.2 g SO_4^{2-} L^{-1} mixed with domestic wastewater achieved a biological sulfate removal rate efficiency of 99%.

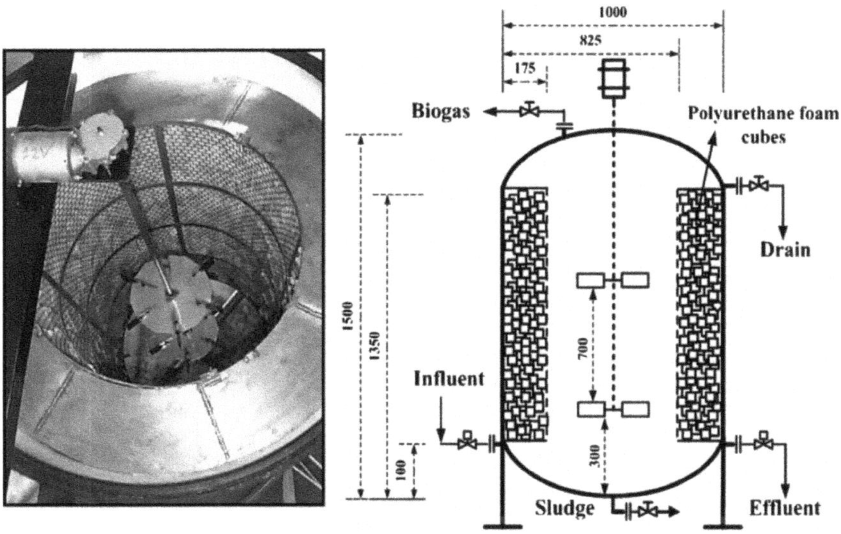

(a) (b)

Figure 7.12. Pilot-scale AnSBBR used to treat domestic wastewater: (a) image and (b) schematic diagram, with all dimensions in mm. The carrier (polyurethane foam cubes) is held in an outer chamber, with the wastewater able to pass freely from the central mixed zone to the biofilm on the carriers via the perforated wall (Sarti *et al.*, 2007). Reproduced with permission of Elsevier Publishing.

Both AnSBR and AnSBBRs have been successfully employed in producing methane and biohydrogen (Sec. 9.1.1).

7.6 Baffled Reactors

The concept of incorporating baffles in anaerobic reactors first occurred early in the 1980s when septic tanks were modified to increase COD removal. Bachman *et al.* (1982; 1985) created the first anaerobic baffled reactor AnBR as a dedicated secondary treatment step rather than a combined settlement/treatment process, such as a septic or Imhoff tank, utilizing anaerobic granulated sludge. The rectangular reactor incorporates a number of vertical baffles, creating chambers where the wastewater passes from one chamber to another under one baffle and over the next, ensuring maximum HRT and biomass contact with the wastewater and minimizing the loss of

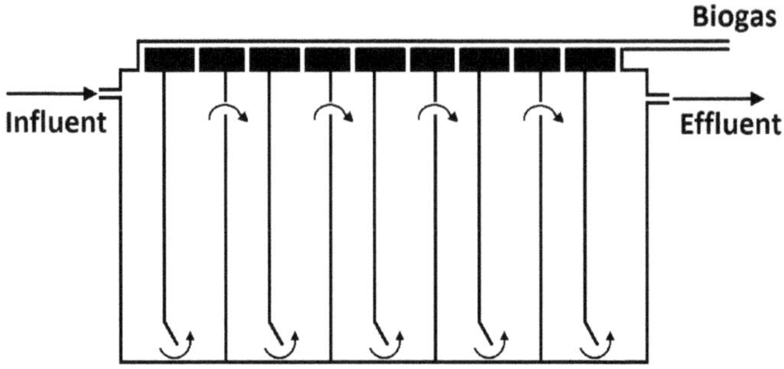

Figure 7.13. Schematic diagram of a five-chambered anaerobic baffled reactor (AnBR).

granules (Fig. 7.13). This is often referred to as a push-flow pattern (Reynaud and Buckley, 2016). There is no GSL separation system as in EGSBs, making them far simpler to operate (Uyanik *et al.*, 2002). Granulation occurs relatively quickly in AnBRs, with a diverse range of bacteria and archaea developing. The plug-flow nature of the reactor caused by multiple chambers ensures that there is a substrate concentration gradient through the system so that biogas production and sludge activity tend to fall off toward the end of the reactor. The series of chambers allows a large degree of separation of acidogenic and methanogenic activity (Zhu *et al.*, 2015). AnBRs are widely used for domestic wastewaters in rural China (Cui and Liang, 2019; Zha *et al.*, 2019), achieving a 62–71% COD removal at loadings ranging from 0.06 to 0.61 kg COD d^{-1} (Da Silva *et al.*, 2017). AnBRs have good biosolids retention, are able to withstand shock and intermittent loading, and have low energy consumption, being easy to operate with low maintenance (Soh *et al.*, 2020).

AnBRs utilizing granular sludge have been successfully used to treat problematic waste streams. For example, Zang *et al.* (2024) were able to successfully treat strong wastewater (COD 4,000 mg L^{-1}) containing 12.47 mg L^{-1} of allicin, a sulfur-containing compound derived from garlic processing, which is highly bactericidal. This was achieved by using a step feed system with 40% loading to chamber 1, 30% to chamber 2, 20% to chamber 3, and 10% to chamber 4, and an effluent recycling rate of 200% to ensure adequate sludge expansion within the chambers as well as diluting the wastewater, thereby reducing toxicity. This achieved an 89 and 90% reduction in the COD and allicin, respectively. The loading and recycling regime controlled the pH to within 6.5–8.0, which was optimal for methanogens.

Packed anaerobic baffled reactors (PABRs) include fixed media in the up-flow channels to increase biomass concentration, increasing COD removals to 90% (Renuka *et al.*, 2016; Feng *et al.*, 2018) (Fig. 7.14). Zheng *et al.* (2023) tested a laboratory-scale PABR to compare performance under summer (32 ± 1°C) and winter temperatures (16 ± 1°C), using different HRTs of 36, 48, 60, 72, and 90 hours. The study showed the greater the HRT, the better the removal of COD, reaching 85 and 52% during summer and winter conditions, respectively (Fig. 7.15), recommending a minimum HRT of 72 hours. As with AnBRs, the COD removal rates fell in the different chambers as the wastewater passed through. The authors observed variations in the bacterial and archaeal flora between chambers. In the four-chambered system, the abundance of hydrogen trophic methanogens decreased progressively with flow direction, while the

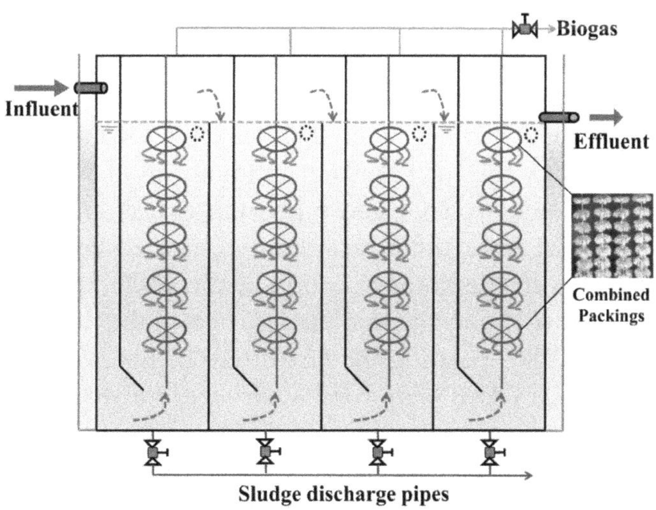

Figure 7.14. Schematic diagram of the laboratory-scale packed anaerobic baffled reactors (PABR) used by Zheng *et al.* (2023). Reproduced with permission of Elsevier Publishing.

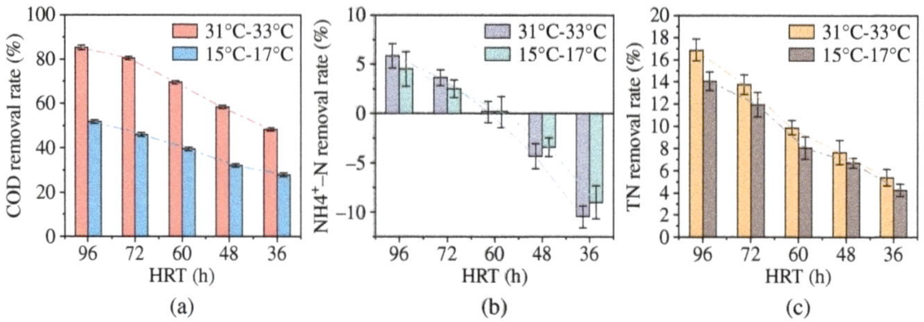

Figure 7.15. Effect of temperature and HRT on the removal of: (a) COD, (b) ammonia (NH_4^+-N), and (c) total nitrogen (TN) in experimental PABRs (Zheng *et al.*, 2023). Reproduced with permission of Elsevier Publishing.

abundance of acetic acid trophic methanogens progressively increased with flow direction (Fig, 7.16). Overall, the most abundant phyla in the PABR were *Chloroflexi* (53.2%), *Bacteroidota* (9.8%), *Proteobacteria* (9.3%), *Actinobaciota* (6.8%), *Firmicutes* (6.3%), *Pattescibateria* (3.7%), *Acidobacteriota* (1.5%), and *Caldisericota* (1.3%). Similar microbial results were reported by Zha *et al.* (2019), who used a PABR to treat black water (i.e., water only from toilets) and found the optimum HRT after testing over a range of 24–72 hours to be 48 hours, with mean reductions of 94% COD, 2.9% TN, 14% NH_4^+-N, and 33% TP (total phosphorus) over the sampling period of 112 days.

A wide range of carriers are used either suspended as bunches, on rods, or as strings (Fig. 7.17). The optimum number of chambers should be kept to between 4 and 6 per unit (Xu *et al.*, 2016), with a width ratio of 1:3 for the downward and upward flow sections of each chamber, with the bottom section of the baffle inclined at 45°. Renuka *et al.* (2016) only incorporated carriers in the final two chambers of their PABR, operating the first three chambers as an AnBR.

The potential of AnBRs for rural wastewater treatment and biogas production offers huge opportunities for sustainable sanitation and energy production at the local level, with research also showing the potential of AnBRs for hydrogen production from specific wastewaters rich in starch, such as potatoes and tapioca (Sec. 9.1.1).

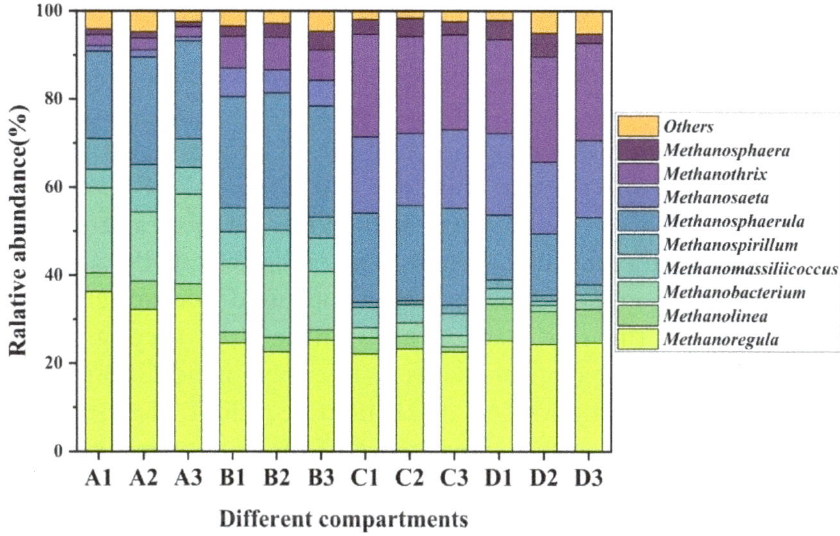

Figure 7.16. Structure of the archaeal community in the four different compartments in an experimental PABR (Fig. 7.14). Biomass samples were taken from the top middle and base of each compartment and mixed to give a single sample. Three replicates (1–3) were taken from each compartment (A–D) (Zheng *et al.*, 2023). Reproduced with permission of Elsevier Publishing.

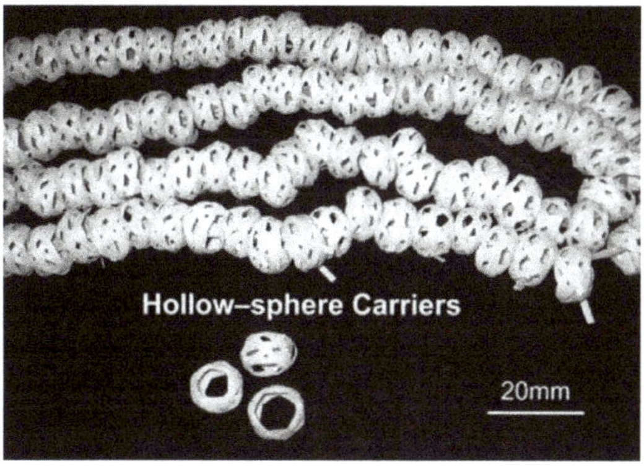

Figure 7.17. A wide range of carriers are used in PABRs, including the standard anaerobic biofilter carrier. However, local natural materials are often employed, such as small spheres woven from bamboo fibers used by Feng *et al.* (2018). Reproduced with permission of Elsevier Publishing.

7.7 Anaerobic Migrating Blanket Reactor

The anaerobic migrating blanket reactor was patented in 1999 and is similar to AnBRs, as it comprises a number of compartments inducing an up-flow down-flow pattern through the tank (Angenent and Dague, 1995; Angenent and Sung, 2001) (Fig. 7.18). Waste is continuously fed to the reactor, with biodegradation and biogas production by granular sludge. Movement is created by the influent wastewater, although mechanical mixers are used intermittently, as required when the flow rate is not sufficient to keep granules expanded within the compartments. While there is no GLS separator, feed distribution system, or effluent recycling, the need for expensive stirrers in each compartment of the reactor makes them more expensive than other designs. Like the baffled reactor, the compartmentalized design with the wastewater entering one end and leaving at the far end means that a substrate gradient is formed so that there is very little COD removal or biogas production in the final compartment, which is often used as a settlement zone preventing biomass loss. Inevitably, granules accumulate in the final compartment. This is overcome by operating the reactor with the hydraulic flow reversed, with the first compartment now operating as the settlement zone. Operationally, this is quite difficult in order to maintain constant removal efficiency, so the feed may be split between the first and penultimate compartment just before reversing the flow in order to condition the granules in the final compartment for higher organic loading. Anaerobic migrating bed reactors require a minimum of three compartments, preferably four or five. Compared

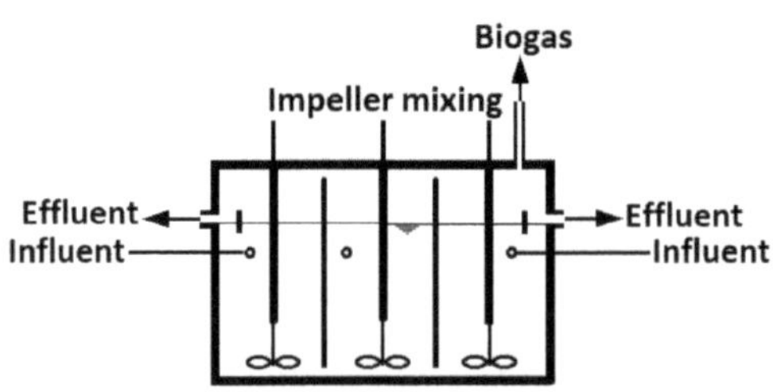

Figure 7.18. Schematic diagram of an anaerobic migrating blanket reactor (Angenent and Sung, 2001). Reproduced with permission of Elsevier Publishing.

to UASBs, the granules that form are smaller and denser, helping to retain them within the tank. Like AnBRs, building an anaerobic migrating bed reactor on site is complex due to the baffles, so pre-fabricated units are easier to construct and install. Some authors use the abbreviation AnMBR for anaerobic migrating bed reactors; however, this is far more widely used for anaerobic membrane reactors (Sec. 9.2.3).

7.8 Hybrid Reactors

A major drawback of UASB reactors is the potential to lose biomass, which has resulted in many innovative GLS separator designs and configurations. The anaerobic hybrid reactor (AnHR) overcomes this to a certain extent by incorporating an anaerobic biofilter within the UASB, creating a hybrid sludge blanket-biofilm reactor (Fig. 7.19). Confusingly, AnHRs are also widely known as hybrid up-flow anaerobic sludge blanket (HUASB) reactors. The AnHB is constructed with the top 20–30% of the reactor volume packed with a carrier, either a modular medium or a standard random biofilter carrier, such as polyurethane foam or plastic units (Figs. 6.10 and 6.12). The medium or carriers provide surface area for biomass development as well as trapping suspended biomass displaced from the lower section. AnHRs are well suited for treating complex industrial wastewaters, especially those that may contain more recalcitrant compounds as well as more readily biodegradable materials. Good COD removals have been reported due to the biofilter section acting as a polishing step as well as adding significantly to the total biomass within the reactor, with excellent biogas yields containing up to 87% methane achievable. A wide range of wastewaters can be successfully treated in AnHRs, with good biogas yields including chemical, petrochemical, pharmaceutical (Mullai *et al.*, 2024), antibiotic production (Mullai *et al.*, 2018), biopesticides (e.g., azadirachtin) (Mullai *et al.*, 2022), pulp paper processing (Hemalatha and Keerthinarayanna, 2017), food processing, brewery, distillery, leather, and heavy metals (Murugesan and Selvarasu, 2017).

Comparing an AnHR with a UASB treating phenolic wastewater, Ramakrishnan and Surampalli (2012) observed that the hybrid

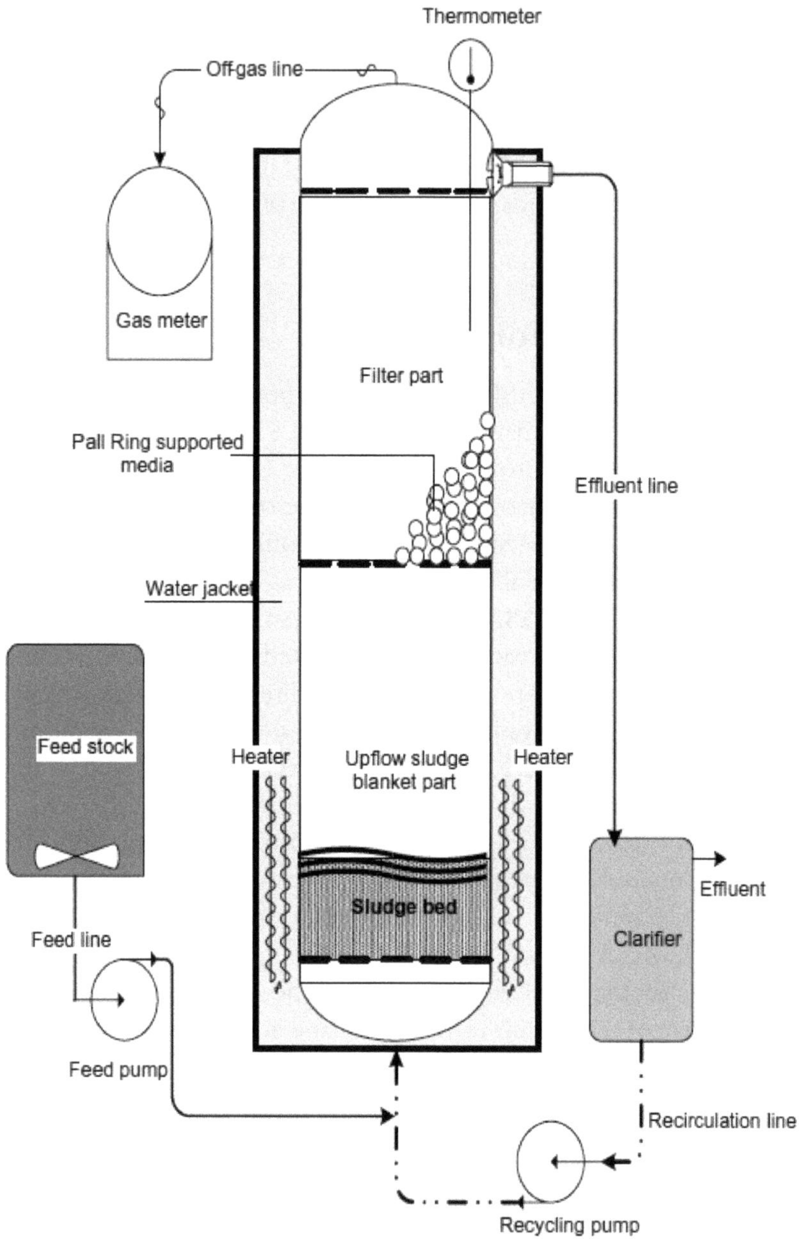

Figure 7.19. Schematic design of an experimental anaerobic hybrid reactor used to treat pharmaceutical wastewater (Oktem *et al.*, 2007). Reproduced with permission of Elsevier Publishing.

reactor was able to achieve effective granulation within 45 days compared to 58 days in the UASB and was able to withstand far greater shock loadings. The AnBR achieved a slightly higher percentage of COD removal at 88% (85% UASB) and phenolics removal at 93% (91% UASB) but similar biogas production rates of 0.33 m^{-3} kg^{-1} COD removed with 67 and 60% methane content in the AnHR and UASB, respectively. Although AnHRs operate well under normal temperatures (12–20°C), many studies on industrial wastewaters have recorded optimal performance in the mesophilic range (30–35°C), especially for biogas production, and so where wastewaters are not already at the correct temperature, heating of the lower chamber may be advantageous.

Treating fermentation-based antibiotic wastewater (COD 15,000 to 25,000 mg L^{-1} and BOD 5,000 to 9,020 mg L^{-1}) in a laboratory-scale AnHR, Mullai *et al.* (2018) recorded optimum biogas production at an organic loading rate of 16 kg COD m^{-3} d^{-1}, with COD removal of 83% at an HRT of 6 hours. The authors successfully used the Stover–Kincannon and Van der Meer and Heertjes models to predict biogas production in full-scale systems. A loss of granulation was noted in an experimental AnHR treating olive oil processing wastewater due to phenolic compounds present, although a reduced performance was maintained by the biomass carriers in the upper section of the reactor, with the hydrogenotroph *Methanobacterium beijingense* the dominant methanogen (Zakoura *et al.*, 2022). In contrast, a UASB using flocculent biomass performed well using the same wastewater.

The major phyla in the AnHR were *Bacteroidetes* (8.4–57.9%), *Firmicutes* (0.7–34.9%), *Chloroflexi* (0.5–40.5%), *Proteobacteria* (2.3–13.0%), *Planctomycetes* (0.6–30.2%), *Actinobacteria* (0.3–6.3%), and *Synergistetes* (0.2–14.8%), with *Prevotella* and *Longilinea*, the dominant allicin resistant genera. Using a laboratory scale AnHR, Oktem *et al.* (2007) describe the acclimatization procedure to treat pharmaceutical wastewater (6,000 mg L^{-1}). Optimum treatment and gas production was achieved at an OLR of 8 kg COD m^{-3} d^{-1}, at an HRT of 2 days, and a reactor up-flow velocity of <12 m d^{-1} producing 72% COD removal and 200 L CH_4 kg^{-1} VS d^{-1}. They observed that COD removal fell with an increased percentage of pharmaceutical wastewater added, with 99, 96, 91, and 85% COD removals at 10, 30, 70, and 100% pharmaceutical wastewater treated at an OLR of 3 kg COD $m^{-3}d^{-1}$ and a reactor up-flow velocity of 4.8 m d^{-1}.

References

Akil, K. and Jayanthi, S. (2012) Anaerobic sequencing batch reactors and its influencing factors: An overview. *Journal of Environmental Science and Engineering*, **54**, (2), 317–322. https://pubmed.ncbi.nlm.nih.gov/24749385/

Angenent, L.T. and Dague, R.R. (1995) A laboratory-scale comparison of the UASB and ASBR processes. In *Proceeding of 50th Purdue Industrial Waste Conference* (eds. C.S. Dalton and R.F Wukasch). Ann Arbor Press, Chelsea, MI, USA, pp. 365–377.

Angenent, L.T. and Sung, S. (2001) Development of anaerobic migrating blanket reactor (AMBR), a novel anaerobic treatment system. *Water Research*, **35**, 1739–1747. https://doi.org/10.1016/S0043-1354(00)00447-4

Bachman, A., Beard, V.L. and McCarty, P.L. (1982) Comparison of fixed-film reactors with a modified sludge blanket reactor. In *Proceedings of the First International Conference on Fixed-film Biological Processes* (eds. Y.C Wu and E.D Smith). Noyes Publications, Park Ridge, USA, pp. 1192–1211.

Bachman, A., Beard, V.L. and McCarthy, P. (1985) Performance characteristics of the anaerobic baffled reactor. *Water Research*, **19**, 99–106. https://doi.org/10.1016/0043-1354(85)90330-6

Berthe, S. and Leta, S. (2023) Anaerobic co-digestion of agro-industrial wastes using anaerobic sequencing batch reactor for bio-energy recovery: Focus on process performance and stability of the methanogenic step. *Journal of Water Process Engineering*, **54**, 103993. https://doi.org/10.1016/j.jwpe.2023.103993

Borzacconi, L., Ottonello, G., Castelló, E., Pelaez, H., Gazzola, A. and Viñas, M. (1999) Denitrification in a carbon and nitrogen removal system for leachate treatment: Performance of a upflow sludge blanket (USB) reactor. *Water Science and Technology*, **40**, (8), 145–151.

Brummeler, E., Hulshoff-Pol, L.W., Dolfing, J., Lettinga, G. and Zehnder, A.J.B. (1985) Methanogenesis in an upflow anaerobic sludge blanket reactor at pH 6 on an acetate-propionate mixture. *Applied and Environmental Microbiology*, **49**, 1472–1477.

Chang, J.E. (1989) Treatment of landfill leachate with an upflow anaerobic reactor combining a sludge bed and a filter. *Water Science and Technology*, **21**, (4–5), 133–143.

Cui, B.J. and Liang, S.X. (2019) Monitoring opportunistic pathogens in domestic wastewater from a pilot-scale anaerobic biofilm reactor to reuse in agricultural irrigation. *Water*, **11**, (6), 1283. https://doi.org/10.3390/w11061283

Da Silva, J.A., Da Silva, G.H.R. and Sarti, A. (2017) Treatment of low-strength sanitary sewage using an anaerobic/aerobic baffled reactor in pilot scale. *Engenharia Sanitaria Ambiental*, 22, (3), 539–549. https://doi.org/10.1590/S1413-41522016144668 Erratum https://doi.org/10.1590/S1413-41522017022000000030

Debik, E. and Coskun, T. (2009) Use of static granular bed reactor (SGBR) with anaerobic sludge to treat poultry slaughterhouse wastewater and kinetic modelling. *Bioresource Technology*, **100**, 2777–2782. https://doi.org/10.1016/j.biortech.2008.12.058

Debik, E., Park, J. and Ellis, T.G. (2005) Leachate treatment using the static granular bed reactor. In *Water Environment Federation 78th Annual Technical Exhibition and Conference*, Washington, DC, pp. 4003–4011. https://www.accesswater.org/publications/proceedings/-291864/leachate-treatment-using-the-static-granular-bed-reactor

Del Nery, V., Damianovic, M.H.Z. and Barros, F.G. (2001) The use of upflow anaerobic sludge blanket reactors in the treatment of poultry slaughterhouse wastewater. *Water Science and Technology*, **44**, (4), 83–88.

Deng, Z., Sierra, J.M., Ferreira, A.L.M., Cerqueda-Garcia, D., Spanjers, h. and van Lier, J.B. (2024) Effect of operating parameters on the performance of an anaerobic sequencing batch reactor (AnSBR) treating protein-rich wastewater. *Environmental Science and Ecotechnology*, **17**, 100296. https://doi.org/10.1016/j.ese.2023.100296

Driessen, W. and Yspeert, P. (1999) Anaerobic treatment of low, medium and high strength effluent in the agro-industry. *Water Science and Technology*, **40**, (8), 221–228.

Dutta, K., Tsai, C.-Y., Chen, W.-H. and Li, J.-G. (2014) Effect of carriers on the performance of anaerobic sequencing batch biofilm reactor treating municipal wastewater. *International Biodeterioration and Biodegradation*, **95**, (A), 84–88. https://doi.org/10.1016/j.ibiod.2014.04.021

Evans, K.M. (2004) Fundamentals of the static granular bed reactor. Ph.D. Thesis, Iowa State University, IA, USA.

Fang, H.H.P., Guohua, L. and Jinfu, Z. (1990) Treatment of brewery effluent by UASB process. *Journal of Environmental Engineering, American Society of Civil Engineering*, **116**, (3), 454–460.

Fdz-Polanco, F., Hidalgo, M.D., Fdz-Polanco, M. and García Encina, P.A. (1999) Anaerobic treatment of polyethylene terephthalate (PET) wastewater from lab to full scale. *Water Science and Technology*, **40**, (8), 229–236.

Feng, H.J., Hu, L.F., Mahmood, Q., Qiu, C. Fang, C. and Shen, D. (2008) Anaerobic domestic wastewater treatment with bamboo carrier anaerobic baffled reactor. *International Biodeterioration and Biodegradation*, **62**, 232–238. https://doi.org/10.1016/j.ibiod.2008.01.009

Fernández, J.M., Méndez, R. and Lema, J.M. (1995) Anaerobic treatment of Eucalyptus fibreboard manufacturing wastewater by a hybrid USBF lab-scale reactor. *Environmental Technology*, **16**, (7), 677–684.

Gray, N.F. (2023) *Activated Sludge: Developments and Sustainable Solutions.* World Scientific, Singapore.

Grotenhuis, J.T.C., Smit, M., Plugge, C.M., Yuansheng, X., van Lammeren, A.A.M., Stams, A.J.M. and Zehnder, A.J.B. (1991) Bacteriological composition and structure of granular sludge adapted to different substrates. *Applied and Environmental Microbiology*, **57**, 1942–1949.

Harmsen, H.J.M., Kengen, H.M.P., Akkermans, A.D.L., Stams, A.J.M. and de Vos, W.M. (1996) Determination and localization of syntrophic propionate-oxidizing bacteria in granular sludge by *in situ* hybridisation using 16S rRNA-based oligonucleotide probes. *Applied and Environmental Microbiology*, **62**, 1656–1663.

Hemalatha, D. and Keerthinarayana, S. (2014) Anaerobic treatment of pulp and paper mill wastewater using hybrid upflow anaerobic sludge blanket reactor (HUASBR). *International Journal of Innovative Research in Science, Engineering and Technology*, **3**, (4), 11576–11584. http://www.ijirset.com/upload/2014/april/93_Anaerobic.pdf

Hulshoff-Pol, L.W., de Zeeuw, W., Velzeboer, C.T.M. and Lettinga, G. (1983) Granulation in UASB reactor. *Water Science and Technol*ogy, **15**, (8–9), 291–304. https://doi.org/10.2166/wst.1983.0172

Hulshoff-Pol, L.W., Dolfing, J., de Zeeuw, W., and Lettinga, G. (1982) Cultivation of well adapted pelletized methanogenic sludge. *Biotechnology Lett*ers, **4**, 329–332. https://doi.org/10.1007/BF00132835

Hulshoff-Pol, L.W., van de Worp, J.J.M., Lettinga, G. and Beverloo, W. A. (1986) Physical characterization of anaerobic granular sludge. In *Proceedings EWPCA Water Treatment Conference, AQUARECH '86.* Amsterdam, The Netherlands, pp. 98–101.

Imai, T., Ukita, M., Sekine, M., Fukagawa, M. and Nakanishi, H. (2000) Fact-finding survey of actual garbage discharged from dormitory and its biological anaerobic-aerobic treatment. *Water Science and Technology*, **41**, (3), 129–135.

Irizar, I., Roche, E., Beltran, S., Aymerich, E. and Esteban-Gutierrez, M. (2018) Model-based design of software sensor for real-time diagnosis of the stability conditions in high-rate anaerobic reactors — Full-scale application of internal circulation technology. *Water Research*, **143**, 479–490. https://doi.org/10.1016/j.watres.2018.06.055

Jeison, D. and Chamy, R. (1999) Comparison of the behaviour of expanded granular sludge bed (EGSB) and upflow anaerobic sludge blanket (UASB) reactors in dilute and concentrated wastewater treatment. *Water Science and Technology*, **40**, (8), 91–97. https://doi.org/10.2166/wst.1999.0393

Jimenez, J., Latrille, E., Harmand, J., Robles, A., Ferrer, J., Gaida, D., Wolf, C., Mairet, F., Bernard, O., Alcaraz-González, V., Mendez-Acosta, H., *et al.* (2015) Instrumentation and control of anaerobic digestion processes: A review and some research challenges. *Reviews in Environmental Science and Biotechnology*, **14**, 615–648. https://doi.org/10.1007/s11157-015-9382-6

Kato, M., Field, J.A., Kleerebezem, R. and Lettinga, G. (1994) Treatment of low strength wastewater in upflow anaerobic sludge blanket (UASB) reactors. *Journal of Fermentation and Bioengineering*, 77, 679–685.

Kumar, V., Thakur, C. and Chaudhari, P.K. (2022) Anaerobic biological treatment of dye bearing water in anaerobic sequencing batch reactor: Performance and kinetics studies. *Journal of the Indian Chemical Society*, **99**, 100673. https://doi.org/10.1016/j.jics.2022.100673

Laguna, A., Ouattara, A., Gonzalez, R.O., Baron, O., Famá, G., El Mamouni, R., Guiot, S., Monroy, O. and Macarie, H. (1999) A simple and low cost technique for determining the granulometry of upflow anaerobic sludge blanket reactor sludge. *Water Science and Technology*, **40**, (8), 1–8. https://doi.org/10.1016/S0273-1223(99)00602-2

Lauzurique, Y., Meza, A., Huilinir, C., Salazar-Gonzalez, R. and Vargas-Morales, G. (2023) Evaluation of the anaerobic digestion of winery wastewater: Effect of fly ash in anaerobic sequencing batch reactor.

Journal of Environmental Chemical Engineering, **11**, 110997. https://doi.org/10.1016/j.jece.2023.110997

Lettinga, G. and Holshoff-Pol, L. (1991) UASB-process design for various types of wastewaters. *Water Science and Technology*, **24**, (8), 87–107. https://doi.org/10.2166/wst.1991.0220

Lettinga, G., Field, J., van Lier, J., Zeeman, G. and Hulshoff Pol, L.W. (1997) Advanced anaerobic wastewater treatment in the near future. *Water Science and Technology*, **35**, (10), 5–12. https://doi.org/10.1016/S0273-1223(97)00222-9

Lettinga, G., Velsen, A. F. M. van, Hobma, S. W., Zeeuw, W. de, and Klapwijk, A. (1980). Use of the upflow sludge blanket (USB) reactor concept for biological waste treatment, especially for anaerobic treatment. *Biotechnology and Bioengineering*, **22**, 699–734.

Lutze, R. and Engelhart, M. (2020) Comparison of CSTR and AnMBR for anaerobic digestion of WAS and lipid rich digestion sludge from the dairy industry. *Water Resources and Industry*, **23**, 100122. https://doi.org/10.1016/j.wri.2019.100122

Mach, K.F. (2000) Development of the static granular bed reactor. Master Thesis, Iowa State University, IA, USA.

MacLeod, F.A., Guiot, S.R. and Costerton, J.W. (1990) Layered structure of bacterial aggregates in an upflow anaerobic sludge bed and filter reactor. *Applied and Environmental Microbiology*, **56**, 1598–1607.

Manjunath, N.T., Mehrotra, I. and Mathur, R.P. (2000) Treatment of wastewater from slaughterhouse by DAF-UASB system. *Water Research*, **34**, (6), 1930–1936. https://doi.org/10.1016/S0043-1354(99)00337-1

Mao, C., Feng, Y., Wang, X. and Ren, G. (2015) Review on research achievements of biogas from anaerobic digestion. *Renewable and Sustainable Energy Reviews*, **45**, 540–555. https://doi.org/10.1016/j.rser.2015.02.032

Martínez, J., López, I., Giani, L. and Borzacconi, L. (2001) Blanket development in a malting wastewater anaerobic treatment. *Water Science and Technology*, **44**, (4), 57–62. https://doi.org/10.2166/wst.2001.0178

Morris, G.G. and Burgess, S. (1984). Two phase anaerobic wastewater treatment. *Water Pollution Control*, **83**, 514–520.

Mullai, P., Vishali, S. and Dharmalingam, K. (2024) Residence time distribution experiments in a hybrid anaerobic blanket reactor treating real pharmaceutical wastewater. *Journal of the Taiwan Institute of Chemical Engineers*, 105381. https://doi.org/10.1016/j.jtice.2024.105381

Mullai, P., Vishali, S. and Sabarathinam, P.L. (2018) Biogas production kinetics in an anaerobic multiphase hybrid reactor treating antibiotic industry wastewater. *Desalination and Water Treatment*, **122**, 247–253. https://doi.org/10.5004/dwt.2018.22831

Mullai, P., Vishali, S. and Sobiya, E. (2022) Experiments and adaptive-network-based fuzzy inference system modelling in a hybrid up-flow anaerobic sludge blanket reactor to assess industrial azadirachtin effluent quality. *Bioresource Technology*, **358**, 127395. https://doi.org/10.1016/j.biortech.2022.127395

Murugesan, M.P. and Selvarasu, P. (2017) Experimental investigation of HUASB reactor for treatment characteristics of leather effluent with varying of different operating parameters. *International Journal of ChemTech Research*, **10**, 478–484. https://sphinxsai.com/2017/ch_vol10_no6/2/(478-484)V10N6CT.pdf

Oktem, Y.A., Ince, O., Sallis, P., Donnelly, T. and Ince, B.K. (2007) Anaerobic treatment of a chemical based pharmaceutical wastewater in a hybrid upflow anaerobic sludge blanket reactor. *Bioresource Technology*, **99**, 1089–1096. https://doi.org/10.1016/j.biortech.2007.02.036

Puñal, A. and Lema, J.M. (1999) Anaerobic treatment of wastewater from a fish-canning factory in a full-scale upflow anaerobic sludge blanket (UASB) reactor. *Water Science and Technology*, **40**, (8), 57–62. https://doi.org/10.1016/S0273-1223(99)00609-5

Puñal, A., Loenzo, A., Roca, E., Hernández, C. and Lema, J.M. (1999) Advanced monitoring of an anaerobic pilot plant treating high strength wastewaters. *Water Science and Technology*, **40**, (8), 237–244. https://doi.org/10.1016/S0273-1223(99)00631-9

Ramakrishnan, A. and Surampalli, R.Y. (2012) Comparative performance of a UASB and anaerobic hybrid reactors for the treatment of complex phenolic wastewater. *Bioresource Technology*, **123**, 352–359. http://dx.doi.org/10.1016/j.biortech.2012.07.072

Rebac, S., Ruskova, J., Gerbens, S., van Lier, J., Stams, A.J.M. and Lettinga, G. (1995) High rate anaerobic treatment of wastewater under psychrophilic conditions. *Journal of Fermentation and Bioengineering*, **80**, 499–506.

Renuka, R., Mohan, S.M., Sowmiya, B. and Raj, S.A. (2016) Performance evaluation of panelled anaerobic baffle-cum-filter reactor in treating municipal wastewater. *Ecological Engineering*, **97**, 1–12. https://doi.org/10.1016/j.ecoleng.2016.07.020

Reynaud, N. and Buckley, C.A. (2016) The anaerobic baffled reactor (ABR) treating communal wastewater under mesophilic conditions: A review. *Water Science and Technology*, **73**, (3), 463–478. https://doi.org/10.2166/wst.2015.539

Sarti, A., Garcia, M.L., Zaiat, M. and Foresti, E. (2007) Domestic sewage treatment in a pilot-scale anaerobic sequencing batch biofilm reactor (ASBBR). *Resources, Conservation and Recycling*, **57**, (1), 237–247. https://doi.org/10.1016/j.resconrec.2006.09.008

Sarti, A., Silva, A.J., Ziaiat, M. and Foresti, E. (2011) Full-scale anaerobic sequencing batch biofilm reactor for supfate-rich wastewater treatment. *Desalination and Water Treatment*, **25**, 13–19. https://doi.org/10.5004/dwt.2011.1864

Show, K.Y. and Lee, D.J. (2016) Chapter 8. Anaerobic treatment vs. aerobic treatment. In *Biotechnology and Bioengineering: Biological Treatment oof Industrial Effluents* (eds. D.-J. Lee *et al.*). Elsevier, London, pp. 205–230. https://doi.org/10.1016/B978-0-444-63665-2.00008-4

Show, K.-Y., Yab, Y., Yao, H., Guo, H., Li, T., Show, D.-Y., Chang, J.-S. and Lee, D.-J. (2020) Anaerobic granulation: A review of granulation hypotheses, bioreactor designs and emerging green applications. *Bioresource Technology*, **300**, 122751. https://doi.org/10.1016/j.biortech.2020.122751

Soh, Y.N.A., Kunacheva, C., Webster, R.D. and Stuckley, D.C. (2020) Composition and biotransformational changes in soluble microbial products (SMPs) along an anaerobic baffled reactor (ABR). *Chemosphere*, **254**, 126775. https://doi.org/10.1016/j.chemosphere.2020.126775

Sousa, D.Z., Smidt, H., Alves, M.M. and Stams, A.J.M. (2009) Ecophysiology of syntrophic communities that degrade saturated and unsaturated long-chain fatty acids. *FEMS Microbiology Ecology*, **68**, 257–272. https://doi.org/10.1111/j.1574-6941.2009.00680.x

Timur, H. and Özturk, I. (1999) Anaerobic sequencing batch reactor treatment of landfill leachate. *Water Research*, **33**, 3225–3230. https://doi.org/10.1016/S0043-1354(99)00048-2

Uyanik, S., Sallis, P.J. and Anderson, G.K. (2002) The effect of polymer addition on granulation in an anaerobic baffled reactor (ABR): Part I. Process performance. *Water Research*, **36**, 933–942. https://doi.org/10.1016/S0043-1354(01)00315-3

van Lier, J.B. (2008) High-Rate anaerobic wastewater treatment: Diversifying from end-of-the-pipe treatment to resource-oriented conversion techniques. *Water, Science and Technology*, **57**, 1137–1148. http://dx.doi.org/10.2166/wst.2008.040

van Lier, J.B., Lens, P.N.L. and Hulshoff Pol, L.W. (2001) Anaerobic treatment for C and S removal in "zero-discharge" paper mills: Effects of process design on S removal efficiencies. *Water Science and Technology*, **44**, (4), 189–195. https://doi.org/10.2166/wst.2001.0218

Vissier, F.A., van Lier, J.B., Macario, A.J.L. and Conway de Macario, E. (1991) Diversity and population dynamics of bacteria in a granular consortium. *Applied and Environmental Microbiology*, **57**, 1728–1734.

Wang, T.W., Huang, Z., Ruan, W., Zhao, M., Shao, Y. and Miao, H. (2018) Insights into sludge granulation during anaerobic treatment of high-strength leachate via a full-scale IC reactor with external circulation system. *Journal of Environmental Sciences,* **64**, 227–234. http://dx.doi.org/10.1016/j.jes.2017.06.024

Wiegant, W.M. and de Man, A.W. (1986) Granulation of biomass in thermophilic upflow anaerobic sludge blanket reactors treating acidified wastewaters. *Biotechnology and Bioeng*ineering, **28**, 718–727. https://doi.org/10.1002/bit.260280511

Wirtz, R.A. and Dague, R.R. (1996) Enhancement of granulation and start-up in anaerobic sequencing batch reactor. *Water and Environmental Research*,**68**,883–892.https://doi.org/10.2175/106143096X127893

Wu, W., Hu, J., Gu, X., Zhao, Y. and Zhang, H. (1987) Cultivation of anaerobic granular sludge with aerobic activated sludge as seed. *Water Research*, **21**, 789–799. https://doi.org/10.1016/0043-1354(87)90154-0

Xu, M., Ding, L., Xu, K., Geng, J. and Ren, H. (2016) Flow patterns and optimization of compartments for the anaerobic baffled reactor. *Desalination and Water Treatment*, **57**, 345–352. https://doi.org/10.1080/19443994.2014.970580

Xue, Y., Ma, H. and Li, Y.-Y. (2023) Anammox-based granulation cycle for sustainable granular sludge biotechnology from mechanisms to strategies: A critical review. *Water Research*, **228**, 119353. https://doi.org/10.1016/j.watres.2022.119353

Yan, Y.G. and Tay, J.H. (1996) Brewery wastewater treatment in UASB reactor at ambient temperature. *Journal of Environmental Engineering, American Society of Civil Engineering,* **122**, (6), 550–553.

Zakoura, M., Kopsahelis, A., Tsigkou, K., Ntougias, S., Ali, S.S. and Kornaros, M. (2022) Performance evaluation of three mesophilic upflow anaerobic sludge blanket bioreactors treating olive mill wastewater: Flocculent and granular inocula tests, organic loading rate effect and anaerobic consortia structure. *Fuel*, **313**, 122951. https://doi.org/10.1016/j.fuel.2021.122951

Zang, H., Wang, M., Li, H., Yan, B. and Peng, H. (2024) Anaerobic baffled reactor (ABR) coupled step-feeding and effluent recycling for treating high concentration allicin wastewater with granulation of sludge. *Journal of Water Process Engineering*, **65**, 105834. https://doi.org/10.1016/j.jwpe.2024.105834

Zha, X., Ma, J. and Tsapekos, P. (2019) Evaluation of an anaerobic baffled reactor for pretreating black water: potential application in rural China. *Journal of Environmental Management*, **251**, 109599. https://doi.org/10.1016/j.jenvman.2019.109599

Zheng, H., Chai, H.-X., Zhao, L., Cao, X., Feng, L. and Ji, F. (2023) Pretreatment of septic tank wastewater by packed baffled reactor: Pollutant degradation and microbial community succession in different compartments. *Environmental Research*, **223**, 115475. https://doi.org/10.1016/j.envres.2023.115475

Zhu, G.F., Zou, R., Jha, A.K., Huang, X., Liu, L. and Liu, C. (2015) Recent developments and future perspectives of anaerobic baffled bioreactor for wastewater treatment and energy recovery. *Critical Reviews in Environmental Science and Technology*, **45** (12), 1243–1276. https://doi.org/10.1080/10643389.2014.924182

Ziels, R.M., Beck, D.A. and Stensel, H.D. (2017) Long-chain fatty acid feeding frequency in anaerobic co-digestion impacts syntrophic community structure and biokinetics. *Water Research*, **117**, 218–229. https://doi.org/10.1016/j.watres.2017.03.060

Zoutberg, G.R. and de Been, P. (1997) The biobed EGSB (Expanded Granular Sludge Blanket) system covers shortcomings of the UASB reactor in the chemical industry. *Water Science and Technology*, **35**, (10), 183–188. https://doi.org/10.2166/wst.1997.0381

Section D

Biogas Production Systems

8

Biogas Production and Cleaning

8.1 Methane Production and Collection

There are three products of anaerobic digestion: the digested sludge, waste liquor, and a mixture of gases. Digested sludge is different from either primary or secondary sludge in a number of ways. It is pathogen-free, stabilized, and far less offensive, having a tarry smell and drying to an inert friable condition, thus making it ideal for disposal to agricultural land. The waste liquor from anaerobic digesters has a high suspended solids (500–1,000 mg L^{-1}) and biochemical oxygen demand (BOD) concentration (400–800 mg L^{-1}) due to all the soluble organics present. However, the BOD of the liquor can be very high indeed, reaching up to 10,000 mg L^{-1}, although up to 60% of the BOD is due to the suspended solids fraction. Because of the degradation of organic nitrogen, the liquor may have high concentrations of soluble nitrogen present. The characteristics and strength of the liquor make it difficult to dispose of or treat separately, and it is returned to the works inlet, where it is diluted by the incoming sewage and treated in admixture aerobically.

Sludge or digester gas is more commonly referred to as biogas. Gas production is the most direct and sensitive measure of the rate of anaerobic digestion, a decrease in production being the first indication that the process is unstable. Apart from trace amounts of water vapor, hydrogen sulfide, hydrogen, nitrogen, unsaturated hydrocarbons, and other gases, biogas is predominantly a mixture of carbon dioxide and methane, the exact proportions of which determine its calorific value.

A typical biogas contains between 65% and 70% methane by volume, with the remaining being largely carbon dioxide. Although hydrogen is an important precursor of methane formation in digestion, the concentration of hydrogen, which can be measured within the reactor, is usually very low, which suggests an immediate uptake by the methanogens. When burnt, biogas produces water, which complicates the determination of its calorific value. The net calorific value for combustion is where the water formed remains in the vapor phase, whereas the gross calorific value is where the water formed is condensed. The calorific values are calculated using the percentage of methane (M) present as the net calorific value = $(334 \times M)$ kJ m^{-3}, or the gross calorific value = $(370 \times M)$ kJ m^{-3}, both at 15.5°C and 1 atmosphere pressure. For normal biogas, the gross calorific value is between 24,000 and 26,000 kJ m^{-3}, with the net value being some 10% less.

The production of biogas can be related to the amount and type of organic matter utilized. For example, the removal of 1 kg of chemical oxygen demand (COD) yields 0.35 m^3 of biogas at standard temperature and pressure (STP), whereas 1 kg of organic carbon would yield 1.87 m^3 at STP. The exact gas yield per kg of volatile solids removed depends on the composition of the waste but is approximately 1.0 m^3. Some food-processing industries produce a pure substrate waste with 80–95% of the organic (volatile) matter removed, whereas only 40–50% of the organic matter in sewage sludge, and even slightly less in animal slurries, will be utilized with a proportionately lower gas yield in terms of m^3 of gas produced per kg of substrate supplied. Therefore, for a mixed primary and secondary sludge from a purely domestic sewage treatment works, the gas production will be of the order of 0.5 $m^3\,kg^{-1}$ organic (volatile) matter, or about 0.375 $m^3\,kg^{-1}$ of total (dry) solids, added to the digester. The composition and quantity of gas produced by complete digestion can be theoretically determined using the equation:

$$\begin{aligned} C_cH_hO_oN_nS_s &+ 1/4(4c - h - 2o + 3n + 2s)H_2O \\ &= 1/8(4c - h + 2o + 3n + 2s)CO_2 + 1/8(4c + h - 2o - 3n - 2s)CH_4 \\ &\quad + nNH_3 + sH_2S \qquad (8.1) \end{aligned}$$

where *c*, *h*, *o*, *n*, and *s* are the number of atoms of carbon, hydrogen, oxygen, nitrogen, and sulfur, respectively. This equation replaces the earlier equation developed by Buswell and Mueller (1952), which ignored the nitrogen and sulfur components of the waste that are largely utilized within the digester.

Thus, by neglecting the very small volume of other gases formed and using the simplified formula, carbohydrates will produce 73% CO_2 and 27% CH_4, lipids 52% CO_2 and 48% CH_4, and protein 73% CO_2 and 27% CH_4. The total volume of biogas produced from each of these basic substrates is 0.75, 1.44, and 0.98 $m^3\,kg^{-1}$ dry matter, respectively (Buswell and Mueller, 1952). Actual values of gas yield and composition from the basic substrates have been given by Konstandt (1976). Lipids are only slowly degraded anaerobically, and the measurement of their removal provides a very useful quality control parameter for the process.

A more accurate estimation of methane production per unit time can be obtained by using the equation:

$$G = 0.35\ (L - 1.42\ S_t) \tag{8.2}$$

where G is the volume of methane produced ($m^3 d^{-1}$), L is the mass of ultimate BOD removed (kg d^{-1}), and S_t is the mass of volatile solids accumulated (kg d^{-1}). S_t can be estimated as:

$$S_t = aL/1 + bt_s \tag{8.3}$$

where a is the mass of volatile solids synthesized per kg of ultimate BOD removed, b is the endogenous respiration constant, and t_s is the solids retention time (Speece and McCarty, 1964). Methane is the most valuable byproduct of anaerobic wastewater treatment, and it is useful to express the yield as $m^3\,CH_4\,kg^{-1}$ organic matter removed, or $m^3\,CH_4\,kg^{-1}$ COD removed. Typical methane yields from mesophilic digesters operated at 35°C are 0.86 $m^3\ kg^{-1}$ organic matter removed for mixed sewage sludge, 0.55 $m^3\,kg^{-1}$ for dairy manure, and 0.42 $m^3\,kg^{-1}$ for pig manure. Under continuous reactor operation, a mesophilic digester will produce between 12 and 16 L of methane per capita per day for primary sludge, rising to about 20 L $ca^{-1}\ d^{-1}$ for combined primary and secondary sludges.

The simplest use of biogas is to use it on-site either to produce heat directly for the anaerobic reactors, which normally operate in the mesophilic range using gas boilers, or to generate electricity using modified diesel (dual fuel) engines to drive alternators, with the cooling water from these engines used to heat the reactors, a system known as combined heat and power (CHP) generation. Some plants now use biogas for vehicle fuel, either in the form of biomethane or LPG, while larger plants and those treating organic wastewaters, especially food and agricultural wastes, can generate enough biogas to make it viable for it to be cleaned to the high standard required for supply to the natural gas grid for domestic and industrial use. Biomethane is different from biogas methane, as it requires further processing, as discussed below. Hydrogen can also be produced via digestion and can be used as a vehicle fuel or for household heating (Sec. 9.1.1). Smaller digestion units are used widely in the Third World for cooking and lighting.

With the increase in the size of anaerobic digesters and the centralization of sludge treatment, large quantities of biogas can be generated enough for sale, either to national gas networks for home and industrial use or for the large-scale production of biomethane for vehicle fuel. However, the drawback is that the biogas contains a wide range of contaminants that need to be removed, as well as increasing the methane content to match the calorific value of natural gas. The precise allowable concentrations of contaminants allowable in biomethane for injection into the gas network vary slightly from country to country. Gas Networks Ireland, which is the key distributor of natural gas in Ireland, has produced a technical specification data sheet for biomethane suppliers who wish to sell their gas to the network (Table 8.1). There are a number of European Standards for biomethane. While standard EN16726-1 provides gas quality standards for natural gas, EN16723-1 deals specifically with biomethane to be injected into natural gas networks, and EN16726-2 deals with its use as vehicle fuel (BS EN, 2016; 2017).

Gas supplied to the grid must have an odor in order to identify leaks. While this is normally added by the supplier, even after the most effective desulfurization, trace amounts of hydrogen sulfide will remain, and with an odor threshold of 0.0047 ppm, odor detection is unlikely to be a problem.

Table 8.1. Technical specifications for maximum allowable contaminant concentrations for biomethane supplied to Gas Networks Ireland.

Parameter	Units	Limit value
Hydrogen sulfide content	mg m^{-3}	≤5
Total S content including H_2S	mg m^{-3}	≤50
Hydrogen content	% mol mol^{-1}	<0.1
Oxygen content	% mol mol^{-1}	≤1.0
Hydrocarbon dewpoint	°C	≤−2
Water (moisture) content	mg m^{-3}	≤50
Wobbe index (high limit)	Mj m^{-3}	51.41
Wobbe index (low limit)	Mj m^{-3}	47.2
Incomplete combustion factor	—	<0.48
Sooting index	—	<0.6
Gross calorific value (CV high limit)	Mj m^{-3}	42.3
Gross calorific value (CV low limit)	Mj m^{-3}	36.9
Carbon dioxide content	% mol mol^{-1}	<2.5
Organo-halides	mg m^{-3}	<1.5
Radioactivity	Becquerels g^{-1}	<5
Ethane	% mol mol^{-1}	<12
Nitrogen	% mol mol^{-1}	≤5
Ammonia content	mg m^{-3}	10
Siloxanes content	mg m^{-3}	0.3
Amines	mg m^{-3}	10
Carbon monoxide	% mol mol^{-1}	0.1

8.2 Hydrogen Sulfide Control and Removal

Any sulfate present in the waste stream will be readily converted to sulfide by sulfur reducing bacteria (SRB) during anaerobic treatment (Fig. 8.1). Sulfate is common in many wastewaters from concentrations between 20 and 200 mg L^{-1} in domestic and up to 35,000 mg L^{-1} in industrial wastewaters. The sulfide can be present in anaerobic reactors in a number of forms, either in solution (S^{2-}, HS^-, and H_2S) or gaseous (H_2S) form, which contaminates the biogas (Jung *et al.*, 2022). The

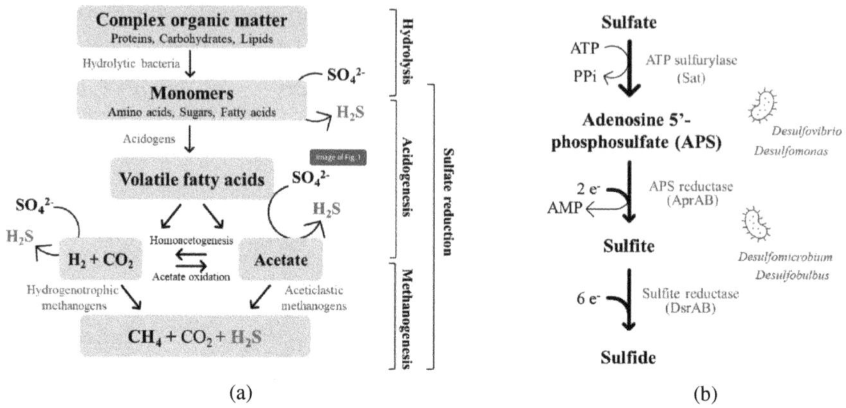

Figure 8.1. Sulfide is formed during anaerobic biodegradation by sulfate reducing bacteria through the dissimilatory sulfate-reduction pathway. (a) The stages at which sulfate is formed, and (b) the key enzymes and bacteria involved in sulfide formation (Jung *et al.*, 2020). Reproduced with permission of Elsevier Publishing.

form in which the sulfide is found is dependent largely on the pH and temperature but is toxic to methanogens at concentrations between 100 and 800 mg total dissolved sulfide L^{-1} (Krayzelova *et al.*, 2015; Wang *et al.*, 2020). Hydrogen sulfide is a hazardous gas with a wide range of environmental and health effects as well as corrosion problems, which are explored in detail in Secs. 3.3.2 and 3.3.3, respectively.

Biogas contains, on average, 0.2% H_2S (typical range 0.08–5.5%) and there are a number of different approaches for directly removing it from the biogas, such as metal precipitation, chemical oxidation, micro-aeration, and most commonly, adsorption and absorption (Table 8.2), although preventing hydrogen sulfide production during anaerobic treatment (i.e., *in-situ* technologies) should be part of good operational practice (Skerman *et al.*, 2012; Jung *et al.*, 2022). Comparative cost analysis of the different desulfurization methods is given by Allegue and Hinge (2014).

The control of hydrogen sulfide is critical for stable anaerobic operation and is increasingly important, as biogas is used off-site for energy generation and other commercial uses. Post-treatment is relatively expensive, especially when sulfur-rich substrates are used as feed to digesters. There is also real potential in recovering not only elemental sulfur (S^0) but also metals as sulfide compounds (Fu and Wang, 2011). In theory, this is most economically done by

Table 8.2. List of current desulfurization methods for biogas (Kapłan *et al.*, 2021). Reproduced under the Creative Commons Attribution (CC BY) license.

Method	Process or technique	Application	Range of biogas desulfurization efficiency
Physical	Absorption (wet)	– scrubbing water — differences in solubility of CO_2 and CH_4 in polar solvents are used; – absorbent: (a) methanol (Rctisol), (b) poly(ethylene glycol) dimethyl ether (Selexol, Genosorb), (c) propylene carbonate (Fluor), (d) sulfolane (Sulfinol).	>94%
	Absorption (dry)	– adsorbent: (a) carbon (activated carbon), (b) mineral, (c) mineral and carbon.	up to 95%
	Absorption (wet)	– dissolution of H_2S in water or organic solvents: (a) Genosorb, (b) Rectisol, (c) Ifpex.	>96%
	Cryogenic separation	– very low temperatures down to –100°C and high pressure of about 40 bar.	up to 99%
Chemical	Absorption	– iron compounds: (a) bog iron ore (porous sedimentary rock).	up to 90%
	Membrane (porous or diffusive)	– chemical potential of solutions or mixtures.	up to 99%
	Absorption (wet)	– caustic soda solution; – aqueous alkanolamine solutions; – ferrous chloride solution.	(95 ÷ 99)%
Biological	Oxidation	– supplying air to biogas.	over 95%
	Microbiological reaction	– bacteria of the family *Thiobacillus* producing elemental sulfur.	(80 ÷ 99)%

controlling sulfur loadings to digesters, by mixing different waste streams, or by *in-situ* treatment (Table 8.3), with many new technologies, such as electrochemical oxidation, iron cycling, and sulfide oxidation coupled with denitrification and stimulation of electric-syntropy in development (Table 8.4). Post-treatment of sulfides is generally more expensive although relatively simple and effective (Jung *et al.*, 2022). All anaerobic treatment methods seem amenable to at least one or more *in-situ* sulfide control technologies.

Table 8.3. The advantages and challenges of using conventional *in-situ* desulfurization methods (Jung *et al.*, 2022). Reproduced with permission of Elsevier Publishing.

Technology	Advantages	Challenges
Metal sulfide precipitation	High sulfide removal efficiency	Need for continuous supply of chemicals
	Simple operation	Loss of digester working volume by precipitates
	Low toxicity	Increase in salinity
Chemical oxidation	High sulfide removal efficiency	Need for continuous supply of chemicals
	Low energy consumption	Toxicity of chemical oxidants to methanogens
	Simple operation	Non-selective oxidation of organic matter
Microaeration	High sulfide removal efficiency	Disturbance of anaerobic metabolism
	Enhanced substrate hydrolysis	Biogas dilution (i.e., reduced energy content)
	Applicable with high sulfide loads	Energy consumption for aeration
	Recovery of elemental sulfur	Risk of explosion
Adsorption	High sulfide removal efficiency	Limited adsorption capacity
	Availability of low-cost adsorbents	Non-selective adsorption of organics or ions
	Process simplicity	Deactivation of adsorbents by biofilm formation
		Need for periodic replacement or regeneration of adsorbents

Table 8.3. (*Continued*)

Technology	Advantages	Challenges
Suppression of sulfidozenesis	Reduced formation of toxic hydrogen sulfide Relief of competitive inhibition of methanogens	Transient suppression of sulfate-reducing activity Potential inhibition of beneficial microorganisms
	Reduced burden on biogas upgrading process	Additional competitive reactions to methanogenesis Possible formation of undesirable compounds

Using different substrates, Peu *et al.* (2012) were able to develop a methodology to determine the hydrogen sulfide content of biogas during anaerobic digestion. A wide range of sulfur contents were recorded, ranging from 1 to 29.6 mg S kg^{-1} TS (total solids) (Table 8.5), and using the biochemical methanogenic potential, biochemical biogas potential, and the sulfur content of the substrates, tested a model that was developed to predict the hydrogen sulfide content of the raw biogas. From this, a clear relationship between the C:S ratio and the concentration of hydrogen sulfide in biogas was established. Two methods were used to calculate the predicted H_2S–S biogas (PSB) content (%): PSB*max* and PSB*bio* (Fig. 8.2), with the former overestimating the actual hydrogen sulfide concentration in biogas while the latter closer to the actual measured values. However, it is recommended that using PSB*max* is a better safety indicator for digester operation and biogas management and is simpler to calculate. The model showed that hydrogen sulfide in biogas could be kept to <2% (v/v) if the C:S ratio was maintained at 40 or above.

8.2.1 *Adsorption*

Desulfurization of biogas is generally achieved by adsorption onto hydrated ferric oxide, although a wide range of other adsorbents is available. Suitable adsorbents have a high specific surface area and can operate at low temperatures, such as activated carbon, which is highly effective and widely used but has high operation costs. Currently, research is being carried out into low-cost and natural alternatives, including wood, coal, peat, and lignite or industrial byproducts such as fly

Table 8.4. The advantages and challenges of using advanced *in-situ* desulfurization methods (Jung *et al.*, 2022). Reproduced with permission of Elsevier Publishing.

Technology	Advantages	Challenges
Electrochemical oxidation	Chemical-free and cost-effective Faster than chemical or biological oxidation High energy efficiency	Passivation of electrode surfaces by S° Regeneration of sulfide from S° by electrode biofilms Need to develop sustainable and cost-effective electrode materials
Iron cycling	Possibility of elemental sulfur recovery Easy regeneration and reuse of iron granules Increased removal capacity of regenerated granules Additional sulfide removal by FeS precipitation Suppression of sulfate-reducing activity	Lack of study in organic-rich, active methanogenic environments Little understanding of how iron granules influence microbial redox reactions and sulfur metabolism
Sulfide oxidation coupled with denitrification	Simultaneous removal of sulfide, nitrogen, and carbon Mitigation of inhibitory effects of sulfide and nitrate Minimal material requirements Simple operation	Need to remove S° from digesters continually to prevent sulfide regeneration High sensitivity to environmental variations Low biomass yield
Stimulation of electric syntrophy	Possibility of recovering S° Possible enhancement of methanogenesis Simple operation	Very limited information Little understanding of the underlying mechanism Unpredictable complexity of electro-syntrophic associations in methanogenic systems

ash, slag, bauxite sludge, and sludge ash (Żarczyński *et al.*, 2014). However, ferric oxide and activated carbon remain the commonest desulfurization absorbents.

Ferric oxide is widely available as a commercially prepared media using $Fe_2O_3.H_2O$. The media reacts with the biogas, which is usually moist, to remove the hydrogen sulfide as ferric sulfide:

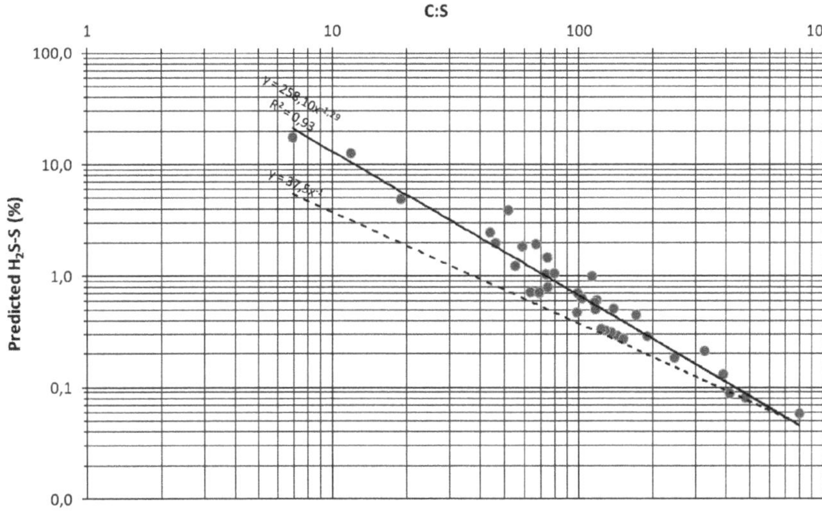

Figure 8.2. Predicted H_2S–S biogas (PSB) content (%) over a range of different C:S ratios (g/g) using two different methods, PSB*max* (solid line) and PBS*bio* (dotted line) (see Table 8.5), based on a wide range of different substrates tested by Peu *et al.* (2012). Reproduced with permission of Elsevier Publishing.

$$Fe_2O_3.H_2O + 3H_2S \rightarrow Fe_2S_3 + 3H_2O \quad (8.4)$$

Any light mercaptans (methyl and ethyl) present are also removed:

$$Fe_2O_3.H_2O + 6RSH \rightarrow Fe_2(RS)_3 + 3H_2O \quad (8.5)$$

The optimum temperature for desulfurization using ferric oxides is 25–50°C at pH 8.0–8.5, and as the reaction requires water, the biogas should not be too dry. Stoichiometrically, 1 kg $Fe_2O_3.H_2O$ absorbs 0.64 kg H_2S, although, in practice, the process is about 90% efficient. Biogas hydrogen sulfide concentrations of <1 ppm are achievable, with 99.98% removal possible even at raw hydrogen sulfide gas concentrations of 1,000 ppm. Once exhausted, the absorbent is regenerated using air or oxygen to oxidize the Fe_2S_3 back to its original state, although this is a highly exothermic reaction:

$$2\,Fe_2S_3 + 3O_2 \rightarrow Fe_2O_3.H_2O + 6S \quad (8.6)$$

Eventually, adsorption sites will be fully taken up with elemental S, and the absorbent will then need to be replaced. Commercial oxide

Table 8.5. Characteristics of 37 different substrates (feedstocks) for anaerobic digestion used in the calculation of potential hydrogen sulfide biogas content, with details of total solids content (TS), volatile solids content (VS), total sulfur, total carbon, crude proteins (CP), biochemical methanogenic potential (BMP), biochemical biogas potential (BBP), carbon biodegradability (CB), carbon to sulfur ratio (C/S), and predicted H_2S–S biogas content (PBS), with PBS*max* measured using total S analysis and biogas production and PBS*bio* determined using the molar ratio between total S and total C using the method of Peu *et al.* (2012). Reproduced with permission of Elsevier Publishing.

	TS (g/kg fd)	VS (g/kg fd)	Total S (g S/ kgTS)	Total C (g C/ kg TS)	CP (g/kg TS)	BMP ($_N$LCH_4/ kg VS)	BBP ($_N$L Biogas/ kg fd)	CB (%)	C/S (g/g)	PBS*max* (%)	PBS*bio* (%)
Feedstock from urban wastewater treatment plant											
Primary sludge	52.8	44.5	4.5	339.7	nd*	nd	nd	nd	74.7	—	—
Biological sludge	57.0	45.8	8.6	505.3	493.4	255.5	18.6	34.6	58.6	1.9	0.6
DAF sludge	47.6	36.3	8.4	576.5	154.0	806.1	39.4	76.9	68.7	0.7	0.5
Feedstock from agricultural resources											
Cover crops											
Camelina crops	213.0	178.0	4.0	394.4	132.9	234.3	84.2	53.7	98.6	0.7	0.4
White mustard crops	228.0	194.0	3.3	390.4	100.9	222.5	86.4	52.0	117.6	0.6	0.3
Brown mustard crops	173.0	154.0	3.7	514.5	98.3	264.1	87.5	52.7	137.9	0.5	0.3
Radish crops	301.0	126.0	3.4	245.8	62.3	236.7	67.6	49.0	73.2	1.0	0.5
Soybean crops	249.0	84.0	2.4	152.6	78.8	305.0	58.2	82.0	63.3	0.7	0.6
Animal wastes											
Farm manure	225.2	209.9	1.2	297.3	71.3	271.5	103.0	82.4	245.3	0.2	0.2
Dairy cow slurry (1)	31.0	26.0	2.8	477.4	141.1	326.9	13.4	48.5	170.5	0.5	0.2
Dairy cow slurry (2)	106.9	79.2	3.8	429.3	140.9	238.7	28.3	33.0	112.7	1.0	0.3

Pig faeces (1)	362.0	280.0	3.1	360.0	147.8	304.0	135.1	55.5	115.4	0.6	0.3
Pig faeces (2)	390.0	320.0	3.9	398.0	124.8	328.0	166.6	57.5	102.4	0.6	0.4
Pig slurry (1)	43.2	30.4	8.0	349.3	172.2	230.0	9.8	34.8	43.6	2.5	0.9
Pig slurry (2)	38.5	25.0	8.0	411.9	204.0	140.2	5.5	18.6	51.5	3.9	0.7
Feedstock from agri-food facilities											
Industrial wastewater treatments											
Biological sludge PS**	60.3	47.8	8.3	553.0	130.3	259.5	18.0	28.9	66.6	1.9	0.6
Biological sludge DMI***	69.2	40.0	7.0	319.4	445.2	282.8	16.9	41.0	45.8	2.0	0.8
Biological sludge PI****	61.6	41.3	4.7	348.4	456.8	205.9	13.7	34.2	74.2	1.5	0.5
Screening refusal PS	328.6	310.8	5.1	500.9	338.7	545.7	247.7	80.6	97.9	0.5	0.4
Screening refusal CS*****	176.2	163.2	2.8	531.4	227.5	490.2	119.2	68.2	189.5	0.3	0.2
Fatty effluents											
Grease trap waste CS	237.7	231.6	1.0	765.9	42.6	863.4	271.7	79.9	797.9	0.1	0.0
Grease trap waste PI	361.3	350.0	1.5	702.3	67.7	1011.5	454.8	96.0	478.9	0.1	0.1
Grease trap waste PS	557.1	552.6	1.5	609.2	102.5	902.9	648.3	102.3	412.0	0.1	0.1
Vegetable residues											
Carrot pulp	177.1	166.7	1.1	447.2	93.1	330.6	107.7	72.8	389.0	0.1	0.1
Onion pulp	211.5	194.6	3.0	434.5	122.5	416.7	153.1	89.2	142.6	0.3	0.3
Shallot pulp	226.7	212.7	3.4	456.0	144.6	413.3	169.9	88.0	134.8	0.3	0.3

(Continued)

Table 8.5. (*Continued*)

	TS (g/kg fd)	VS (g/kg fd)	Total S (g S/kgTS)	Total C (g C/kg TS)	CP (g/kg TS)	BMP ($_{N}LCH_4$/kg VS)	BBP ($_{N}$L Biogas/kg fd)	CB (%)	C/S (g/g)	PBS*max* (%)	PBS*bio* (%)
Others											
Rumen content CS	166.3	155.4	1.5	493.6	124.2	341.1	82.7	54.0	325.6	0.2	0.1
Meat waste PI	330.3	293.9	4.9	626.5	332.2	871.0	346.9	89.8	127.3	0.3	0.3
Bovine blood CS	155.8	145.1	6.8	505.1	925.0	454.9	92.3	62.8	74.5	0.8	0.5
Pig mucus PS	190.5	169.1	7.7	526.0	695.1	579.5	141.6	75.7	68.6	0.7	0.5
Pig bristles PS	285.5	278.4	26.0	492.5	904.2	277.3	105.5	40.2	18.9	4.9	2.0
Fish viscera	418.0	380.0	4.6	568.6	269.6	657.9	403.0	90.8	122.6	0.3	0.3
Feedstock from municipal waste											
Canteen refusals	271.3	257.5	3.6	545.1	244.2	570.8	249.6	90.4	151.4	0.3	0.2
Dehydrated canteen refusals	975.4	827.1	4.0	464.2	305.0	395.3	535.9	63.4	116.2	0.5	0.3
Harvested red seaweeds	200.9	120.2	23.3	276.2	136.9	133.1	25.6	24.7	11.9	12.8	3.2
Harvested green seaweeds	222.0	128.0	29.6	203.4	141.0	132.8	26.0	30.8	6.9	17.7	5.5
Lawn mowing	238.2	187.5	5.5	438.2	220.4	289.6	86.7	44.5	79.2	1.1	0.5

*Not determined.
**Pig slaughterhouse.
***Dairy milk industry.
****Pig industry.
*****Cattle slaughterhouse.

pellets are widely available and can be used in a variety of different desulfurization reactor designs. Wood or ceramic granules or pellets are coated with Fe_2O_3 or Fe_3O_4 and used in low-pressure vessels, with the biogas passing through in a down-flow direction. Ceramic-based pellets are possibly better due to their stability and uniform size, optimizing gas flow contact and ease of handling. Examples of commercial desulfurization adsorbents are SulfaTreat 410-HP, Sulfa-Bind, Sulfur-Rite, and BioBond. Regeneration is generally achieved by blowing air through the media, in which case at least two units will be required for continuous operation as one unit is taken out of operation (i.e., batch regeneration). Alternatively, a small amount of air (1%) can be added to the biogas so long as the hydrogen sulfide concentration is not too high (i.e., continuous regeneration). Where high hydrogen sulfide concentration is present, batch regeneration is required. Some media are non-flammable, such as Sulfa-Bind, eliminating the risk of combustion during the regeneration phase.

Dry scrubbers are used for small to medium biogas flows and hydrogen sulfide contents, with the media housed in compact vertical cylindrical vessels made out of HDPE or stainless steel. Chemical or biological scrubbers are generally used for higher biogas flow rates. The dry scrubbing process is often referred to iron sponge scrubbing, and some basic design criteria are given in Table 8.6 (Zicari *et al.*, 2003). Some research has suggested that the spent media could be used as a soil amendment or in brick manufacture or road construction. Currently, the main disposal route is landfill.

Activated carbon is extremely effective in removing hydrogen sulfide by adsorption while also allowing its oxidation to elemental sulfur or sulfate by exposing the carbon to air or oxygen.

$$2H_2S + O_2 \rightarrow 0.25S_8 + 2H_2O \tag{8.7}$$

Normal practice is to inject a small amount of oxygen into the biogas stream, although the activated carbon can be periodically removed and left exposed to the air for oxidation to occur. Biogas must be free from dust and moisture before treatment. Improved efficiency in desulfurization by activated carbon can be achieved by using alkaline or oxide coatings, which can increase hydrogen sulfide removal capacity from 10 to 20 kg H_2S m^{-3} for normal carbon to 120–140 kg H_2S m^{-3} for coated carbon. However, it is not feasible to regenerate all coated activated carbon medium, which has to be landfilled. SULFUSORB® GAC, manufactured by Calgon Carbon Corporation, is impregnated with copper oxide, and like other metal-oxide-coated granular activated carbon (GAC), it is possible to regenerate the medium by heat treatment.

Table 8.6. Basic design criteria for an iron sponge scrubber adapted from (Zicari *et al.*, 2003).

Design parameter	Recommendation
Tower/column	Stainless steel construction; two columns in series preferred
Biogas feed	Downward flow to maintain bed moisture Use column with the most contaminated medium first
Residence time	Biogas residence time >60 seconds
Temperature	18–40°C to optimize reaction kinetics; prevent medium drying
Tower/column height	Minimum 3 m for H_2S removal; 6 m for mercaptans
Superficial gas velocity	Liner velocity 0.6–3 m min^{-1}
Surface mass loading rate	<10 g S min^{-1} m^{-2}
Humidity	Optimum 40% (range 25–55%)
pH	8–10; use sodium carbonate supplement
Pressure	Minimum 140 kPa

8.2.2 *Absorption*

Hydrogen sulfide can also be removed from biogas by absorption in a solvent, most commonly water, but also methanol, propylene carbonate, and ethers of polyethylene glycol, preferably under alkaline conditions. Two columns are operated in series: the absorber or scrubber, followed by the regeneration vessel. Here, the hydrogen sulfide is stripped from the biogas into the solvent, usually water, which then requires it to be treated, which makes it ideal for use at wastewater treatment plants. High biogas flow rates can be treated at 100–10,000 m^3h^{-1}, with up to 99% H_2S removal. The resultant contaminated waste stream may contain up to 20–30 g m^{-3}. Where water scrubbing is

Figure 8.3. Water scrubber media is designed to maximize hydrogen sulfide and carbon dioxide absorption into water. Shown here is the Tellerette® 2R medium (70 × 25 mm). Reproduced with permission of Verantis Environmental Solutions Group, OH, USA. https://www.verantis.com/products/tellerette-packing/tellerette-packing

used, the hydrogen sulfide is removed with pressurized water, which also removes the carbon dioxide present. Water is ideal as a solvent, as both hydrogen sulfide and carbon dioxide are more soluble than methane. To ensure maximum contact between the biogas moving upward through the column and the water moving downward, the column is packed with a medium, ensuring a large surface area for gaseous absorption to take place. For example, Tellerette™ media come in a range of sizes from 46 mm (width) × 19 mm (height) up to 105 mm × 56 mm (Fig. 8.3) and is designed to create numerous droplets that collide with the next piece of media, causing it to burst, creating new surfaces and maximizing water gas contact, hence absorption. The structural design of the media is more important than the actual surface area, which is more important in biofilm reactors. If the water is to be reused, which is often not the case, where the final effluent is used at wastewater treatment plants (WWTPs), then it must go through an air or steam stripper to remove the contaminants.

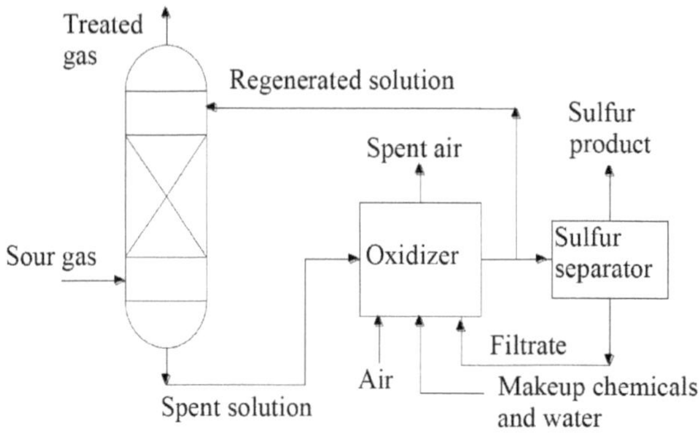

Figure 8.4. Schematic layout of the LO-CAT® system, which removes hydrogen sulfide from biogas using chemical absorption by catalytic oxidation using a chelated iron salt solution. https://merichemtech.com/sulfur-recovery-with-lo-cat/

8.2.3 *Catalytic Scrubbing*

Chelated iron solutions are widely used in a range of patented liquid redox systems to convert hydrogen sulfide to elemental sulfur. They can operate at low temperatures and employ continuous regeneration, with removal efficiencies >99.9%. The system uses a proprietary catalyst solution, which is then regenerated — an example of which is the LO-CAT®system from Merichem Technologies. Iron ions are bound to organic-chelating agents, which remove hydrogen sulfide by the reduction of ferric chelated iron (Fe^{3+}) into ferrous chelated iron (Fe^{2+}), both of which are in aqueous solution. The ferrous iron is subsequently re-oxidized to ferric chelated iron in the regeneration stage by air stripping (Fig. 8.4).

$$\textit{Purification}: H_2S + 2[Fe^{3+}] \rightarrow 2[Fe^{2+}] + S + 2H^+ \qquad (8.8)$$

$$\textit{Regeneration}: 2[Fe^{2+}] + 0.5O_2 + 2H^+ \rightarrow 2[Fe^{3+}] + H_2O \qquad (8.9)$$

8.2.4 *Biological Oxidation*

Where hydrogen sulfide is in low concentrations, then it can be removed *in situ* by bacterial action using microaeration. Sulfur oxidizing bacteria (SOB) are predominantly chemoautotrophs, Gram-negative, and of the genera *Thiobacillus, Thiothrix, Sulfolobus,*

Thermothrix, and *Beegiatoa*, which oxidize hydrogen sulfide to elemental sulfur, which is then deposited into their cells, forming characteristic granules that are easily seen using phase contrast microscopy.

$$H_2S + 0.5O_2 \rightarrow S^0 + H_2O \ \Delta G^{0\prime} = -218 \text{ kJ mol}^{-1} \quad (8.10)$$

The energy produced by this reaction is used to fix the carbon dioxide required for cell growth. SOBs are microaerophilic, which means that they are found in areas of very low oxygen tension, ensuring abundant reduced compounds but just enough oxygen for metabolism. Most favorable growth conditions mimic their natural habitat, such as marsh sediments. SOB species vary in their optimum pH and temperature and so can work successfully in many different treatment scenarios. Filamentous forms include *Beggiatoa* and *Thiothrix*, while the rod-shaped SOBs belong to the genus *Thiobacillus*. In 2000, a reclassification of the species belonging to the genus *Thiobacillus* was carried out, with some reclassified into the new genera *Acidithiobacillus* and *Halothiobacillus* (Kelly and Wood, 2000), so *Thiobacillus thiooxidans* is now *Acidithiobacillus thiooxidans*. While all SOB are obligate aerobes, *Acidothiobacillus denitrificans* is actually a facultative anaerobe, which can replace oxygen with nitrate as its terminal electron acceptor to convert hydrogen sulfide to sulfate.

Biological desulfurization falls into three different treatment processes: biofilters, biotrickling filters, and bioscrubbers (Fig. 8.5), with major advantages over other treatment desulfurization methods, including lower operational costs and reduced use of chemicals (although nutrients are sometimes required), no regeneration step, and no polluting end products requiring disposal. There are a number of different biological outcomes using SOB, with the oxygen saturation concentration controlling the production of elemental sulfur or sulfate ions (Pudi *et al.*, 2022):

$$H_2S + H_2O \rightarrow HS^- + H_3O^+ \quad (8.11)$$

$$H_2S + 2O_2 \rightarrow SO_4^{2-} + 2H^+ \quad (8.12)$$

$$HS^- + 0.5O_2 \rightarrow S^0 + H_2O \quad (8.13)$$

$$S^0 + H_2O + 0.5O_2 \rightarrow SO_4^{2-} + 2H^+ \quad (8.14)$$

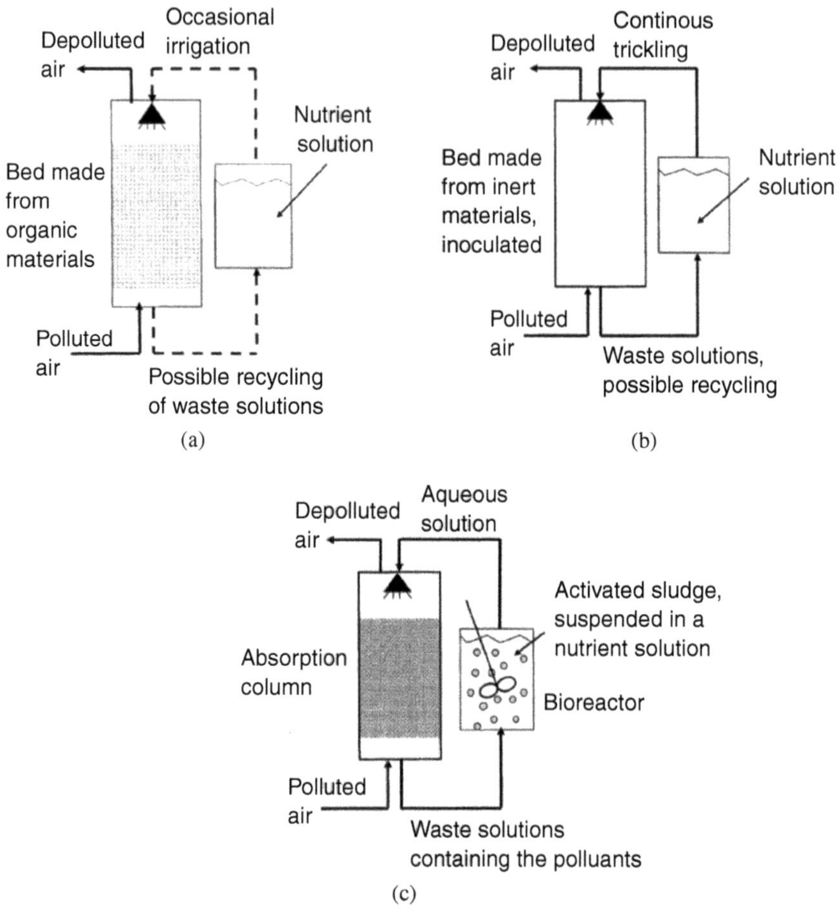

Figure 8.5. Schematic view of biofiltration systems used for the desulfurization of biogas: (a) biofilter, (b) biotrickling filter, and (c) bioscrubber (Pudi *et al.*, 2022). Reproduced with permission of Elsevier Publishing.

Biofiltration

Both biofilters and biotrickling filters employ a random or modular medium on which a biofilm develops, comprised primarily of SOB. Air must be made available as the bacteria are aerobic, and a controlled amount is normally injected into the biogas stream to give 4% dissolved oxygen saturation. Biofilters use organic media, which provides the necessary nutrients (Fig. 8.5a), whereas biotrickling filters use plastic media, requiring nutrients to be added separately in the recycled process water (Fig. 8.5b). A wide variety of other materials has been used as media, including compost, wood chips, charcoal, larva, zeolites, and ceramics (Pokorna *et al.*, 2015).

As the hydrogen sulfide is broken down into sulfate, H_2SO_4 is also formed. While the reactor is generally operated under acidic conditions (pH 1–3), some buffering is required, which, in the case of biofilters, is achieved by adding alkaline material to the bed or using a medium that is alkaline. In biotrickling filters, pH control is easier as the recycled process water can be buffered by chemical addition. Among the commercial units available is the Biopuric® process from Biothane and various systems from BioGlasclean, such as QSR®(quick sludge remover), MBR (moving bed reactor), and MUW® (make-up water). Generally, the removal efficiency of hydrogen sulfide by biofilters and biotrickling filters is between 90 and 99%. The majority of the hydrogen sulfide is oxidized to elemental sulfur, which builds up in the biomass and requires the medium to be flushed every few months. To reach the required standards for biomethane, the treated biogas may need polishing by passing through an activated carbon adsorption vessel (Allegue and Hinge, 2014).

Bioscrubbing

A bioscrubber consists of an absorption tower where hydrogen sulfide is absorbed into water or alkaline solution, followed by an aerobic treatment phase, where the sulfur is removed with the effluent recycled back to the absorption tower. The key advantages are that there is no injection of air or oxygen into the biogas and that bioscrubbing has high hydrogen sulfide removal efficiency (99%) (Fig. 8.5c).

The Biogasclean MBR™ is a moving biological reactor where the gas bubbles ensure that the floating medium is kept in suspension and that the digested wastewater used as the liquid phase is kept fully mixed. The biogas is compressed, cooled, and injected with air and released via diffusers at the base of the reactor, where it mobilizes the media-rich in SOB, oxidizing the hydrogen sulfide to elemental sulfur. The process is specifically designed for treating biogas from ethanol distilling, which has a high hydrogen sulfide concentration of between 20,000 and 30,000 ppm and uses significantly less oxygen

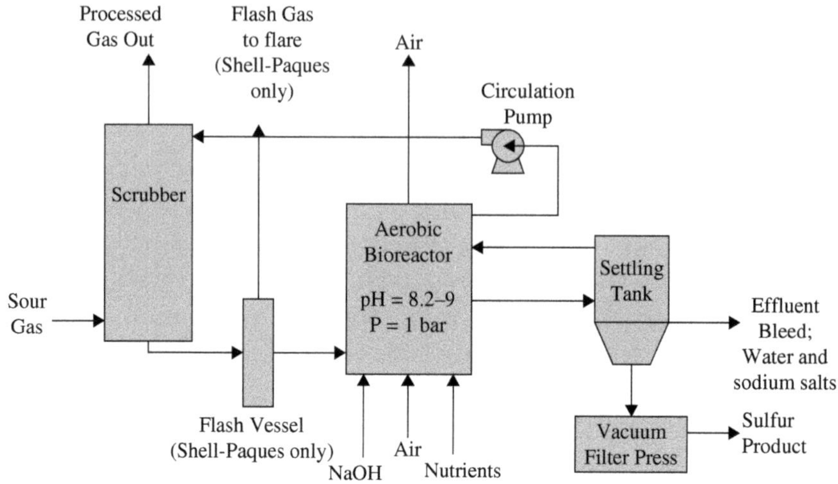

Figure 8.6. (a) Layout of a chemical scrubber to remove the hydrogen sulfide using a counter flow of alkaline wash water. (b) Example of a chemical scrubber using a counter flow of sodium hydroxide. Reproduced with permission of Biogas Products Ltd. www.biogasproducts.co.uk

(25%) than biofilters. Biological desulfurization techniques and the microbiology of SOB have been reviewed by Pudi *et al.* (2022).

Caustic scrubbers remove the hydrogen sulfide using a counter flow of alkaline wash water (pH 8.2–9.0). The biogas enters the bottom of a tower packed with medium to ensure maximum liquid-gas contact, while recirculated NaOH is sprayed downward through the medium, absorbing the hydrogen sulfide (Fig. 8.6a). The wastewater leaving the scrubber enters the completely mixed biological reactor, where sulfur bacteria convert the dissolved sulfide into elemental sulfur and sodium hydroxide, thus restoring the caustic nature of the wash water (Fig. 8.6b).

$$\text{Scrubber: } NaOH + H_2S \rightarrow NaHS + H_2O \tag{8.15}$$

$$\text{Bioreactor: } NaHS + 0.5O_2 \rightarrow S_0 + NaOH \tag{8.16}$$

A second reaction also occurs in the bioreactor, with <5% of the NaHS converted to sodium sulfate:

$$2NaHS + 4O_2 \rightarrow NaHSO_4 \rightarrow Na_2SO_4 + H_2SO_4 \tag{8.17}$$

This requires a small addition of sodium hydroxide to maintain the optimum alkaline pH. An example of a commercial system is THIOPAQ™ by Paques.

8.2.5 *Metal Sulfide Precipitation*

Hydrogen sulfide in biogas can be employed to remove and recover metals from waste streams as highly insoluble salts even at very low concentrations, offering the potential not only for enhanced environmental protection but also the recovery of currently lost valuable materials.

Metal wastewaters are among the most difficult to treat, having low pH, low COD, and high concentration of metals, requiring expensive physicochemical treatment, principally ion exchange, adsorption, and membrane filtration, with chemical precipitation, coagulation-filtration, electrolysis, flotation, or reverse osmosis also used (Fu and Wang, 2011). Therefore, they are often discharged into the sewer, with the resultant metals ending up in the sludge. Metals are also a common component of sewage often associated with personal care products. There is increasing interest in the recovery of metals from a wide range of wastewaters, especially those with high concentrations, as part of the circular economy approach to waste management.

The commonest metals in wastewater are zinc, copper, nickel, lead, chromium, cadmium, and mercury. While copper and zinc are both essential trace elements for human health, excessive amounts are toxic. All these metals can cause severe health problems, with nickel carcinogenic, mercury a neurotoxin, and lead causing damage to the central nervous system. They all have both acute and chronic toxicities to a wide range of aquatic species.

Compared to the physicochemical treatment options, sulfide precipitation using SRB offers a range of advantages, including the ability to form insoluble metal sulfides at low pH (2.5–3.0), high settling velocities of the sulfides, excellent thickening and dewatering of metal sludges, and low capital and operational costs. Selective metal recovery is possible by controlling the pH (Table 9.7), although this is generally used when treating acid mine drainage rather than municipal wastewaters (Bijmans *et al.*, 2009; Sampaio *et al.*, 2009). SRB use sulfate ions as terminal electron acceptors under anaerobic conditions, utilizing simple organic compounds (CH_2O) as an electron donor (Eq. (9.3)). Sulfides are produced, which react with metals (M^{2+}) present to form insoluble metal sulfides (MS), which are very stable (Eq. (9.4)). During these reactions, alkalinity is released into the wastewater (Eq. (9.5)).

It is important that metal sulfides are fully recovered from waste streams and not allowed to enter the watercourse, where they can be environmentally damaging. The sulfide precipitate is made up of very small particles, which, like nanoparticles, have high specific surface areas and are very reactive (Sec. 9.2.2).

Sulfate reduction and metal precipitation can be carried out, either in single reactors or multi-stage reactors where these steps are separated. The latter is more efficient as direct contact between metals and the biomass is avoided. Down-flow fluidized beds and air lift reactors are most efficient, with removal efficiencies >98% for most metals (Table 9.9) (Kumar *et al.*, 2021).

8.2.6 *In-situ Sulfide Control*

Metal salts (chlorides, phosphates, and oxides) can be added to reactors, mainly digesters, to control hydrogen sulfide in biogas. They react with the hydrogen sulfide to form insoluble iron sulfides, thereby removing a significant portion of soluble hydrogen sulfide (Kiilerich *et al.*, 2017). Although zinc and copper can be used, iron salts are predominately employed and, in particular, ferrous chloride ($FeCl_2$), which is highly effective and has a low toxicity:

$$\underset{\text{Ferrous chloride}}{FeCl_2} + H_2S \rightarrow FeS + 2HCl \tag{8.18}$$

$$\underset{\text{Ferric chloride}}{2FeCl_3} + 3H_2S \rightarrow 2FeS + S + 6HCl \tag{8.19}$$

$$\underset{\text{Ferric hydroxide}}{2Fe(OH)_3} + H_2S \rightarrow 2Fe(OH)_2 + S + 2H_2O \tag{8.20}$$

$$\underset{\text{Ferrous hydroxide}}{Fe(OH)_2} + H_2S \rightarrow FeS + H_2O \tag{8.21}$$

$$\underset{\text{Ferrous oxide}}{FeO} + H_2S \rightarrow FeS + H_2O \quad (8.22)$$

$$\underset{\text{Cupric oxide}}{CuO} + H_2S \rightarrow CuS + H_2O \quad (8.23)$$

$$\underset{\text{Zinc oxide}}{ZnO} + H_2S \rightarrow ZnS + H_2O \quad (8.24)$$

Zero-valent iron has also been found to be highly effective (Al Mamam and Torii, 2015; Su *et al.*, 2015). However, metal salts used *in situ* cannot reduce hydrogen sulfide concentrations low enough to meet current standards for national grid or vehicle specifications on their own, so an additional desulfurization step will be required. The constant addition of metal salts is expensive and an additional operational factor, although savings can be made by reducing the amount of hydrogen sulfide that ultimately ends up in the biogas, requiring further treatment (Table 8.3).

There is growing interest in iron dosing of the sewer network or inlet works of WWTPs that reduces hydrogen sulfide formation, enhances phosphate removal, and improves sludge dewaterability. In a major study, Kulandaivelu *et al.* (2020) compared the effects of dosing the wastewater at the WWTP inlet (Trail 1) and within the sewer network (Trail 2) with ferrous chloride. They found that COD removal in the primary settlement phase was increased by 49 and 47% in Trials 1 and 2, respectively, with 76 and 53% increase in phosphate removal in the biological nutrient removal activated sludge system, respectively. Hydrogen sulfide was reduced by 36 and 45% during anaerobic digestion in Trials 1 and 2, respectively, with sludge dewaterability improved by 10% in both trials. The head space in the sewer network showed significant reductions in hydrogen sulfide concentration. The idea of using iron salts for multiple functions within the wastewater treatment cycle rather than simply for desulfurization for biogas upgrading is an example of the modern sustainability approach to WWTP operation and management.

Air or oxygen dosing to the biogas digester headspace or gas storage tank allows biological aerobic oxidation of the hydrogen sulfide to occur, with 80–99% reductions recorded. The process is known as microaeration and has already been explored above in conjunction with other systems. Oxygen acts as an electron acceptor for the oxidation of hydrogen sulfide to elemental sulfur, sulfate, or thiosulfate:

$$H_2S + O_2 \rightarrow S^0 + 2OH^- \tag{8.25}$$

$$H_2S + 2O_2 \rightarrow SO_4^{2-} + 2H^+ \tag{8.26}$$

$$4H_2S + 5O_2 \rightarrow 2S_2O_3 + 4H_2O \tag{8.27}$$

Thiosulfate is the main product of anaerobic sulfide oxidation, which can subsequently be reduced to sulfide again by sulfate reducing bacteria. This tends to occur at high hydrogen sulfide concentrations due to the inhibition of SOB. Elemental sulfur is the main product of biotic activity by SOB occurring at hydrogen sulfide concentrations of <5 mM. Microaeration is an effective technology, and although five times more air is required than pure air, using air increases the risk of explosion when mixed with methane. Mixing is very important to ensure maximum contact between oxygen and the sulfide, so the location of the air or oxygen inlet in the reactor is critical (Jung *et al.*, 2022). However, residual oxygen levels of 0.5 to 1.8% remain, which may exceed required biomethane standards, with the methane content of biogas also diluted due to the presence of oxygen. Oxygen is a well-documented inhibitor of methanogenesis as well as increasing the risk of explosion.

Other less successful methods of *in-situ* treatment include adsorption and the suppression of SOB by, for example, operating at a more alkaline pH (Table 8.3), but neither are as reliable or successful as the previous methods. Current research on alternative *in-situ* methods of desulfurization has focussed on electrochemical oxidation, iron cycling, and sulfide oxidation coupled with denitrification and stimulation of electric syntrophy (Table 8.4), all of which have been reviewed by Jung *et al.* (2022).

8.3 Carbon Dioxide Removal Technologies

Biogas contains, on average, 35% carbon dioxide (typical range 14–48%), with some removed during desulfurization. However, to optimize the calorific value of the biogas, most of the carbon dioxide must be removed to <2.5% (Table 8.1). There are numerous carbon dioxide removal methods, including membrane filtration, pressure swing adsorption (PSA), absorption, both with and without chemical reaction, cryogenic removal, and by adding propane (Hagen *et al.*, 2001).

Membrane separation takes advantage of the differences in permeability of methane and carbon dioxide as well as hydrogen sulfide, which can also be removed by a membrane, usually in the form of hollow fibers made from polymers, such as acetate-cellulose. Operating at a partial pressure of between 25 and 40 bar, the retained enriched methane is also at high pressure, usually eliminating the need for further compression before supply to the gas grid. The permeability of hydrogen sulfide is 60 times greater than methane, while carbon dioxide is 20 times higher, making membrane filtration ideal for both desulfurization and carbon dioxide removal. However, as methane purity increases, the methane yield falls. Methane purity can be increased by using a number of membranes in series, but this also coincides with methane permeating through the membrane (Fig. 8.7). To a certain extent, this can be minimized by recirculating a portion of the permeate (i.e., the carbon-dioxide-enriched gas) through the last membrane for a second time. Alternatively, only one or two membranes can be used, and the lower quality is brought up to standard by the addition of propane.

PSA employs four adsorber tanks filled with zeolites or granular activated carbon, with each tank operating alternating cycles of (i) adsorption, (ii) depressurization regeneration (desorption under vacuum), and (iii) repressurization. The biogas enters the vessel at the base, and as it passes through the medium, carbon dioxide, oxygen, and nitrogen are all adsorbed, and the biogas collected from the top has methane purity in excess of 97%. Before purity levels fall, the adsorber is closed, and the untreated biogas enters the next vessel, where the medium has already been regenerated and repressurized. The first vessel is now depressurized to atmospheric pressure and then evacuated by a vacuum pump, regenerating the medium. The vessel is then repressurized, ready to commence the cycle once again. By using tanks in series, a continuous process can be maintained (Fig. 8.8).

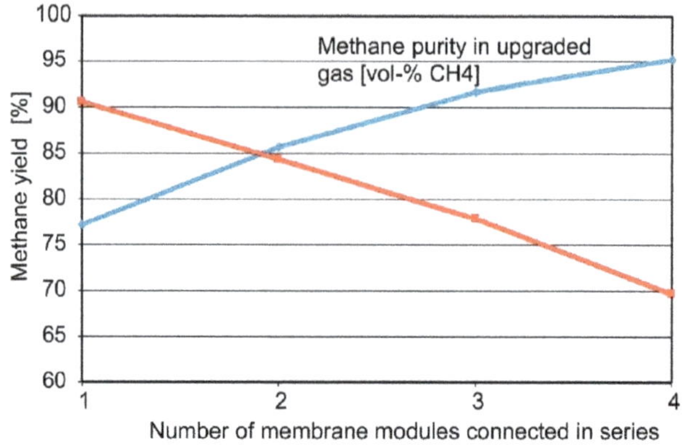

Figure 8.7. The effect of using membranes in series to improve methane purity (blue line) on methane yield (red line) using recirculation (Hagen *et al.*, 2001). Reproduced with permission of the Swedish Gas Center, Malmo, Sweden. info@sgc.se/Joakim.lundgren@ltu..se

The most common absorption method is using a counterflow water scrubber packed with a random medium to ensure maximum gas–water contact. The gas is pressurized before entering the base of the scrubber to increase the solubility of the carbon dioxide; however, methane is also partially soluble in pressurized water, so some will be lost in the wash water. This is recovered using a flash

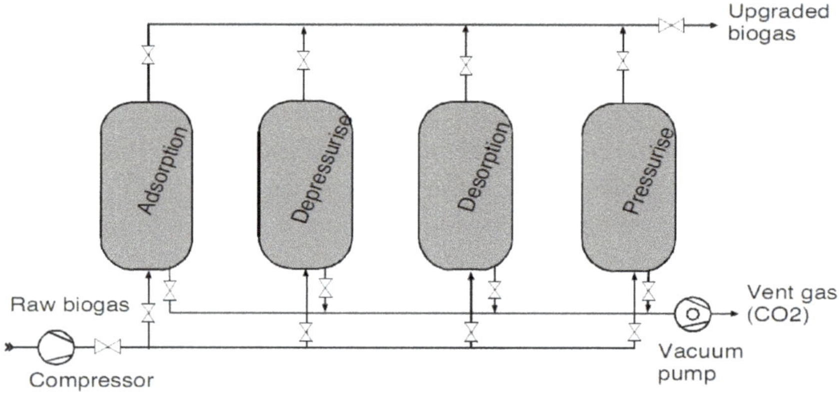

Figure 8.8. Pressure swing adsorption (PSA) is widely used for the removal of carbon dioxide from biogas.

tank, where the wash water is partially depressurized, so that when the methane comes out of the solution, it can be returned to the incoming biogas stream. The wash water is then fully depressurized to atmospheric pressure and passed downward through the desorption (regeneration) column, with air pumped up from the base, which strips out the carbon dioxide. The water is then reheated to the required temperature before recycling back to the absorption column (Fig. 8.9). Desulfurization absorption using water also removes carbon dioxide, but the reverse only works with low hydrogen sulfide concentrations and requires a different operational approach to regeneration due to the build-up of sulfur on media.

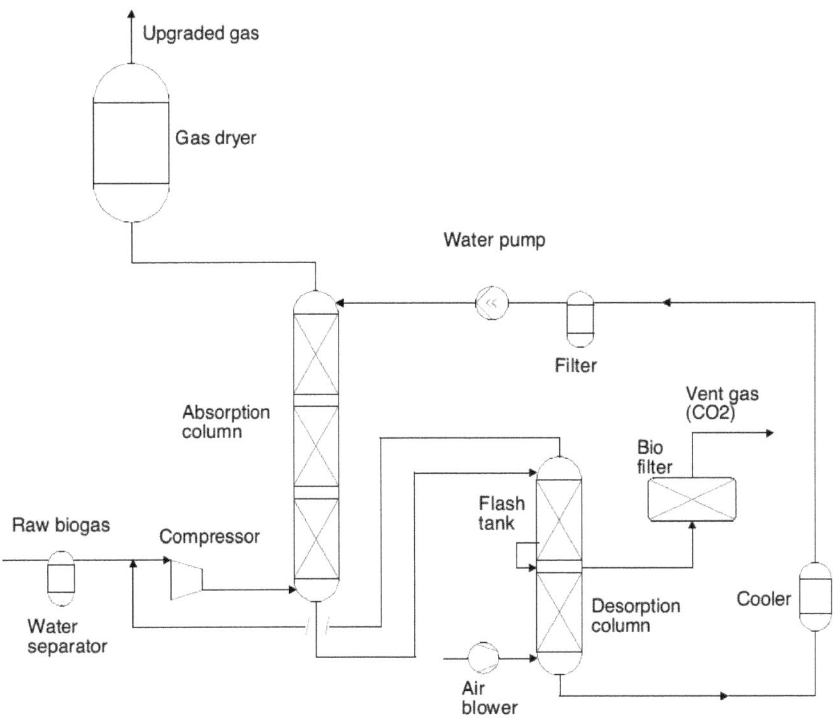

Figure 8.9. Water scrubbers are used to reduce carbon dioxide concentration in biogas. The system comprises an absorption column, followed by a regeneration column comprising a flash separator to recover any dissolved methane before the desorption section, where carbon dioxide is recovered or vented (Hagen *et al.*, 2001). Reproduced with permission of the Swedish Gas Center, Malmo.

One advantage of the process is that treated effluent can be used instead of clean water without the need to treat and recirculate. This can be advantageous in some treatment processes where recirculation is normally employed. However, the methane removed by the pressurized water during absorption will have to be recovered by depressurization in a flash tank before return to the WWTP. Carbon dioxide can also be removed by chemical action, or cryogenically (Hagen *et al.*, 2001).

8.4 Removal of Water

Biogas is saturated with water vapor, approximately 5% at 35°C, and much of this condensates in the pipework as it cools to ambient temperature. It can be collected by ensuring the pipework has a very small slope (10 mm m^{-1}), allowing the condensate to fall back into drip traps. If moisture remains a problem, then chilling may be required, lowering the dew point to 0.5–1.0°C. Hydrogen sulfide is also partially removed in the condensate. There are alternative adsorption and absorption methods using drying agents, but these tend to be expensive and are not always successful.

8.5 Removal of Other Impurities

Other gases found in biogas are hydrogen (0–5%), carbon monoxide (0–2.1%), nitrogen (0.6–7.5%), and oxygen (<0.1%) although generally found in trace amounts and so require no removal unless they exceed required thresholds. However, ammonia, volatile organic compounds (VOCs), including terpenes, and siloxanes may need to be removed when biogas is upgraded to biomethane for off-site energy use (Table 8.1). Particulate matter is removed using simple filters.

Ammonia: Ammonia is generally removed during desulfurization using adsorption or a water scrubbing process. Activated carbon is widely used for ammonia removal, but large-scale industrial units rely on acid washing with weak sulfuric or nitric acid.

Oxygen: Oxygen is normally only found in trace amounts in biogas, although it can be introduced during desulfurization. If it is a problem, it can be removed by membrane separation or PSA.

Terpenes: The removal efficiency of terpenes (Sec. 3.5.2) varies significantly between methods, with PSA achieving 100% removal, while water scrubbers achieve higher removal efficiencies (83–96%) than amine scrubbers (61–85%) (Arrhenius *et al.*, 2017).

Siloxanes: Currently, siloxane (Sec. 3.6) is removed from the biogas using adsorption, usually activated carbon, although a wide range of other absorbents has been used, such as metal oxides, zeolites, and polyurethane foams (Paglini *et al.*, 2022; Werkneh, 2022). For high biogas volumes, or high siloxane concentrations, adsorption can be improved by employing an initial refrigeration/condensation step before adsorption, with efficiencies of >99% removal possible, allowing for concentrations <1.0 mg m^{-3} (de Arespacochaga *et al.,* 2015). Some siloxanes are also removed during biogas chilling to control the moisture content.

References

Al Mamun, M.R. and Torii, S. (2015) Removal of hydrogen sulphide from biogas using zero-valent iron. *Journal of Clean Energy Technologies*, **3**, (6), 428–432. https://doi.org/10.7763/JOCET.2015.V3.236

Allegue, L.B. and Hinge, J. (2014) Biogas upgrading: Evaluation of methods for H_2S removal. Danish Technological Institute, Arras, Denmark. https://www.teknologisk.dk/_/media/60599_Biogas+upgrading.+Evaluation+of+methods+for+H2S+removal.pdf

Arrhenius, K., Holmqvist, A., Carlsson, M., Engelbrektsson, J., Jansson, A., Rosell, L., Yaghooby, H. and Fischer, A. (2017) Terpenes in biogas plants digesting food wastes: Study to gain insight into the role of terpenes. Report: 2017:350. Energiforsk AB, Sweden. https://energiforskmedia.blob.core.windows.net/media/22208/terpenes-in-biogas-plants-digesting-food-wastes-energiforskrapport-2017-350.pdf

Bijmans, M.F.M., van Helvoort, P.J., Buisman, C.J.N. and Lens, P.N.L. (2009) Effect of the sulfide concentration on zinc bio-precipitation in a single stage sulfidogenic bioreactor at pH 5.5. *Separation and Purification*

Technology, **69**, 243–248. https://doi.org/10.1016/j.seppur.2009.07.023.

BS EN. (2016) Natural gas and biomethane for use in transport and biomethane for injection in the natural gas network. BS EN 16723-1:2016, British Standards Institution, London. https://knowledge.bsigroup.com/products/natural-gas-and-biomethane-for-use-in-transport-and-biomethane-for-injection-in-the-natural-gas-network-specifications-for-biomethane-for-injection-in-the-natural-gas-network/standard

BS EN. (2017) Natural gas and biomethane for use in transport and biomethane for injection in the natural gas network — Automotive fuels specification. BS EN 16723-2:2017, British Standards Institution, London. https://www.en-standard.eu/bs-en-16723-2-2017-natural-gas-and-biomethane-for-use-in-transport-and-biomethane-for-injection-in-the-natural-gas-network-automotive-fuels-specification/

Buswell, A.M. and Mueller, H.F. (1952). Mechanisms of methane fermentation. *Industrial Engineering Chemistry*, **44**, 550–552.

de Arespacochaga, N., Valderrama, C., Raich-Montiu, J., Crest, M., Mehta, S. and Cortina, J.L. (2015) Understanding the effects of the origin, occurrence, monitoring, control, fate and removal of siloxanes on the energetic valorization of sewage biogas — a review. *Renewable and Sustainable Energy Review*, **52**, 366–381. https://doi.org/10.1016/j.rser.2015.07.106

Fu, F. and Wang, Q. (2011) Removal of heavy metal ions from wastewaters: A review. *Journal of Environmental Management*, **92,** (3), 407–418. https://doi.org/10.1016/j.jenvman.2010.11.011

Hagen, M., Polman, K., Jensen, J., Myken, A., Jonsson, O. and Dahl, A. (2001) Adding gas from biogas to the grid. Report SGC 118, Swedish Gas Center, Malmo, Sweden. http://sgc.camero.se/ckfinder/userfiles/files/SGC118.pdf

Jung, H., Kim, D., Choi, H. and Lee, C. (2022) A review of technologies for *in-situ* sulfide control in anaerobic digestion. *Renewable and Sustainable Energy Reviews*, **157**, 112068. https://doi.org/10.1016/j.rser.2021.112068

Kapłan, M., Klimek, K., Syrotyuk, S., Konieczny, R., Jura, B., Smoliński, A., Szymenderski, J., Budnik, K., Anders, D. and Dybek, B. (2021) Raw biogas desulfurization using the adsorption-absorption technique for a pilot production of agricultural biogas from pig slurry in Poland. *Energies,* **14,** (18), 5929. https://doi.org/10.3390/en14185929

Kelly, D.P. and Wood, A.P. (2000) Reclassification of some species of *Thiobacillus* to the newly designated genera *Thiobacillus* gen.nov., *Halotriobacillus* gen. nov. and *Thermithiobacillus* gen. nov. *International Journal of Systematic and Evolutionary Microbiology*, **50**, 511–516. https://doi.org/10.1099/00207713-50-2-511

Kiilerich, B., Van de Ven, W., Nielsen, A.H. and Vollertsen, J. (2017) Sulfide precipitation in wastewater at short timescales. *Water*, **9**, (9), 670. https://doi.org/10.3390/w9090670

Konstandt, H.G. (1976). *Microbial Energy Conversion* (eds. H.G. Schlegel and J. Barnea). Erich Goltze, Gotingen.

Krayzelova, L., Bartacek, J., Díaz, I., Jeison, D., Volcke, E.I.P. and Jenicek, P. (2015) Microaeration for hydrogen sulfide removal during anaerobic treatment: A review. *Reviews in Environmental Science and Bio/Technology*, **14**, 703–725. https://doi.org/10.1007/s11157-015-9386-2

Kulandaivelu, J., Shrestha, S., Khan, W., Dwyer, J., Steward, A., Bell, L., McPhee, P., Smith, P., Hu, S., Yuan, Z. and Jiang, G. (2020) Full-scale investigation of ferrous dosing in sewers and a wastewater treatment plant for multiple benefits. *Chemosphere*, **250**, 126221. https://doi.org/10.1016/j.chemosphere.2020.126221

Kumar, M., Nandi, M. and Pakshirajan, K. (2021) Recent advances in heavy metal recovery from wastewater by biogenic sulphide precipitation. *Journal of Environmental Management*, **278**, 111555. https://doi.org/10.1016/j.jenvman.2020.111555

Paglini, R., Gandiglio, M. and Lanzini, A. (2022) Technologies for deep biogas purification and use in zero-emission fuel cells systems — a review. *Energies*, **15**, (3551), 1–30. https://doi.org/10.3390/en15103551

Peu, P., Picard, S., Diara, A., Girault, R., Béline, F., Bridoux, G. and Dabert, P. (2012) Prediction of hydrogen sulphide production during anaerobic digestion of organic substrates. *Bioresource Technology*, **121**, 419–424. http://dx.doi.org/10.1016/j.biortech.2012.06.112

Pokorna, D. and Zabranska, J. (2015) Sufur-oxidizing bacteria in environmental technology. *Biotechnology Advances*, **33**, 1246–1259. http://dx.doi.org/10.1016/j.biotechadv.2015.02.007

Pudi, A., Rezaei, M., Signorini, V., Andersson, M.P., Baschetti, M.G. and Mansouri, S.S. (2022) Hydrogen sulfide capture and removal technologies: A comprehensive review of recent developments and emerging trends. *Separation and Purification Technology*, **298**, 121448. https://doi.org/10.1016/j.seppur.2022.121448

Sampaio, R.M.M., Timmers, R.A., Xu, Y., Keesman, K.J. and Lens, P.N.L. (2009) Selective precipitation of Cu from Zn in a pS controlled continuously stirred tank reactor. *Journal of Hazardous Materials*, **165**, 256–265. https://doi.org/10.1016/j.jhazmat.2008.09.117

Skerman, A., Collman, G., Sohn, J.H. and Pott, L. (2012) Options for biogas cleaning and use on-farm. RIRDC Publication No. 12/056, Rural Industries Research and Development Corporation, The Commonwealth of Australia. https://era.daf.qld.gov.au/id/eprint/2280/1/RIRDC_FinalReport12-056_Skerman.pdf

Speece, R.E. and McCarty, P.L. (1964). Nutrient requirements and biological solids accumulation in anaerobic digestion. In *Advances in Water Pollution Research. Proceedings of the First International Conference on Water Pollution Research* (ed. W.W. Ekenfelder). Pergamon, Oxford, pp. 305–333.

Su, L., Zhen, G., Zhang, L., Zhao, Y., Niu, D. and Chai, X. (2015) The use of the core–shell structure of zero-valent iron nanoparticles (NZVI) for long-term removal of sulphide in sludge during anaerobic digestion. *Environmental Sciences: Processes and Impacts*, **17**, 2013–2021. https://pubs.rsc.org/en/content/articlelanding/2015/em/c5em00470e

Wang, H., Li, J., Yi, Y., Nobu, M., Narihiro, T. and Tang, Y. (2020) Response to inhibitory conditions of acetate-degrading methanogenic microbial community. *Journal of Bioscience and Bioengineering*, **129**, (4), 476–485. https://doi.org/10.1016/j.jbiosc.2019.10.006

Werkneh, A.A. (2022) Biogas impurities: Environmental and health implications, removal technologies and future perspectives. *Heliyon*, **8**, 10929. https://doi.org/10.1016/j.heliyon.2022.e10929

Żarczyński, A., Rosiak, K., Anielak, P. and Wolf, W. (2014) Practical methods of cleaning biogas from hydrogen sulphide. Part 1, Application of solid sorbents. *Acta Innovations*, **12**, 24–34. https://www.proakademia.eu/gfx/baza_wiedzy/255/nr_12_24-34_2.pdf

Zicari, S., Graf, K., Ma, J., Minott, S.J., Pryor, S.W. and Saikkonen, K. (2003) Removal of hydrogen sulphide from biogas using cow-manure compost. Thesis presented to the Faculty of the Graduate School of Cornell University in Partial Fulfilment of the Requirements for the Degree of Master of Science. Cornell University, New York. http://www.green-trust.org/Al%20Rutan/MS-Thesis-Steve-Zicari.pdf

Section E

The Future

9

Sustainability and Future of Anaerobic Wastewater Treatment

9.1 Sustainability and the Circular Economy

Achieving sustainable wastewater treatment is one of the most important engineering challenges facing society, as the need to mitigate the dual problems of population growth and global warming becomes critical. The traditional roles of protecting surface and ground waters, preserving biodiversity, and preventing pathogen transfer have all become more challenging as wastewaters become more complex with new chemicals that are increasingly inhibitory or toxic to treatment. To fulfill the needs of sustainability, circularity has become a defining layer on top of our existing environmental and health priorities, with carbon harvesting, recovery of metals and other useful products, and, of course, reuse of water — all achievable expectations. Traditionally, wastewater treatment has relied on separation and aerobic processes, with a large portion of the carbon in the wastewater lost as carbon dioxide through aerobic metabolism, with anaerobic treatment occasionally used for sludge stabilization. In order to achieve sustainability, the emphasis of treatment has to change to primarily anaerobic treatment, with aerobic treatment used for effluent polishing where filtration technologies are unable to achieve the final effluent quality, although many natural treatment options are available (Gray, 2022).

Traditionally, anaerobic digestion was a slow process, operated at low microbial concentrations, requiring large reactors, and subject to many operational problems. But today's modern designs and a greater understanding of the microbial basis of anaerobic treatment, especially of the methanogenic archaea, have seen a rapid development of compact, rapid, and highly efficient processes, many utilizing granular sludges.

Modern wastewater treatment has embraced the circular economy ideology, becoming increasingly sustainable. Among the huge range of valuable byproducts that can be generated by fermentation are biomethane, biohydrogen, bioplastics, biodiesel, and volatile fatty acids (VFAs). VFAs are a valuable raw material used in the food, chemical, and pharmaceutical sectors as well as the carbon source for a wide range of bio-based compounds (Atasoy *et al.*, 2018).

9.1.1 *Biohydrogen and Dark Fermentation*

While hydrogen is an ideal alternative to fossil fuels, its success in terms of carbon footprint depends on how it is produced. Thermochemical processes can be a source of carbon and other pollutants, so the possibility of using anaerobic fermentation to produce biohydrogen is extremely attractive. Key operational factors that affect the maintenance of hydrogen-producing bacterial communities include pH, temperature, organic loading rate, substrate concentration, and characteristics. Show *et al.* (2007) operated a laboratory-scale anaerobic continuously stirred tank reactor (AnCSTR) using a granular sludge to successfully produce biohydrogen, with butyrate, acetate, and ethanol, the major metabolic byproducts. Using glucose as a substrate, the optimum solids loading rate (SLR) was between 0.6 and 0.8 g glucose g^{-1} VSS h^{-1}, with a maximum biohydrogen production of 3.26 L L^{-1} h^{-1}, a yield of 1.84 mol H_2 mol^{-1} glucose at an organic loading rate of 20 g glucose L^{-1} h^{-1}, with a hydraulic retention time (HRT) of 0.5 hours. The relationship between HRT and influent substrate concentration on biohydrogen production is shown in Fig. 9.1, with a yield affected at concentrations >20 g L^{-1} or an HRT <0.5 hours.

Food crops such as potatoes and tapioca are a vital part of the diet of billions of people, with starch products resulting in significant

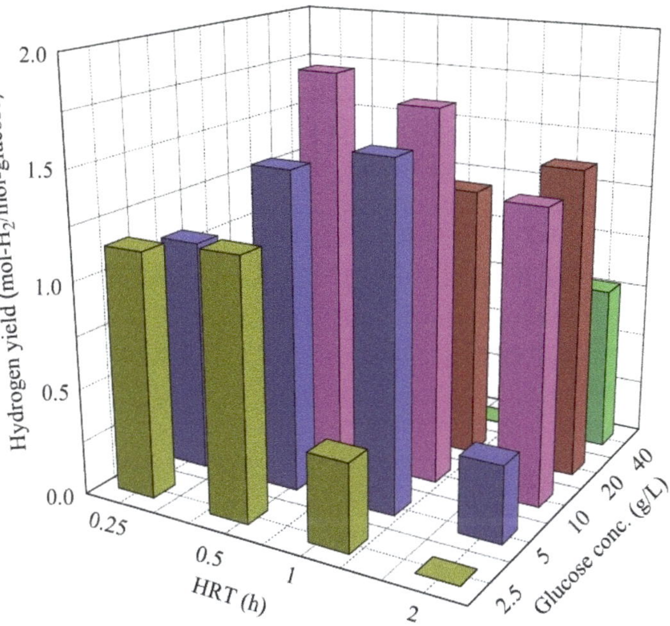

Figure 9.1. The effect of hydraulic retention time and influent substrate concentration on hydrogen yield (mol H_2 mol^{-1} glucose) (Show *et al.*, 2007). Reproduced with permission of Elsevier Publishing.

water pollution in many countries but particularly in Asia (Thanwised *et al.*, 2012). Traditional treatments, such as stabilization ponds including anaerobic lagoons, result in significant odor production as well as the uncontrolled release of huge quantities of greenhouse gases (Rajbhandari and Annachhatre, 2004). Closed anaerobic treatment of such wastes by contact treatment systems can harvest the carbon in these wastewaters to produce biogas. Many of the starch products manufactured need to be dried, making them energy intensive; therefore, using anaerobic treatment processes such as anaerobic baffled reactors (AnBRs), anaerobic fixed film reactors (AnFFRs), and up-flow sludge blanket reactors (UASBs) close the production loop, resulting in the biogas produced being able to offset energy requirements by up to 75%. For example, tapioca production results in 20 m^3 of wastewater from each ton of starch produced, and with an average chemical oxygen demand (COD) of 16,000 mg L^{-1}, it can yield up to

10 m^3 of biogas per m^3 of wastewater treated with significant COD reductions. Starch-rich wastewaters have also been used to generate biohydrogen for energy use (Wang *et al.*, 2008). For example, the treatment of wastewaters from tapioca, cassava, and molasses processing resulted in biohydrogen yields of 13, 54, and 46 L H_2 kg^{-1} COD, respectively, under ideal operating conditions using an AnBR, a UASB, and a different AnBR system, although COD removal was poor (>30%) (Thanwised *et al.*, 2012). Some methane can be generated, which is minimized by controlling the HRT, with up to 70–80% of the biogas comprising carbon dioxide, so a lot of the embedded carbon is lost.

Dark fermentation

Hydrogen is an alternative to fossil fuels, which can be used as a substitute for natural gas or as a replacement fuel for propane and natural gas household boilers, many of which are already able to use hydrogen with minor modifications. Biohydrogen (Bio-H_2) offers a sustainable manufacturing pathway, producing a zero-carbon, combustion-high gravimetric energy density (120 kJ g^{-1}) fuel. It has been mainly generated using hydrocarbons or gas, and there are increasing opportunities to use anaerobic fermentation to produce Bio-H_2. Currently, photo-fermentation or direct or indirect photolysis is the major method for producing Bio-H_2, although an exciting alternative is dark fermentation. Unlike the former methods, dark fermentation does not require light nor a wide range of nutrients and is able to be produced by a range of standard anaerobic treatment systems (e.g., anaerobic completely stirred reactor (AnCSTR), anaerobic membrane reactor (AnMBR), upflow anaerobic sludge bed reactor (UASB), and anaerobic sequencing batch reactor (AnSBR)) using mixed cultures of both facultative and obligate anaerobes, with a wide range of potential substrates, both liquid and solids.

Dark fermentation can be carried out over a wide temperature range of 20 to 80°C, and, like normal anaerobic breakdown, carbon dioxide and hydrogen sulfide are also produced. The Bio-H_2 is produced primarily from glucose molecules in carbohydrates with VFAs

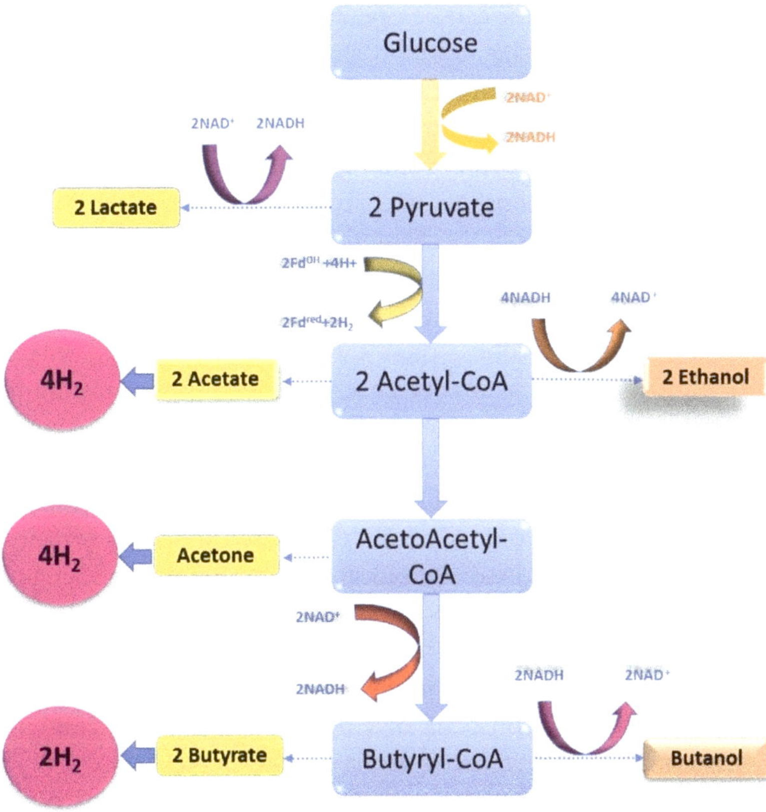

Figure 9.2. Main breakdown pathways of glucose during dark fermentation (Ahmed *et al.*, 2024). Reproduced with permission of Elsevier Publishing.

or alcohols produced as byproducts. The key operational factors are pH, HRT, and partial pressure of the gas — the latter directing the metabolic pathway and electron flows so that acetate, butyrate, propionate, and alcohols are the main byproducts, along with hydrogen (Fig. 9.2). The main pathways are acetate and butyrate:

$$\text{Acetate pathway: } C_6H_{12}O_6 + 2H_2O \rightarrow 2CH_3COOH + 2CO_2 + 4H_2\ \Delta G^0\ -126\ \text{kJ mol}^{-1} \quad (9.1)$$

$$\text{Butyrate pathway: } C_6H_{12}O_6 \rightarrow CH_3CH_2CH_2COOH + 2CO_2 + 2H_2\ \Delta G^0\ -214\ \text{kJ mol}^{-1} \quad (9.2)$$

The anaerobic reactor can be operated to control how the Bio-H_2 is generated, although the acetate pathway is preferred as it produces twice as much Bio-H_2 as butyrate, while lactic acid bacteria suppress Bio-H_2 production. Methanogenesis must be suppressed to improve Bio-H_2 production and prevent methane from being produced.

Unlike methanogenesis, the microbial consortia involved in hydrogen production are constantly changing, leading to a variation in Bio-H_2 production and process instability. Hydrogen-producing species fall into three broad groups, the main group comprising spore-forming obligate anaerobes, then the non-spore-forming obligate anaerobes, and, finally, facultative anaerobes. The spore-forming obligate anaerobes are classed as main hydrogen producers and include the most efficient hydrogen-producing species, with *Clostridium* spp., the best-known genera producing 1.5–3.0 mol H_2 mol^{-1} hexose. Although they are sensitive to low pH ($pH < 5.0$), they are abundant in anaerobic reactors and can also be used as a biocatalyst. In theory, hydrogen yield is restricted to just 4 moles per mol of glucose.(Eq. (9.1)) This is known as the Thauer limit and restricts the potential output from dark fermentation. However, exciting new research by Ergal *et al.* (2020), who used an artificial consortia of *Clostridium acetobutylicum* and a facultative anaerobe *Enterobacter aerogenes* to increase the yield beyond the Thauer limit to 5.6 mol mol^{-1} glucose. Non-spore-forming obligate anaerobes are classed as *auxiliary functioning species* and include *Ethanoligenens harbinense* (Pylum Bacillota), which produces ethanol as well as Bio-H_2 and carbon dioxide, *Acetanaerobacterium elongatum* (Order: Clostridiales) with Bio-H_2 yields of 1.55 mol mol^{-1} glucose, *Megasphaera* spp. (Phylum: Firmicutes) that utilize lactate to produce hydrogen, *Acidaminococcus* sp. (Phylum: Firmicutes), and *Prevotella* sp. (Phylum: Bacteroidetes), which is able to use a wide range of substrates and actively reduce methane production by diverting the hydrogen flow in glycolysis away from methanogenesis. The final group comprises the facultative anaerobes, which are classed as *challenged producers* and include a wide range of genera, including *Citrobacter* sp., *Klebsiella* sp., *Enterobacter* sp., *Bacillus* sp., and also *Shewanella oneidensis* and *Pseudomonas stutzeri* (Dahiya *et al.*, 2021).

Supplements have been shown to improve dark fermentation, with Fe, Ni, and Mg ions playing an important role in the formation of cofactors for enzymes, being inhibitory at low concentrations. Nanoparticles (NPs) have been widely used experimentally to enhance hydrogen-producing microorganisms during dark fermentation in terms of production rate and efficiency by improving metabolic activity. For example, NPs reduce the dependence on NADH and improve the formation of formate hydrogen lactase, an enzyme used to decompose formate and improve electron adsorption and transport between hydrogen-producing bacteria and hydrogenase. There have been many studies on the effect of NPs on dark fermentation, the majority of which show a large increase in Bio-H_2 production; for example, FeO NPs specifically enhance acetate synthesis and reduce ethanol formation, thereby increasing Bio-H_2 yield. These studies and the role of supplements to enhance dark fermentation have been reviewed by Ahmed *et al.* (2024).

The concept of a closed-loop integrated biorefinery able to deal with a wide range of liquid and solid wastes has been proposed by Venkata Mohan *et al.* (2020), with waste VFAs from acidogenesis being used for the production of a wide range of products from alcohols to bioplastics (Fig. 9.3). However, dark fermentation faces many challenges before it can be fully scaled-up to become a major industrial producer of hydrogen (Sarkar *et al.*, 2021; Dahiya *et al.*, 2021).

9.1.2 *VFA Production*

In terms of resource recovery, there are many waste streams that offer opportunities to produce valuable byproducts. One of the most valuable waste streams is that produced by the dairy industry, being rich in carbohydrates, proteins, and fats with 2,250,000 m^3 produced worldwide daily. Using the Eurostat website (https://ec.europa.eu/eurostat/web/main/home), it is estimated that 9.15 Mt acetate, 5.39 Mt butyrate, and 6.47 Mt propionate could be recovered from that waste stream alone, far in excess of current global demand. Of the three, acetic acid is the most widely used, being a key building block in the manufacture of paint, rubber, plastics, synthetic fibers, textile finishes, pesticides, polymer emulsions, and paper coatings, and in the food industry as a flavor, acidity regulator, and preservative (Cheryan, 2009). Current manufacturing costs for VFAs vary, as does

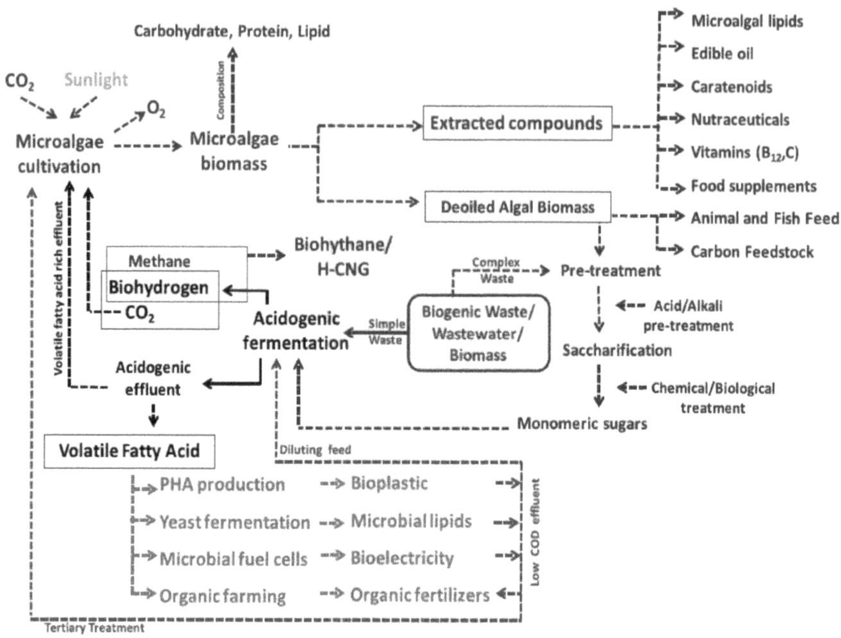

Figure 9.3. Details of a closed-loop biorefinery incorporating dark fermentation for the production of biohydrogen (Dahiya *et al.*, 2021). Reproduced with permission of Elsevier Publishing.

demand (Table 9.1). VFAs are easy to transport and store, have a high market value, and have a very low carbon footprint when produced using anaerobic fermentation.

Currently, the vast majority of VFAs (90%) are produced synthetically from petrochemical derivatives (Agnihotri *et al.*, 2022). While VFAs can also be produced by fermentation, the development of large-scale biorefineries is still some way away, but with increasing markets and the move away from fossil fuels, it is inevitable that the change will be made with wastewater, a key substrate (Bhatia and Yang, 2017; Venkata Mohan *et al.*, 2019). Within all anaerobic treatment processes, VFA can be promoted by inhibiting methanogens. While, in theory, this could be done using a methanogenic inhibitor, the normal practice is to reduce the HRT and solids retention time (SRT) and adjust the pH to <6.0 or >8.0, which effectively inhibits

Table 9.1. Uses, production methods, and market for volatile fatty acids (Atasoy *et al.*, 2018). Reproduced with permission of Elsevier Publishing.

VFAs	Chemical formula	Market size (kton/year)	Market price (€/ton)	Usage/application	Production methods	References
Acetic acid	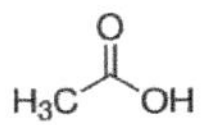	14,000–17,000	400–800	Vinyl acetate monomer (polymers, adhesives, dyes), food additive, solvent, vinegar, ester production, chemicals	Chemical synthesis (carboxylation of methanol) and microbial fermentation (oxidative and anaerobic)	Bhatia and Yang (2017)
Butyric acid	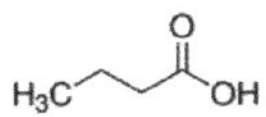	90–105	1,500–1,650	Animal and human food additive, chemical intermediate, solvent, flavoring agent	Chemical synthesis (oxidation of butyraldehyde), extraction from butter, microbial fermentation	Zigová and Šturdík (2000)
Propionic acid	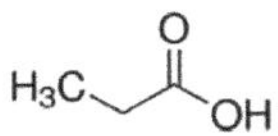	350–470	2,000–2,500	Esters used food industry as aroma additive, food additive, flavoring, pharmaceuticals, animal feed supplement, fishing bait additive	Chemical Synthesis (ethylene hydro formylation, carboxylation of ethylene, direct oxidation of hydrocarbons), by product of acetic acid manufacturing, microbial fermentation	Cheryan (2009)

the methanogens, allowing fermentative bacteria to flourish. Maximum VFA production is achieved under alkaline conditions (pH 10). The key bacterial species employed for VFA production varies; for example, for propionic acid, *Propionibacterium freudenreichii, P. acidipropionici, P. thoenii, Propionibacterium shermanii, and P. jensenii*; for acetic acid fermentation, *Acetobacter spp. Gluconacetobacter spp.*, and *Gluconobacter spp.*; and for butyric acid fermentation, *Clostridium butyricum, C. kluyveri, C. beijerinckii, C. barkeri, C. acetobutylicum, C. thermobutyricum, C. thermopalmarium, C. pasteurianum, Butyribacterium spp., Sarcina spp., Megasphaera spp., Fusobacterium nucleatum, Peptococcus asacelarolyticus, Butyrivibrio fibrisolvens, Pseudobutyrivibrio ruminis,* and *Eubacterium limosum*. Atasoy *et al.* (2020) used cheese-processing wastewater in both a batch reactor and an AnSBR to produce VFAs with yields of 0.97 and 0.95 g^{-1} VS COD g^{-1} SCOD, respectively. While a wide range of VFAs was produced, the dominant type was lactic acid (48%) in the batch and propionic acid (80%) in the AnSBR. Certain bacterial families were correlated with different VFAs, for example, butyric acid (Gracilibacteraceae, Desulfovibrionaceae), propionic acid (Desulfovibrionaceae, Synergistaceae), and lactic acid (Pseudomonadaceae, Rhodocyclaceae) (Zigová and Šturdĺk, 2000). A similar study using cheese whey as a substrate and an AnSBR compared performance using a range of HRT and SRT scenarios and observed optimum process efficiency at longer SRTs and shorter HRT operating conditions (Lagora-Costa *et al.*, 2020). The degree of acidification rose from 0.73 to 0.83 when the SRT was increased from 5 to 15 days, with hydrolytic bacteria replaced by acidogenic bacteria, and the HRT reduced from 3 to 1 days (Table 9.2). Other waste streams, such as chicken manure, have huge potential for energy recovery, with globally approximately 3.85 billion egg-laying hens and 50 billion broilers producing hundreds of millions of tonnes of manure annually (Sakar *et al.*, 2009). Using an AnMBR, Yin *et al.* (2022) were able to produce a high yield of VFA of 0.90 g VFA g^{-1} VS from chicken manure without any pH control, with acetic acid the major VFA produced (>50%). Another untapped waste stream for acidogenic fermentation is olive mill wastewater, with acetic acid the dominant VFA produced. Yarimtepe *et al.* (2017)

Table 9.2. Performance of an AnSBR treating cheese whey (a) at different solids residence times (SRTs) of of 5, 10, and 15 days and (b) at different hydraulic residence times (HRTs) of 1, 2, and 3 days (Lagora-Costa *et al.*, 2020). Reproduced with permission of Elsevier Publishing.

(a)

	SRT = 5d	SRT = 10d	SRT = 15d
Operation time (d)	51	34	38
Feed concentration (g COD L^{-1})	11.90 (± 0.53)	12.06 (± 0.44)	11.54 (± 0.35)
HRT (d)	2	2	2
OLR (g COD L^{-1} d^{-1})	6.01 (± 0.32)	6.03 (± 0.22)	5.77 (± 0.17)
VSS (g VSS L^{-1})	4.19 (± 0.39)	5.36 (± 0.52)	7.38 (± 0.66)
F/M ratio (g COD g VSS^{-1} d^{-1})	1.46 (± 0.14)	1.13 (± 0.09)	0.81 (± 0.04)
Total VFA (g COD-VFA L^{-1})	9.72 (± 0.66)	9.44 (± 0.40)	10.23 (± 0.34)
Acetic acid (%)	37.39 (± 2.95)	25.83 (± 3.36)	35.71 (± 3.43)
Propionic acid (%)	20.34 (± 3.69)	7.52 (± 1.46)	10.99 (± 1.42)
Butyric acid (%)	22.13 (± 4.35)	59.62 (± 2.71)	48.40 (± 4.10)
Valeric add (%)	4.11 (± 2.24)	5.46 (± 2.20)	4.90 (± 1.32)
Lactic acid (%)	16.03 (± 7.37)	1.57 (± 1.86)	0.00
DA[a]	0.73 (± 0.05)	0.75 (± 0.03)	0.83 (± 0.04)
$Y_{VFA/S}$ (g COD-VFA g COD-Lactose^{-1})[b]	0.78 (± 0.05)	0.78 (± 0.03)	0.87 (± 0.04)
Productivity (g COD-VFA L^{-1} d^{-1})	4.45 (± 0.29)	4.51 (± 0.20)	4.54 (± 0.48)

[a]DA: Degree of acidification.
[b]$Y_{VFA/S}$: Acidification yield.

(b)

	HRT = ld	HRT = 2d	HRT = 3d
Operation time (d)	51	43	66
Feed concentration (g COD L^{-1})	5.58 (± 0.17)	11.54 (± 0.35)	17.13 (± 0.79)
OLR (g COD L^{-1} d^{-1})	5.58 (± 0.17)	5.77 (± 0.17)	5.71 (± 0.26)
SRT (d)	15	15	15
VSS (g VSS L^{-1})	7.99 (± 0.56)	7.38 (± 0.66)	7.18 (± 0.44)
F/M ratio (g COD g VSS^{-1} d^{-1})	0.69 (± 0.05)	0.81 (± 0.04)	0.80 (± 0.06)
Total VFA (g COD-VFA L^{-1})	5.15 (± 0.36)	10.23 (± 0.34)	14.93 (± 0.66)
Acetic acid (%)	27.97 (± 3.08)	35.71 (± 3.43)	23.33 (± 2.98)

(*Continued*)

Table 9.2. (*Continued*)
(b)

	HRT = ld	HRT = 2d	HRT = 3d
Propionic acid (%)	8.84 (± 4.88)	10.99 (± 1.42)	6.83 (± 2.10)
Butyric acid (%)	53.97 (± 7.93)	48.40 (± 4.10)	60.91 (± 4.24)
Valeric acid (%)	8.20 (± 4.06)	4.90 (± 1.32)	7.90 (± 2.28)
Lactic add (%)	1.02 (± 0.87)	0.00 (± 0.00)	1.04 (± 1.66)
DA[c]	0.83 (± 0.03)	0.83 (± 0.04)	0.79 (± 0.05)
$Y_{VFA/S}$(g COD-VFA g COD-Lactose^{-1})[d]	0.90 (± 0.05)	0.87 (± 0.03)	0.86 (± 0.06)
Productivity (g COD-VFA L^{-1} d^{-1})	4.65 (± 0.15)	4.54 (± 0.48)	4.53 (± 0.34)

[c]DA: Degree of acidification.
[d]$Y_{VFA/S}$: Acidification yield.

successfully operated a laboratory-scale AnSBR using olive mill wastewater, pre-treated by electrocoagulation to reduce its toxicity and solids content, at a pH of 5.5 with an HRT of 2 days, with an acidification rate of 68% at a maximum effective loading of 20 g sCOD L^{-1} d^{-1}.

The greatest challenge for commercial anaerobic fermentation production is the recovery of VFAs and, more specifically, separating the individual VFAs. There are a number of methods currently employed, including gas stripping with absorption, adsorption, electrolysis, solvent extraction, nanofiltration, reverse osmosis, and PTFE membranes. These have been reviewed by Atosoy *et al.* (2018).

9.2 Biotechnology and the Future of Anaerobic Wastewater Treatment

9.2.1 *Granulation*

Under certain operating conditions, including plug flow conditions, high shear stress, and long sludge ages, aerobic and anaerobic bacterial flocs can develop into dense, much larger and rounded granules. Granules rapidly separate from suspension, ensuring high biomass retention while offering much higher rates of biodegradation activity, including methane production. There are a number of theories as to how granules are formed, although the role of quorum sensing (QS) and signal molecules are clearly critical in their formation, especially in promoting and accelerating the synthesis of extra cellular polymers (ECPs), more recently referred to as extracellular polymeric substances (EPSs), cellular metabolism, and improving the surface attachment capability of granules (see the section *Quorum Sensing* below). Apart from rapid settlement, granules offer a number of advantages over flocs, which include protection against toxic and inhibitory compounds, defense against extreme pH, and high flow rates, ensuring close contact between the various groups of anaerobic microorganisms and rapid breakdown from hydrolysis through to methanogenesis, high concentrations of microorganisms, and the ability to withstand shock and variable loadings. Each granule matrix supports a diverse syntrophic community of bacteria and archaea, creating what is essentially an independent treatment unit.

Starting up a granular sludge from scratch can take up to four to six months, which is why biofilm carriers are so widely used as an alternative. Seeding directly with granules from another reactor is a more rapid option, if available, but can be expensive due to transportation and does not always work due to sudden changes in food sources and other operating conditions. There is a wide range of physicochemical factors that affect granulation, such as temperature, pH, substrate type, HRT, up-flow liquid velocity, organic loading rate, availability of trace elements, and nutrients. These have been evaluated by Tay *et al.* (2006) and summarized in Table 9.3.

The development, structure, and microbiology of aerobic granular sludge have been discussed at length in an earlier book in this series (Gray, 2023).

Theory and life cycle

The formation of granules is not accurately understood, and to explain the process, many models have been proposed, which include both physicochemical and structural models. Physicochemical models include the inert nuclei model, selection pressure model, multivalence positive ion-bonding model, EPS bonding model, synthetic and natural polymer-bonding model, secondary minimum adhesion model, local dehydration and hydrophobic interaction model, and the

Table 9.3. Optimum conditions to support and improve granulation adapted from Tay *et al.* (2006) and Show *et al.* (2020). Reproduced with permission of Elsevier Publishing.

	Key operating factors
Biological	• Seed sludge for a start-up should be acclimatized with the same wastewater to be treated. If this is not possible, then the seed must comprise as much diversity of bacterial populations as possible. • Microbial loading for a start-up should not be excessive (<0.6 kg COD kg^{-1} VSS d^{-1}).
Chemical	• Divalent compounds should be supplemented in the feed. • Granulation will progress sooner on unacidified wastewaters than on acidified feeds. The pH of the reactor should be kept close to neutral. • Chemicals such as caustic soda, soda ash, or sodium bicarbonate can be added to maintain alkalinity and buffering capacity in the reactor. • Essential growth nutrients such as N, P, S and trace elements (Fe, Ni, Co) should be present in sufficient amounts.
Physical	• Reactor design and operating temperature can affect the liquid mixing, the liquid, and gas up-flow velocities and even the size and shape of the granule. • Temperature also affects the biochemical reaction rates. The optimal temperatures for treatment are in the range of mesophilic 30–38°C and thermophilic 50–60°C.
Feed	• Lower concentration of wastewater feed should be used. It was found that the lower the feed concentration, the faster the granule will be developed. The concentration should, however, be sufficient to sustain bacterial growth. A minimum concentration was suggested at 500 mg COD L^{-1}. • The presence of dispersed suspended solids in the feed may retard or even deter granulation. These substances should be avoided or be removed in a pretreatment. • Toxic substances should be avoided. If toxic wastewater is to be treated, the inoculum microbes should be well acclimatized.

surface tension model. Structural models comprise the Capetown's model, spaghetti theory, syntrophic microcolony model, multi-layer model, cellular automaton model, and the cell-to-cell communication model. All these models, including the proton translocation-dehydration theory, have been fully explored by Liu *et al.* (2004) and more recently by Show *et al.* (2020).

Granulation comprises a number of stages — it shares the first two with flocculation, while the last two require specific operating conditions, relating in particular to the feeding regime and sheer stress as described above.

Initially, bacteria have to make contact with each other by passive or active movement and become attached by physicochemical forces (e.g., electrostatic, van der Waals forces, hydrophobic effect, etc.), which are reversible. As bacteria are generally negatively charged, they have a hydrophobic surface. They also possess fimbriae or attachment pili, which are different from flagella, which are used for locomotion, being both thinner (3–8 nm) and shorter, and also very numerous (up to 1,000 per cell). Fimbriae are also important in adhesion to conditioned inert surfaces as well as in the formation of biofilms (Lettinga *et al.*, 1980; Hulshoff-Pol, 1989), probably by overcoming the electrostatic repulsion between the cell wall and medium surface. Their role in floc formation is thought to be similar to attaching to inert particles or fibers, as well as other cells. However, the surface properties of individual bacterial cells, such as the presence of EPS, lipo-polysaccharides, proteins, and mycolic acid, all increase hydrophobicity and hence adhesion potential (Wilschut and Hoekstra, 1984).

EPSs are not food reserves, like poly-β-hydroxybutrate, and are not easily decomposed. EPS is a polymer that can be composed of a number of organic compounds, such as polysaccharides, amino polysaccharides, and proteins (Sato and Ose, 1980). Lipids may also be present with minor amounts of nucleic acids and other biopolymers, but the exact nature of these flocculating polymers will depend on the species of bacteria. Each polymer will have varying surface properties and charges that will influence not only the settling characteristics but also, importantly, the water-binding properties of the floc (Wilén *et al.*, 2003). The polymer not only gives the floc components cohesion and an open porous structure, but it also allows suspended particles in the wastewater to bind to the floc by adsorption. Cations and anions, including phosphorus and a range of pollutants and toxic compounds, are also adsorbed by such polymers. The EPS matrix is a dynamic system that enables cells in flocs to function in a manner similar to multicellular organisms. They have unique sorption properties and Decho (2000) has suggested that the sorption of heavy metals may be a strategy to protect the bacteria against toxic effects.

The next stage is the irreversible attachment of the bacteria to each other, which is achieved by the secretion of EPS that creates strong chemical bonds between the cells. The bacteria continue to produce EPS, creating a matrix in which bacteria can divide and new bacteria are attracted and attached, thus creating an active floc. It is at this stage that methanogens start to become enmeshed into flocs.

The bonding of particles in solution and the formation of agglomerates can be explained by Derjaguin–Landau–Verwey–Overbeek (DLVO) forces (van der Waals and electrostatic forces), non-DLVO forces (bridging, hydrophobic forces), and physical enmeshment. So, flocs are held together largely by electrostatic forces. At neutral pH, divalent cations, in particular Ca^{2+}, bridge the negatively charged functional groups on EPS produced by the floc-forming bacteria, which are referred to as the divalent cation bridging theory (DCBT), enhancing floc aggregation and structural stability. This bridging also occurs between other materials enhancing enmeshment (Higgins *et al.*, 2004b) (Fig. 9.4).

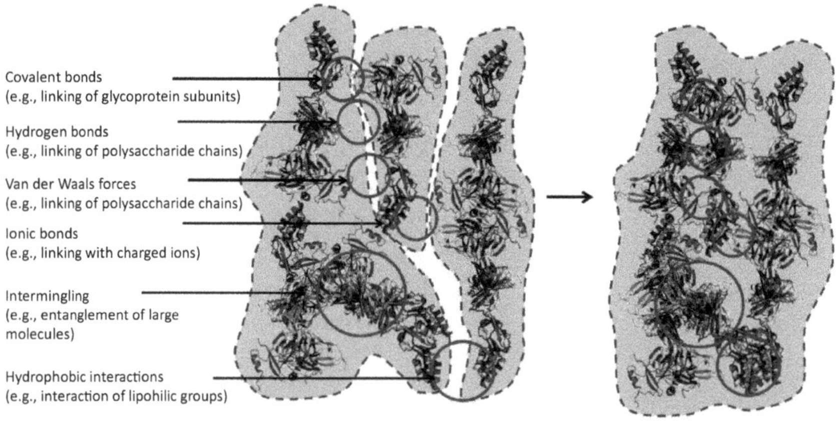

Figure 9.4. Representation of attachment and aggregation processes (Tansel, 2018). Reproduced with permission of Elsevier Publishing.

The importance of divalent cations in floc formation, strength, and settleability is well known with calcium ions, which are preferentially bound to EPS, having an important role in floc formation and size through the construction of calcium bridges (Schmidt and Ahring, 1996; Schmidt and Ahring, 1994; Shen *et al.*, 1993). High concentrations of sodium ions in wastewaters can displace the divalent cations from within the floc, reducing bridging and leading to the weakening of the floc structure (Bruus *et al.*, 1992: Higgins and Novak, 1997a) (Fig. 9.5), with flocculation efficiency that can be

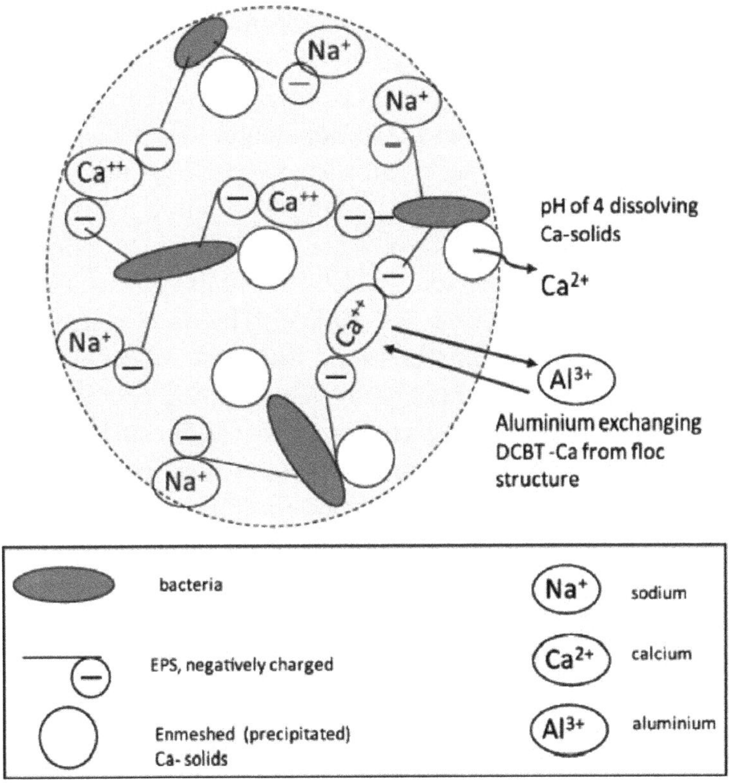

Figure 9.5. Within the activated sludge floc, Ca^{2+} ions create bridges that can be displaced by Na^{+} ions, weakening the floc and leading to possible deflocculation. Within each floc are two pools of calcium, Ca^{2+}, involved in bridging (DCBT-Ca) and precipitated calcium solids. Al^{3+} exchanges with DCBT-Ca while Ca-solids are dissolved at low pH (Peeters *et al.*, 2011). Reproduced with permission of Elsevier Publishing.

estimated by the ratio of monovalent to divalent cations in meq L^{-1}, known as the M+/D++ ratio (Higgins *et al.*, 2004a). The total quantity of Ca^{2+} ions in flocs is split between those required for bridge formation (i.e., DCBT-Ca) and precipitated Ca-solids that become enmeshed within the floc. Total Ca^{2+} in flocs varies over time, with approximately half the DCBT-Ca fraction depending on the actual sludge and wastewater characteristics (e.g., whether the wastewater arises from a hard or soft water catchment) (Bruus *et al.*, 1992). Figure 9.5 shows the nature of divalent bridging in a floc and that at low pH, Ca-solids are leached from flocs. At low pH with Al^{3+} ions present, both exchangeable (DCBT-Ca) and enmeshed (Ca) solids are leached.

Once this initial agglomeration of cells is complete, the floc then begins to go through a series of developmental stages. Early development is often referred to as expansion, with the floc becoming increasingly larger and developing a unique structure. While flocs can still attract new bacteria, both the same and different species, most of their growth is by cell division. Unlike biofilms, the structure that develops is largely spherical and simple. This is not random but depends again on cells communicating with each other through QS. The voids have an important function, allowing nutrients to penetrate from the liquid phase deep into the center of the floc and, at the same time, allowing methane and other gases, as well as biodegradable materials, to be removed. It is assumed that only the bacteria in the outer surface layer actively reproduce, while those within the floc remain viable and actively metabolizing. If this is not the case, then the floc or granule would continue to grow from the inside out, forcing it to break up. Adsorption of organic and inert material takes place both on the EPS matrix as well as on adsorption sites on the bacterial cell walls. In this way, the EPS matrix is creating a digestive system, to quote Flemming and Wingender (2010), retaining extra-cellular enzymes within the floc where they can release metabolized material for absorption by other bacteria and archaea. This is just one example of how the floc community works together, creating a massive advantage over those microorganisms that live as individuals, with up and down-regulation of genes controlling the release of gene products, which govern the actions of individuals or layers of bacteria pivotal in community organization.

Granular sludge comprises much larger, denser, and rounded flocs, referred to as granules. The use of inert particles or nucleating agents to increase the rate of granulation and to increase stability has had some success. These include the addition to the substrate of metal ions (Ca^{2+}, Mg^{2+}, and Fe^{3+}) and carriers, such as biochar and nanoparticles. Biochar stimulates EPS secretion, thus promoting granulation (Zhang *et al.*, 2017b). However, granulation can be achieved without the use of any nucleating agents. The main difference between flocs and granules is the larger amount of EPS produced in granules, which also has a stronger adhesive nature. Under normal operation conditions, the EPS in granules is found as a gel, which is not the case with normal flocs (Seviour *et al.*, 2009), with the hydrogel matrix key to the greater strength and stability of granules.

The precise mechanism of granule formation is not fully understood, although Nancharaiah and Sarvajith (2019) have explored the current theories, including the role of QS, where granulation and granule stability appear related to ATP-dependent N-acyl-homoserine lactone (AHLs). Current research is increasingly highlighting the role of QS and signal molecules in promoting and accelerating the synthesis of extracellular polymers and improving the surface attachment capability of microorganisms (Sarma *et al.*, 2017; Zhang *et al.*, 2019), both critical attributes for successful granulation and granule stability (i.e., the AHL-based granulation strategy).

Steps 3 and 4 are unique to granulation and require specific operating conditions, namely, a feast-famine regime associated with plug flow systems, which encourages the secretion of EPS that creates aggregates and a sufficient up-flow velocity. Granulation requires a separation stage, usually built into reactors, which allows weaker-formed flocs to be lost with the final effluent, while the denser flocs are retained due to the more rapid terminal settling velocities (Qin *et al.*, 2004). High up-flow velocities subject aggregates to a high-shear stress environment within the reactor that helps to create the dense granules. Therefore, correct operational conditions are critical not only for granularization but to sustaining the granules in their optimal condition. The process of granulation and the parameters affecting it are discussed by Hamza *et al.* (2022). The addition of Ca^{2+} mg and Al^{2+} ions has been shown to help granulation in UASBs, although excessive Ca concentrations (>500 mg L^{-1}) have the opposite effect (Yu *et al.*, 2001).

The surface nature and shape of granules alter as the sludge matures. Initially, they have rough, irregular shapes that gradually become smother, increasingly spherical, and develop a fine-grained porous structure. This is shown by Xiao *et al.* (2023), who, by studying granulation in detail using scanning electron microscopy, were able to show the transition from flocs to mature granules. Initially, the seed flocs comprised filaments that became small compact granules after 40 days of operation. By day 190, the granules were larger with discernible pores, allowing biogas to escape — the filamentous bacteria had disappeared after day 40 but returned by day 190 on the surface of the granules, which were largely composed of cocci and rod-shaped cells. The authors speculated that the filaments acted as docking stations for other microbes as well as giving granules structural stability, although their role in adsorption, as seen in filaments, is also a possibility. The fluorescent staining shows that the cells within the granules remained active (Fig. 9.6).

Quorum sensing

Much research has been done on the role of quorum sensing (QS), a molecular communication system, in the creation of complex microbial structures (i.e., biofilms, flocs, and granules) with a high degree of organization and cooperation between bacterial cells. Bacteria can sense the density of other cells close by, and when a threshold number (or quorum) is achieved, they begin to work together, using a range of physicochemical and biological processes to form a floc, biofilm, or granule.

QS is a continuous process where individual bacteria produce signaling molecules or autoinducers (AI) that diffuse through the cellular membrane into the liquid phase. So, the higher the density of bacteria in an area, the higher the concentration of AI molecules

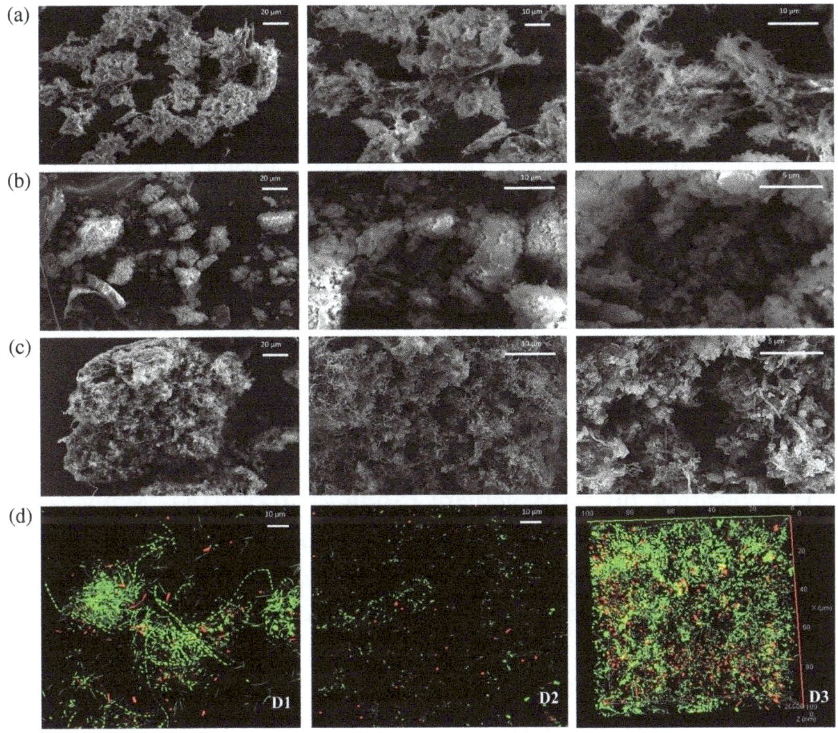

Figure 9.6. Scanning electron microscopy (SEM) and confocal laser scanning microscopy (CLSM) images of anaerobic sludge granules at different stages. (a) Seed sludge. (b) Granular sludge on day 40. (c) Granular sludge on day 190. (d) CLSM images of granular sludge on day 190 stained with live (green) and dead (red) staining kit; D1 — outer layer of the granules; D2 — the inner part of granules; and D3 — the composite 3-D image of granules (Xiao *et al.*, 2023). Reproduced with permission of Elsevier Publishing.

released, which eventually triggers a collective behavioral change in all the bacteria. The secreted AIs are a mixture of homoserine lactones and specific peptides, which control specific repressor or activator sequences in the bacterial DNA responsible for the production of mRNA. This forms new protein, which includes, among others, the genes for the formation of a floc or other structure. Where this gene is absent, then an organized floc is not produced.

Not all bacteria produce AIs or chemical signals responsible for cell-to-cell communication, although over 100 Proteobacteria species are able to produce the QS molecule AHL thought responsible for flocculation, EPS production, and general intraspecies communication (Tan *et al.*, 2014). Details of all the common QS AIs produced by bacteria have been summarized by Maddela *et al.* (2019), including AHLs, autoinducing peptides (AIPs), autoinducers-2 (AI-2), pseudomonas quinolone signal (PQS), diffusible signal factor (DSF), and autoinducers-3 (AI-3). AHLs are primarily produced by Gram-negative bacteria, as are DSF. AIPs are produced by Gram-positive bacteria, while AI-2 signaling molecules are found in both Gram-positive and negative bacteria. The high density of microorganisms in wastewater treatment systems results in high concentrations of these chemicals and of AHLs in particular (Chong *et al.*, 2012), with the highest concentrations found within and around the aggregated bacteria, indicating high QS activity (Song *et al.*, 2014). Many commonly occurring wastewater bacteria are known to produce signaling molecules, for example, *Acinetobacter, Aeromonas, Chitinimonas, Citrobacter, Enterobacter, Klebsiella, Leclercia, Malikia, Raoultella, Pseudomonas,* and *Serratia* (Maddela *et al.*, 2019), including many obligate anaerobic bacteria and archaea such as *Methanosaeta harundinacea* 6Ac (Li *et al.*, 2015).

QS signaling molecules are produced by specific genes, and those responsible for cell aggregation have been identified as BtaI1/BtaR1 (homologs of LuxI/LuxR), which mediate C8-HSL signaling (Chandler *et al.*, 2009). Producing signals are generated by AHLs synthase genes (luxI homolog), while regulatory genes (luxR homolog) perceive and respond to AHLs. The main signaling molecules in anaerobic processes, which are very common, are AHLs AI-2 and DSF, with DSF mainly located within granules and AI-2 in the liquid phase. Feng *et al.* (2014) noted that AHLs with long acyl chains tended to diffuse into granules; those with short to medium acyl chains were found in the liquid. Ma *et al.* (2018) examined ten full-scale industrial anaerobic granular bioreactors and were able to detect four different AHLs, with C8-HSL and C10-HSL both facilitating granulation through EPS synthesis, which they appear to control and were produced by 12 different microorganisms. These two AHLs seem to be universally associated with granulation. Although the ability to produce AHLs is not universal, it may be that the majority of bacteria respond to them. Li *et al.* (2021) studied the recovery of a UASB reactor after a shock loading by increasing the organic loading rate (OLR) by stages and observing a change in microbial community and endogenous AHLs during recovery. In particular, C14-HSL improved aggregation through increasing protein synthesis of EPS. The metabolism of acidogenic fermentation bacteria was improved by C4-HSL, C8-HSL, and 3-oxoC14-HSL. Interestingly, the authors concluded that 3-oxoC6-HSL was a key signal molecule in enhancing methanogenesis.

The role of QS in anaerobic treatment has been reviewed in detail by He and Zhu (2024), with their key functions in the regulation of EPS secretion, promoting biofilm formation, acceleration of granulation, and the promotion of bacterial metabolism and symbiosis (Fig. 9.7). AHLs play a significant role at every stage of digestion from, increasing the activity of hydrolytic acid-producing bacteria, thereby increasing acid production to regulate the expression of methane-related genes in methanogens, increasing methane production. Hue and Zhu use the term "controlling social behavior," and that is what the signaling molecules are doing, controlling the interactions within the four stages of anaerobic digestion. They have also compiled a useful list of QS studies in different anaerobic treatment processes (Table 9.4). QS signals are disrupted by quorum quenching (QQ) molecules, which effectively turn off that particular function (Song *et al.*, 2014). QS molecules have a significant impact on floc and granule structure, composition, and functions, as do QQ molecules with the development of operational strategies to utilize them at the core of making anaerobic processes more efficient and sustainable.

9.2.2 *Role of Metals*

Trace metals

Trace elements and, in particular, metals are essential for optimal functionality of microorganisms in the anaerobic process (Table 9.5a). These and other micronutrients need to be present in sufficient

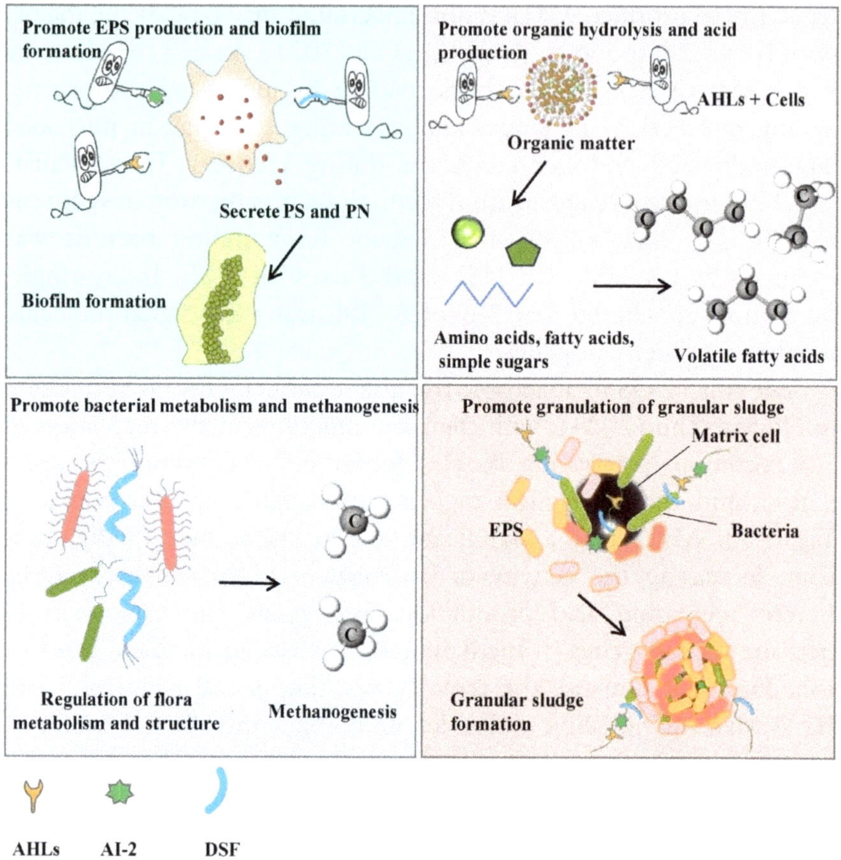

Figure 9.7. Graphical representation of key roles of quorum sensing in anaerobic treatment (He and Zhu, 2024). Reproduced with permission of Elsevier Publishing.

quantities to support both growth and metabolic activity, although it is essential that they are not only present but in a bioassimilable form (Zandvoort *et al.*, 2006). Methanogens are particularly sensitive to metal concentrations being inhibited at either high or low concentrations (Table 9.5b). The essential metals for anaerobes are Fe, Zn, Ni, Co, Cu, Mn, V, Mo, Se, and W (Fermoso *et al.*, 2009).

The most important and essential metal is Fe, which is used as a co-enzyme, oxidation-reduction agent, and terminal acceptor, with a positive correlation between the acetate utilization rate and iron

Table 9.4. Examples of signaling molecules and their specific functions in different systems based on case studies. Details of individual case studies are given by He and Zhu (2024). Reproduced with permission of Elsevier Publishing.

Systems	Function	Regulatory performance	Types of signaling molecules	References
Bacteria	***EPS***	Syntrophic and methanogenic bacteria can bi-directionally regulate their EPS production through the QS system to construct a syntrophic colony and improve the methane fermentation efficiency of propionic acid and butyric acid.	3OC6-HSL, C10-HSL	(Li *et al.*, 2019)
	Vital movement	The methanogenic archaeon *Methanosaeta harundinacea* 6Ac encodes an active community sensing system for regulating cell assembly and carbon metabolic fluxes.	—	(Zhang *et al.*, 2012)
	Metabolism	(1) C14-HSL dominates the phase I of SBOB, and C12-HSL dominates the phase IV of SBOB; C4-HSL dominated the whole growth process of SEOB, and C12-HSL dominated the phase IV of SEOB; C6 dominates phase I and IV of SPOB, and C12-HSL secretion increases from phase I to III; The concentrations of C4-HSL, C6-HSL, and C12-HSL were the highest in the first two phase of AM (phase I and II); C4-HSL, C10-HSL, OXOC10-HSL, and OXOC14-HSL are the highest in the first two phases of HM (phase I and II), and C12-HSL is the dominant signaling molecule in phase III and IV.[a]	(1) C4-HSL, C6-HSL, C10-HSL, 3OC10-HSL, C12-HSL, C14- HSL, 3OC14-HSL	(Chen *et al.*, 2023; Li *et al.*, 2019; Zhang *et al.*, 2019)

(*Continued*)

Table 9.4. (*Continued*)

Systems	Function	Regulatory performance	Types of signaling molecules	References
		(2) Intraspecies regulation was mediated by OXOC6-HSL and C10-HSL for SPOB and HM, respectively. Metabolic coordination between SPOB and AM was bi-directionally regulated under the mediation of the enjoyment signaling molecule, whereas metabolic coordination of SPOB by HM was unidirectionally regulated.[a] (3) C4-HSL can promote the growth of butyric acid-producing bacteria at 15 g L^{-1} NaCl.	(2) 3OC6-HSL, C10-HSL (3) C4-HSL	
Anaerobic granular sludge	***EPS***	Adding C6-HSL, C8-HSL can increase the concentration of TB-PN.	C6-HSL, C8-HSL	(Lv *et al.*, 2018)
	Metabolism	(1) The addition of AHLs changed the microbial community structure of bacteria and methanogens to different degrees, and improved organic matter removal and methanogenesis. (2) The addition of C6-HSL, C8-HSL, and 3OC6-HSL induced the enrichment of actinomycetes and changed the community structure; the addition of C6-HSL, 3OC6-HSL, and C4-HSL promoted the potential methanogenic metabolic pathways of granules. (3) The addition of 3OC10-HSL, C6-HSL, and C8-HSL increased the abundance of acetyl coenzyme A and acetyltransferase in the system, resulting in a more active fatty acid production reaction.	(1) C4-HSL, C6-HSL, C8-HSL, 3OC6-HSL, 3OC8-HSL (2) C4-HSL, C6-HSL, C8-HSL, 3OC6-HSL (3) 3OC10-HSL, C6-HSL, C8-HSL	(Lv *et al.*, 2018; Dang *et al.*, 2022; Li *et al.*, 2021a; 2021b; 2021c)

	Sludge granulation	(1) AI-2 promoted the formation of LE-EPS when alkaline; AI-2 and DSF jointly regulated sludge granulation when acidic, neutral, and weakly alkaline. (2) C4-HSL, C6-HSL, AI-2, and DSF jointly regulated to maintain the stability of sludge particles.	(1) AHLs, DSF, AI-2	(Ding *et al.*, 2015; 2016)
	Calcification of granular sludge	C6-HSL and 3OC6-HSL can improve the calcification of anaerobic granular sludge and reduce the production of calcium carbonate.	C6-HSL, 3OC6-HSL	(Li *et al.*, 2021b; Fu *et al.*, 2022a; 2022b)
UAnMBR Reactor	***Bioabundance***	Ten types of AHLs were detected, with C10-HSL being the predominant AHL, followed by C4-HSL, and C8-HSL. Biostatistical analyses showed that C10-HSL was associated only with bacteria, while the other types of AHLs were associated with both bacteria and archaea.	C4-HSL,3OC4-HSL,C6-HSL, 3OC6-HSL, C8-HSL, 3OC8-HSL, C10-HSL, 3OC10-HSL, C12-HSL, 3OC12-HSL	(Tabraiz *et al.*, 2020)
AnMBR Reactor	***Metabolism***	(1) Starvation may lead to an increase in medium- and short chain AHLs and AI-2 during long-term operation of the AnMBR. (2) Lower concentrations of AI-2 may reduce the occurrence of acetic acid and favor the occurrence of hydrogenated methane, thus reducing the accumulation of VFA and increasing methane production.	(1) AHLs, AI-2 (2) AI-2	(Shah *et al.*, 2022; Xiao *et al.*, 2019)
EGSB Reactor	***Metabolism and Sludge granulation***	C4-HSL and C8-HSL enhance anaerobic sludge granulation by stimulating the secretion of proteins into EPS, and 3OC6-HSL and 3OC14-HSL contribute to methanogenesis.	C4-HSL,C8-HSL,3OC6-HSL, 3OC14-HSL	(Yan *et al.*, 2021b)

(*Continued*)

Table 9.4. (*Continued*)

Systems	Function	Regulatory performance	Types of signaling molecules	References
UASB Reactor	*EPS*	C14-HSL improves protein synthesis in EPS.	C14-HSL	(Li *et al.*, 2021a)
	Sludge granulation	(1) C6-HSL can increase the abundance and metabolic activity of acetate crumb methanogens, promote EPS secretion, and reassemble decomposed sludge into large particles. (2) C8-HSL and C10-HSL were involved in EPS production and granulation process.	(1) C6-HSL (2) C8-HSL, C10-HSL	(Zhang *et al.*, 2022b; Ma *et al.*, 2018a, 2018b)
	Metabolism	(1) C4-HSL, C8-HSL and 3OC14-HSL all promoted the metabolism of acid-producing fermenters; 3OC6-HSL promoted methanogenesis. (2) The dominant signaling molecule C4-HSL, which promotes extracellular polysaccharide synthesis and aspects of acidogenic metabolism, regulates metabolism at low alkalinity (700–1,800 mg L^{-1}); 3OC6-HSL, 3OC12-HSL, and C14-HSL regulate metabolism at alkalinity of 1,800–2,200 mg L^{-1}. (3) The introduction of C6-HSL, C8-HSL, C10-HSL and 3OC6-HSL signalling molecules in UASB increased the overall number of bacteria and archaea and favored methanogenesis.	(1) C4-HSL, C8 HSL, 3OC14-HSL (2) C4-HSL, 3OC6-HSL, 3OC12-HSL, C14-HSL 3 C6-HSL, C8-HSL, C10-HSL (3) OC6-HSL	(Li *et al.*, 2021a; Yan *et al.*, 2021b)

[a]Phase I is the lag phase, Phase II is the logarithmic phase, Phase III is the third phase, and Phase IV is the logarithmic phase. SPOB — syntrophic propionate-oxidizing bacteria, SBOB — syntrophic butyrate-oxidizing bacteria, SEOB — syntrophic ethanol-oxidizing bacteria , HA — homoacetogens, AM — methanogens including aceticlastic methanogens, and HM — hydrogenotrophic methanogens.

Table 9.5. (a) The application of trace metals in anaerobic processes. (b) The optimum, stimulatory, and inhibiting concentrations of trace elements and their corresponding metal-rich enzymes on key anaerobic microorganisms (Nakhate *et al.*, 2023). Reproduced by permission of Elsevier Publishing.

(a)

TEs	Anaerobic digestion	Stimulatory	Inhibitory	Conclusion
Fe	Promotes organic degradation, stabilizes food waste, and reacts with H_2S to form Fe-S	1.4–2.3 g/L	NA	Stimulation for substrate hydrolysis
Ca	Enhances methane production, prevents foaming and granulation of anaerobic sludge	120–200 mg/L	7,000–8,000 mg/L	Helps in maintenance of microbial community
Mg	Enhances methane production and prevents foaming	720 mg/L	300–400 mg/L	Inhibits acetoclastic methanogenesis at a higher concentration
Co	Cofactor of methyltransferase and carbon monoxide dehydrogenate (CODH)	0.02 mg/L Boosts acidogenesis	>250 mg/L Suppress methanogenesis	Supply in trace amount is beneficial
Ni	Component of cofactor F430 required for methyl reductase activity during the last step of methanogenesis	0.5uM –0.1 mg/L VFA oxidation	1 mg/L Hampers acetoclastic methanogenesis	Fix VFA accumulation problem
Ni and Co	Stimulate both biogas production and methane content of biogas	Ni: 0.6 mg/kg and Co: 0.05 mg/kg	Ni: 250 mg/L Co: 250 mg/L	Acetoclastic methanogenesis pathway
Se and W	Tolerate high ammonia concentration and preventing VFA accumulation	NA	6 ng Se/g COD and 14 ng W/g COD	Accelerate microbial metabolism
Cu	Cofactor for microbial enzymes and coenzymes involved in the methane production	10 mg/L	30–50 mg/L	Enhances conversion of organics
Mo	Stable reactor operation at high OLR	0–1.6 mg/L	NA	Essential micronutrient for methnogens

(*Continued*)

Table 9.5. (*Continued*)

(b)

Trace element	Steps	Microbes	Optimum / Stimulatory concentration	Inhibitory concentration	Enzymes
Ni	Acetogenesis	Syntrophs	0.03–27 mg/L	35–1,600 mg/L	CODH, Methylreductase, Hydrogenase
	Methanogenesis	*Methanosarcina barkei*	0.0059 mg/L	NA	MCR/F430
		Methanobacterium smithii	0.059 mg/L	NA	
		Methanobacterium bryantii	0.96 g/L	NA	
Co	Methanogenesis	*Methanococcus voltae*	0.001 g/L	0.1289 g/L	Methyltransferase /Factor III
Fe	Acetogenesis and methanogenesis	NA	<0.33 mg/L	NA	Formylmethanofuran dehydrogenase, CODH, ACS, Hydrogenase
	Methanogenesis	*Methanobacterium thermoautotrophicum*	0.073 g/L	NA	Formylmethanofuran dehydrogenase
Mo	Methanogenesis	*Methanobacterium autotrophicum*	>0.002 g/L	NA	Formylmethanofuran dehydrogenase
		Methanosarcina barkeri	0.048 mg/L	NA	Formylmethanofuran dehydrogenase
		Methanospirillum hungatei	0.0064 g/L	NA	FDH
Se	Methanogenesis	*Methanococcus voltae*	0.64 g/L	NA	NA
		Methanococcus maripaludis	0.0157 g/L	NA	NA

		Methanococcus vannielii	0.79 g/L	NA	NA
W	Methanogenesis	*Methanococcus vannielii*	0.75 g/ L	NA	Formylmethanofuran dehydrogenase
Mo	Methanogenesis	Syntrophs	0.79 mg/L	NA	FDH
Se	—	—	—	—	—
W	—	—	—	—	—

NA: Not Available.

availability, as well as being essential for methanogen growth. Cobalt is often cited as the limiting metal due to its low concentration in most wastewaters and sludges, with all methanogens requiring cobalt, nickel, and iron. The role of cobalt and other major metals are summarized in Table 9.6.

While anaerobic wastewater treatment processes can be inhibited by excessive concentrations of metals, cellular function, and, in

Table 9.6. Role of some essential trace metals in the production of key enzymes required in anaerobic metabolism (Fermoso *et al.*, 2009). Reproduced by permission of Elsevier Publishing.

Element	Functions	Element	Functions
Cu	• Superoxide dismutase • Hydrogenase (facultative anaerobes) • Nitrite reductase • Acetyl-CoA synthase	Ni	• CO-dehydrogenase • Acetyl-CoA synthase • Methyl-CoM reductase (F_{430}) • Urease • Stabilize DNA. RNA • Hydrogenase
Co	• B_{12}-enzymes • CO-dehydrogenase • Methyltransferase	Se	• Hydrogenase • Formate dehydrogenase • Glycin reductase
Fe	• Hydrogenase • CO-dehydrogenase • Methane monooxygenase • NO-reductase • Superoxide dismutase • Nitrite and nitrate reductase • Nitrogenase	W	• Formate dehydrogenase • Formylmethanofuran-dehydrogenase • Aldehyde-oxydoreductase • Antagonist of Mo
Mn	• Stabilize methyltransferase in methane-producing bacteria • Redox reactions	Zn	• Hydrogenase • Formate dehydrogenase • Superoxide dismutase
Mo	• Formate dehydrogenase • Nitrate reductase • Nitrogenase	V	• Nitrogenase • Chloroperoxydase • Bromineperoxydase

particular, biogas production can be affected when they are in short supply or not in an assimilable form. For example, Fermoso *et al.* (2008a) demonstrated how a low assimilable Ni concentration inhibited methanogenic activity in a UASB reactor. They observed that this could be rectified by using a metal supplement. However, this was not the case when Zn was the limiting metal, with a supplement failing to readjust the pH caused by excessive acidification due to reduced methanogenic activity (Fermoso *et al.*, 2008b). For supplements to work effectively, they must be in a bioavailable form, which depends on the metal speciation. This requires the metals to present as free forms or bound to organic ligands, and must be added at a suitable rate so that they are not precipitated or removed by complexation before they can be utilized by the microorganisms (Roussel, 2012). Metal solubility in reactors is affected by the pH, with solubility increasing as the pH falls, as well as affecting the retention of metals in the biofilm or granules. The most used inorganic form of metal supplement is either as a metal chloride or sulfate due to their solubility, but they can be readily precipitated by the hydrogen sulfide present into insoluble metal sulfides. Metals can also be effectively removed by phosphate or carbonate precipitation or biomass chelation. By using a metal-organic compound as a supplement, precipitation by hydrogen sulfide is largely prevented, although the metal is less readily available to microorganisms.

Three chelators (i.e., aminopolycarboxylic chelating ligands) are often used to maximize metal bioavailability: ethylene diamine tetra acetic acid (EDTA), ethylenediamine-N N'-disuccinic acid (EDDS), and nitrilotriacetic acid (NTA) (Chauhan *et al.*, 2015; Zhang *et al.*, 2015). For example, cobalt was found to be limiting in a two-stage anaerobic digestion system by Li *et al.* (2023a). While the acidogenic phase was unaffected, the methanogens were inhibited by the lack of Co^{2+}. The optimum concentration of 2 mg L^{-1} was restored by using the chelating agent EDDS for Co^{2+} addition, which increased the levels of both vitamin B_{12} and the important methanogen coenzyme F_{420}, restoring normal metabolic activity and biogas production.

The abundance of the key methanogens *Methanofollis* and *Methanosarcina* were enhanced.

In order to optimize anaerobic treatment processes, especially in terms of biogas production, optimum nutrient balances must be maintained in reactors, which is especially important with industrial wastewaters, which are generally deficient in one or more micronutrients or trace elements. The behavior of metals in anaerobic digesters has been the subject of an interesting study by Jimmy Roussel at Birmingham University and gives an insight as to how metal supplements work (Roussel, 2012), while Zandvoort *et al.* (2006) give an excellent overview of the importance of trace metals in enzyme reactions, metal speciation, and mobility (Fig. 9.8) and the need for supplement dosing.

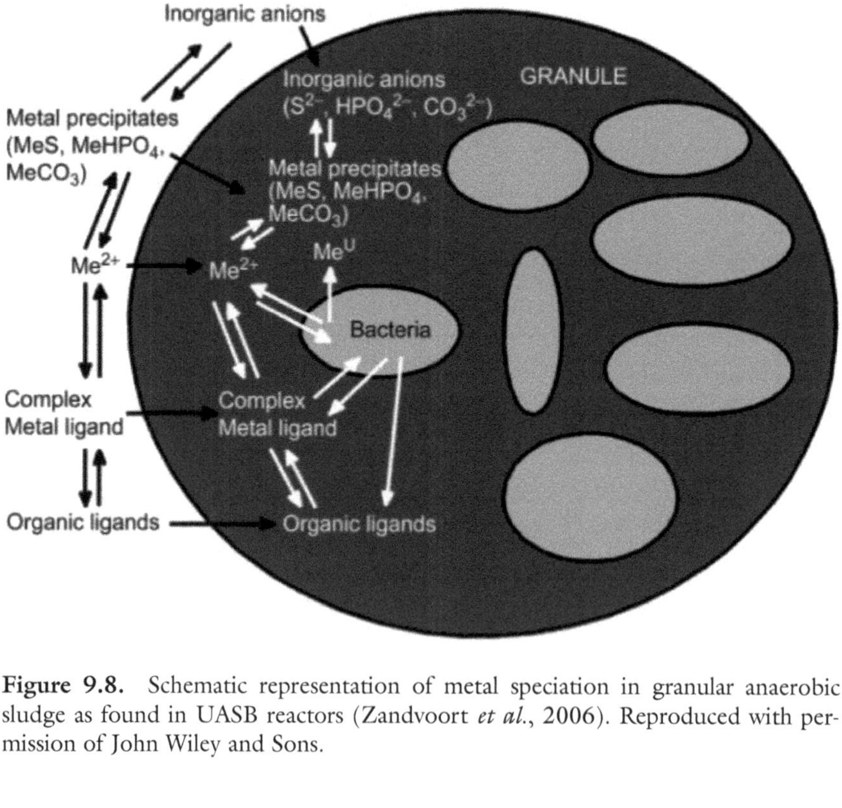

Figure 9.8. Schematic representation of metal speciation in granular anaerobic sludge as found in UASB reactors (Zandvoort *et al.*, 2006). Reproduced with permission of John Wiley and Sons.

Heavy metal recovery

Using sulfate reducing bacteria (SRB) to recover metals from wastewaters is set to play an important role in safeguarding our future supplies of important metals as well as protecting the environment from heavy metal pollution. Recovering and the reuse of useful materials, in this case, metals, from waste streams is a central strand in the circular economy concept. The key advantage to using anaerobic treatment in the recovery of metals is that even at very low concentrations, they can be effectively recovered, producing highly insoluble salts at low capital and operating costs.

SRB use SO_4^{2-} as a terminal electron acceptor under anaerobic conditions to oxidize simple organic compounds (CH_2O), which act as an electron donor:

$$2CH_2O + SO_4^{2-} \rightarrow H_2S + 2HCO_3 \tag{9.3}$$

The hydrogen sulfide produced reacts with dissolved metals (M^{2+}) to form a metal sulfide, which is highly insoluble and so readily precipitates out of solution.

$$H_2S + M^{2+} \rightarrow MS + 2H^+ \tag{9.4}$$

The reaction is completed, causing an increase in alkalinity.

$$HCO_3^- + H^+ \rightarrow CO_2 + H_2O \tag{9.5}$$

The pH is important not only for successful sulfate reduction but also for metal recovery, as different metals precipitate at specific pHs (Table 9.7), leading to selective metal recovery. The effective recovery of the metal sulfide precipitate is also important, as they can naturally contain metal sulfide particles of various sizes, including those that fall into the NP size range. These metallic NPs (MeNPs) are very small, have a large surface area, and are very reactive, making them very toxic to a wide range of organisms. Polycrystalline metal NPs are generally different in their size range for each metal: FeS and PbS <10 nm, CdS and CuS <14 nm, and NiS and ZnS <17 nm, with smallest particles produced when treating acid mine drainage (AMD) 2–10 nm.

Table 9.7. Effective pH for metal recovery by precipitation by hydrogen sulfide.

Metal	pH
Fe	~6
Cu	2.8 and 3
Zn	5.5–6.0
Ni	7.5–8.0
Cd	4.5–5.0
Pb	7.5–8.5
Mn	9.42 and 10.34

However, metal sulfide NPs have a wide range of potentially very useful uses, including the manufacture of electronics (e.g., batteries, LEDs, solar cells), the use in environmental remediation, biomedicine (e.g., drug therapy, fluorescent probes), antifungal agents in agriculture, electroplating, and catalytic reactions, such as water splitting, to generate hydrogen (Table 9.8) (Kumar *et al.*, 2021; Carmona *et al.*, 2023).

A wide range of reactors can be used, both attached and suspended biomass systems, with sulfate reduction and metal precipitation occurring together in single reactor systems, while in multi-stage systems, the two processes can be separated, making the recovery of metal sulfide precipitation easier and more effective (Fig. 9.9). Kumar *et al.* (2021) compared the various reactor types and found inverse or down-flow fluidized beds (AnIFBs) the best in terms of metal recovery (Table 9.9). In a study by Villa-Gomez *et al.* (2011), a single-stage AnIFB was able to recover between 91 and 99% of the common wastewater metals from an influent containing just 5 mg L^{-1} and at a COD/sulfate ratio of 1 at neutral pH. The problem of single-stage systems is that heavy metals can inhibit the SRB as they can only tolerate short-term exposure.

9.2.3 *Membrane Technology*

One of the drawbacks of anaerobic wastewater treatment is the need for tertiary treatment, either in the form of aerobic treatment and

Table 9.8. The synthesis method, size, and shape and the main applications of metal sulfide nanoparticles (kumar *et al.*, 2021). Reproduced by permission of elsevier publishing.

Metal sulfide Nanoparticles (MeS NPs)	Synthesis methods	Size and shape	Applications
Iron (Fe)	Surfactant assisted hot-injection method (chemical synthesis), high-energy milling (mechanical method)	Cubic shape with good uniformity and size distribution in the range 60–200 nm	Photovoltaic devices, labeling of biological materials, transformer core materials, environmental remediation and deactivation of radioactive heavy metals
Zinc (Zn)	Lewis acid-base reaction (chemical synthesis), Thio-urea hydrolysis (hydrothermal synthesis), metal-organic framework strategy	Spherical crystals with an average diameter of 5.5 nm	*In vitro* imaging, light emitting diodes, bioactive coatings, fuel cells
Copper (Cu)	Biological synthesis, sonochemical synthesis using ultrasonicator, *in-situ* growth by solvothemral method, microwave irradiation, surfactant-based synthesis, enzymatic treatment (green synthetic method)	Different types — spherical particles of average size 4 nm; triangular plates of average size 12 nm; distorted spherical to oval shape with slight aggregations and size 8–14 nm.	Dye degradation, solar cells, immuno-sensors, drug-therapy, agriculture, lithium ion batteries, wound dressings, catalyst (click reaction), air filtration technology
Cadmium (Cd)	Chemical precipitation method, sonochemical method.	Varies from a cubic to hexagonal crystalline structure, with size ranging from 2.4 nm to 6.2 nm	Solar cells, fluorescent probe, water purification
Lead (Pb)	Chemical precipitation method, ultrasonic irradiation.	Uniform nanowires of average diameter 40 nm and length 5μ.	Solar cells, telecommunications
Nickel (Ni)	Hydrothermal synthesis, solvothemral method using a domestic microwave oven.	Sphere-like structure with an average crystallite size of 22 nm	Dye-sensitized solar cells, infrared detectors.

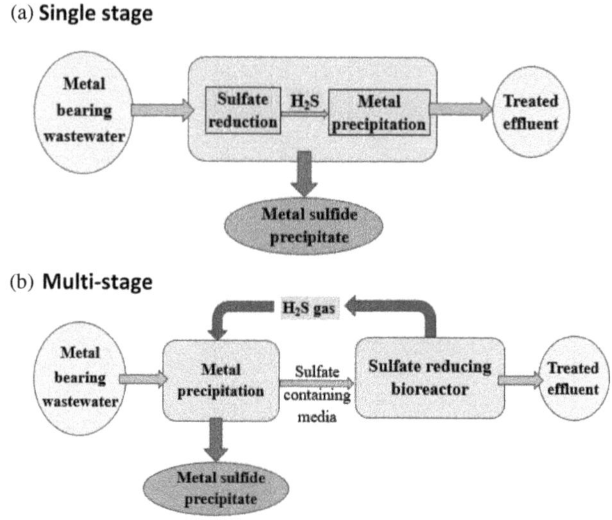

Figure 9.9. Schematic layout of single and multistage systems for metal sulfide recovery from wastewater (Kumar *et al.*, 2021). Reproduced with permission of Elsevier Publishing.

Table 9.9. Advantages and disadvantages of anaerobic reactors for biological sulfate reduction (Kumar *et al.*, 2021). Reproduced by permission of Elsevier Publishing.

Bioreactor type	Advantages	Disadvantages
Continuous stirred tank reactor	Consistent and reliable performance	High solids retention time (SRT) requires a large reactor volume; biomass washout
Gas lift reactor	High mass transfer efficiency, very good mixing	High pressure needed to pump gas inside the reactor
Packed bed reactor	High SRT results in lower reactor volume than CSTR; can be operated in up-flow or down-flow mode	Clogging; high pressure for pumping the flow
Membrane bioreactor	No need of sedimentation basin; high biomass retention results in high substrate degradation rate; avoids direct contact between metals and SRB	High cost to overcome the trans-membrane pressure; requires periodic backwash of the membrane

Table 9.9. (*Continued*)

Bioreactor type	Advantages	Disadvantages
Up-flow anaerobic sludge blanket (UASB) reactor	No clogging; good biomass settling capacity; no requirement of support material	Biomass washout; high susceptibility to influent characteristics
Bubble column reactor	Good mixing; used for turbid medium and shear sensitive cells	Not available (NA)
Fluidized bed reactor	No channeling or clogging, biomass retention is good; biomass activity and treatment efficiency are high	Energy needed for carrier fluidization; shear force can detach biomass; less volume available for biomass compared to the UASB reactor due to the inert biomass carrier
Anaerobic filter reactor	Low shear force; sludge retention time is more than that for CSTR; H_2S stripping is effective in downflow mode	Channeling of flow can be possible; pressure gradients can be large
Anaerobic digester	Excellent biomass holding capacity as compared with the CSTR	Biomass disintegrates into flocs and sludge
Anaerobic hybrid reactor	Less vulnerable to clogging compared with the AFR; sludge removal easier than in the AFR; biomass retention better than in the UASB	Not available
Anaerobic baffled reactor	High sludge retention time; reduced costs due to no biomass carrier; good tolerance toward shocks of hydraulic and organic loading	Not available

sedimentation or filtration. Membrane technology has improved the efficiency of anaerobic treatment by employing membranes to retain solids and biomass within the reactor while also ensuring a high-quality final effluent (i.e., permeate), which is particle-free with extremely low pathogen and antibiotic gene concentrations (Ozgun *et al.*, 2013). By separating the SRT from the HRT, membranes can significantly reduce treatment times, required reactor volume, start-up times, and improve final effluent quality. They can also be used for upgrading biogas. AnMBRs are ideal where slow-growing microorganisms are utilized, especially where municipal wastewaters are treated where its low strength limits growth, and for treating inhibitory contaminants (Ji *et al.*, 2020; Li *et al.*, 2023b).

AnMBRs were developed in the 1980s and employ either ultrafiltration or microfiltration membranes within the bioreactor, often as a submerged hollow fiber membrane or as a crossflow membrane at the effluent outlet; alternatively, both can be used as a separate side stream process (Fig. 9.10). In crossflow systems, the liquid flows

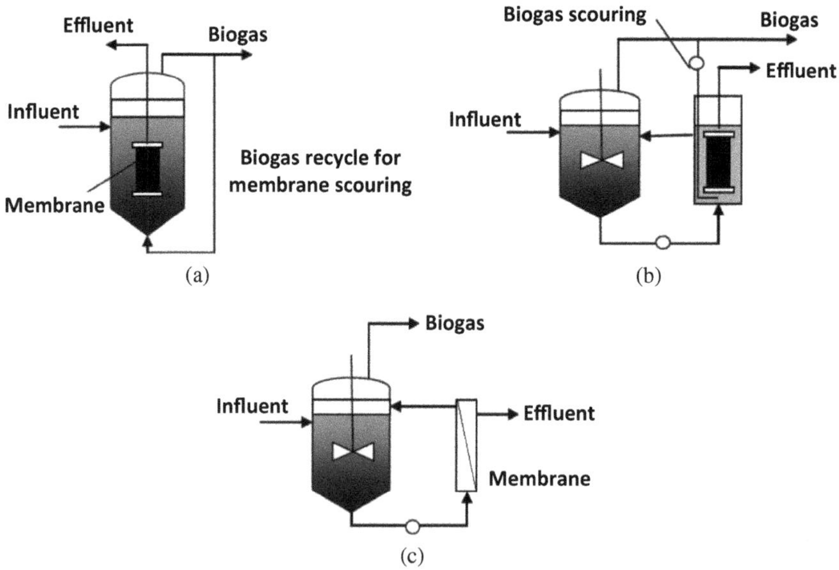

Figure 9.10. There are three common anaerobic membrane reactor configurations: (a) the membrane is submerged within the reactor, (b) external side stream hollow filter membrane, or (c) an external cross flow membrane (Chang, 2014). Reproduced under the Creative Commons Attribution (CC BY) license (https://creativecommons.org/licenses/by/4.0/).

along the membrane surface at high velocity, creating sufficient shear to control fouling and for the permeate to pass the membrane. These are generally used with UASB and AnCSTR reactors and require extra energy for pumping. In submerged systems, the membrane is located within the reactor, with a vacuum used to draw the permeate, which reduces energy consumption. Three membrane types are used: tubular, hollow-fiber, and flat sheet membranes, all of which come in the form of cassettes or units, which are easier to handle. In this case, fouling is controlled by scouring using biogas (Fig. 9.10). The types of membranes available are fully explained in an earlier volume in this series (Gray, 2023).

All wastewater anaerobic bioreactors can employ membrane separation, including sludge digesters treating sewage sludges from conventional aerobic treatment (Liao *et al.*, 2006). However, one of the limiting factors in AnMBR technology is the fouling of the membranes, preventing the passage of the permeate, with increased energy costs as permeability declines. Soluble microbial products and EPS are both key factors in membrane fouling, both becoming more problematic at lower temperatures. While liquid or biogas scouring is generally sufficient, occasional chemical cleaning may be necessary.

While AnMBRs can be operated successfully at ambient temperatures (5–30°C), seasonal temperature fluctuations have been reported as causing operational difficulties, with both COD removal and methane yield falling as temperatures fall. The reported performance of AnMBRs treating wastewater varies, with COD removal between 69 and 93%, a methane yield of 0.07–0.244 L CH_4 g^{-1} COD (Table 9.10a), and when used for sewage sludge digestion, methane yields range from 0.19 to 0.31 L CH_4 g^{-1} VS, with COD removal at 97–99% (Table 9.10b).

Upgrading biogas by converting the carbon dioxide present to methane by introducing hydrogen to improve methanogenesis is considered highly sustainable and can be achieved in AnMBRs by improving hydrogen transfer to the liquid phase. This can be achieved using submerged hollow-fiber or tubular membrane cassettes placed in the digestate, although this method can result in the unwanted accumulation of VFAs in the reactor due to excess hydrogen, so a separate unit is more reliable but more expensive, both in capital and operating costs (Fig. 9.6). This has been reviewed by Sun *et al.* (2022) and Li *et al.* (2023b).

Table 9.10. Performance of anaerobic membrane reactors for (a) wastewater treatment and (b) sludge digestion (Li *et al.*, 2023). Reproduced under the Creative Commons Attribution license.

(a) Wastewater treatment bioreactors

System configuration*	Scale (L)	Temperature (°C)	$COD_{influent}$ (mg/L)	HRT (h)	SRT (d)	Methane yield (L/g-COD)	COD removal efficiency (%)
S-FS-AnMBR	40	18/24	428–477	48–12	50	—	69–89
S-HF-AnMBR	20	15	412	6–24	21–491	0.21–0.23	90.5
S-HF-AnMBR	1,650	25	762	11–21	100	0.2	92
ES-HF-AnMBR	34,400	10–27	755–1,403	25–41	70	0.07–0.17	93
S-HF-AnMBR	5,000	15–25	203–490	8	20–71	0.205–0.244	77–93
ES-TM-AnMBR/UASB	160 + 150	18	892	7–17	—	0.199–0.235	73–90

*S: submerged; ES: external submerged; FS: flat sheet; HF: hollow fiber; TM: tubular membrane.

(b) Sludge treatment systems

System configuration*	Scale (L)	Temperature (°C)	OLR (g-COD/Ld)	HRT (d)	SRT (d)	Methane yield	COD removal (%)
ES-HF-AnMBR/CSTR	15	35	1.72–3.72	15–30	25–60	0.239 L/g-VS	99.0
ES-HF-AnMBR/CSTR	15	55	4.31	15	—	0.31 L/g-VS	96.7
ET-AnMBR	500	24, 35, 55	3.4–3.7	6.9–7.3	22–39	0.19–0.34 L/g-VS	—
ET-HF-AnMBR/CSTR	2.37 + 0.1	37	0.21–0.27	30–33	102–107	0.19–0.24 L/g-VS	—
ET-CM-AnMBR/CSTR	25	35 & 55	4.8–10.4	3–7	30–50	0.213–0.245 L/g-COD	—

*ES: external submerged; HF: hollow fiber; ET: external tubular; CM: ceramic membrane.

9.2.4 *The Future*

Conventional (aerobic) biological wastewater treatment treats the majority of domestic wastewaters and a significant proportion of industrial wastewaters with anaerobic processes, usually limited to sludge digestion. Aerobic treatment has been the vanguard for over a hundred years in the improvement of public health and environmental protection, especially in the improvement of river water quality. However, the high-energy demand for aerobic processes and their comparatively high-carbon footprint, combined with increasing needs to better control nutrients and pathogens and to conserve resources, including water, has led water companies to increasingly look toward anaerobic wastewater treatment to provide a more sustainable alternative. Also, the core change in the concept from a disposal facility to a resource recovery center embracing the concepts of the circular and green economies has forced us as an industry to alter our traditional expectations of what wastewater treatment is. Physical and chemical treatment methods remain core process options, but how we deal with the biological treatment of wastewater, where we employ microorganisms to breakdown the biodegradable fraction of our waste streams, is where the main opportunities lie for the recovery of carbon, nutrients, and many other resources, rather than simply discharging them back into the environment as oxidized and still polluting end-products (e.g., CO_2, NO_2, NO_3, PO_4).

Innovative use of combining different biological processes can yield exciting results. For example, combining dark and photo fermentation can produce high-quality effluents with nutrient removal but capturing most of the available carbon as methane (Elsayad *et al.*, 2024).

A major advancement in anaerobic wastewater treatment is the incorporation of membranes, which has made the treatment of municipal wastewaters anaerobically a feasible option, although limited post-treatment will still be required, especially for nutrient removal. The main problem with AnMBRs is the energy balance, with energy required in particular to control membrane fouling. Energy demand by an AnMBR is estimated at 1–2 kWh m^{-3}, although this varies widely in the literature, depending on reactor configuration and wastewater type. However, methane production will easily cancel out this energy demand, usually with a significant surplus that can be used elsewhere on-site or exported to the grid. Key problems with membrane systems, apart from fouling, is the build-up of inhibitory substances, including salinity (Aslam *et al.*, 2022). One recent innovation to control membrane fouling is to use a low-voltage electric field, a process known as the electrochemical field-assisted anaerobic membrane reactor (AnEMBR), which also improves pollutant removal and overall methane generation (Hu *et al.*, 2024).

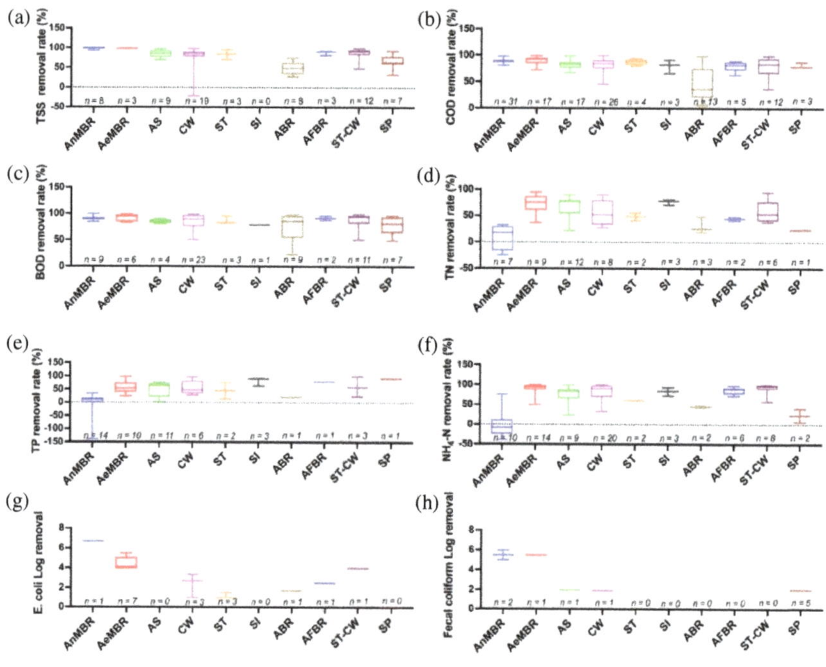

Figure 9.11. Comparison of performance efficiency of decentralized wastewater treatment technologies, in terms of percentage removal. The anaerobic membrane reactor (AnMBR) is compared to aerobic MBRs (AeMBRs), activated sludge (AS), constructed wetland (CW), septic tanks (STs), anaerobic baffled reactors (AnBRs), aerated fixed-bed reactors (AFBRs), septic tank-constructed wetland (ST-CW), and stabilization ponds (SPs) using the following parameters: (a) TSS, (b) COD, (c) BOD, (d) TN, (e) TP, (f) NH4+–N, (g) *E. coli*, and (h) fecal coliforms (Zhang *et al.*, 2023). Reproduced under the Creative Commons Attribution (CC BY) license (https://creativecommons.org/licenses/by/4.0/).

Zhang *et al.* (2023) compared the performance of AnMBRs to a range of other similar aerobic and anaerobic technologies (Fig. 9.11). AnMBRs perform similarly to aerobic MBRs, except that AnMBRs are unable to remove nutrients. In terms of the removal of suspended solids, including pathogens, MBRs are exceptional at achieving removal rates of 98–99%, compared to activated sludge (86%), constructed wetland (77%), septic tanks (84%), AnBRs (50%), aerated fixed-bed reactors (89%), septic tank-constructed wetland (87%), and stabilization ponds (67%). AnMBRs performed equally well as aerobic MBRs in terms of COD (89% and 90%, respectively) and BOD removal (91% and 92%). In terms of the loading rate, an AnMBR can operate at 1.92 Kg COD $m^{-1}d^{-1}$, compared to 0.67, 0.56, and 0.8 for aerobic MBRs, activated sludge, and AnBRs, respectively (Fig. 9.11).

It is estimated that between 3 and 4% of global electricity supplies are consumed by wastewater treatment plants (WWTPs) being primarily used for aeration systems, pumping, and sludge dewatering. The objective of wastewater sustainability is to obtain net zero in terms of energy consumption and, if possible, even become net positive. However, as effluent standards and loadings increase, more energy is required. Carbon neutrality is only possible using natural treatment systems or anaerobic processes, the latter being the only option for centralized wastewater treatment systems. The primary objective of sustainable wastewater treatment is to ensure that greenhouse gases are not released directly from unit processes. For example, carbon dioxide from aerobic processes or methane from anaerobic digestion released to the air, or unstabilized organic matter discharged to water or land, where it subsequently rapidly degrades (i.e., direct carbon emissions). A secondary objective is to reduce carbon emissions derived from energy consumption and other operational activities (i.e., indirect carbon emissions). Zhang *et al.* (2024) define a carbon-neutral anaerobic-based WWTP as being "simplified in process, such as one-stage or integrated design, and has a high removal efficiency for different concentration levels, and should be capable realizing energy self-sufficiency for either high-strength or low-strength wastewater." Among enhancement options suggested by the authors are: (a) *co-digestion* to ensure an optimum C:N ratio. Co-digestion can correct the substrate balance, optimizing pH, alkalinity, and prevent excessive VFA production and free ammonia nitrogen causing inhibition. (b) Ensuring adequate biodiversity of bacteria and archaea, which may require the addition of a co-cultured consortium of microbes, especially where industrial wastewaters are treated. (c) The disadvantage of methane production is that carbon dioxide is produced when it is utilized. This can be overcome by producing hydrogen instead, which, when used, produces only water; and finally, (d) wastewaters often contain recalcitrant organic matter, which can be pre-treated using chemical addition to make them more amenable to anaerobic breakdown by destroying the functional groups within the organic molecules.

Biochar can also be added, which improves microbial activity by providing a stable habitat, allowing high densities of bacteria and archaea to thrive. Biochar also helps regulate the alkalinity within the reactor (Wang *et al.*, 2018) (Fig. 9.12). Conductive materials, both iron and carbon-based, have also been shown to be effective in the treatment of antibiotic wastewaters. Apart from biochar, conductive materials include zero-valent iron, magnetite, activated carbon, graphene, and combinations of iron and organic materials. They work by ensuring rapid adsorption and biodegradation of antibiotics, as well as reducing inhibitory effects and preventing the expression of antibiotic resistant genes (ARGs). All these materials are also electron carriers, which accelerate interspecies electron transfer, facilitating organic matter breakdown and methanogenesis (Song *et al.*, 2024; Zhao *et al.*, 2024).

Anaerobic treatment has many advantages, including the ability to treat high organic loading rates, low production rates of stabilized sludge, excellent potential for bioenergy and byproduct transformation, and more stable and reliable degradation capability as well as requiring a fraction of the energy that a comparable aerobic system would require (Table 9.11). We tend to think of nutrient removal as requiring an aerobic step, but today, both nitrogen and phosphorus can also be effectively treated using anaerobic technologies, including anaerobic ammonium oxidation (anammox) and sulfur-based autotrophic denitrification (SAD) (Zhang *et al.*, 2024) (Fig. 9.12).

Nitrogen removal is normally carried out using a combination of aerobic and anoxic biological treatment and requires extra energy

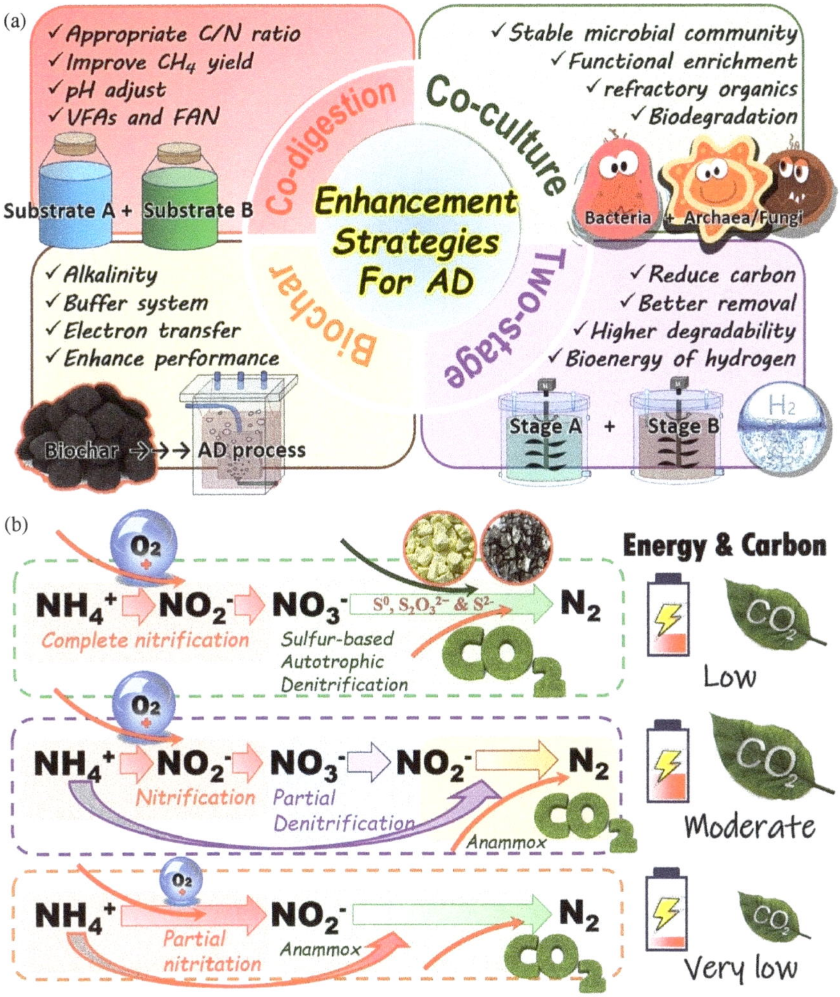

Figure 9.12. Visual summary of (a) enhancement strategies for anaerobic treatment and (b) carbon-neutral biological nitrogen removal technologies (Zhang *et al.*, 2024). Reproduced with permission of Elsevier Publishing.

for the aeration of wastewaters to achieve sufficient oxygen for the nitrifying bacteria. Another source of global warming is caused by incomplete nitrification — denitrification producing nitrous oxide, a greenhouse gas three hundred times more powerful than carbon dioxide. Strong wastewaters associated with anaerobic treatment are often rich in protein amino acids, amines, and amides, all rich in

Table 9.11. Comparison of modern anaerobic wastewater treatment systems treating both low and high-strength wastewaters at a range of different organic loading rates and hydraulic retention times (Zhang *et al.*, 2024). Reproduced with permission of Elsevier Publishing.

Process	High-strength wastewater	COD (mg/L)	OLR (kg COD/m^3/d)	HRT (h)	Removal COD (%)	Methane yield (L CH_4/g COD)
UASB	Swine	4,939–5,124	10.3–20.1	6–12	89.1–93.8	0.03–0.261
	Distillery	21,000	5–33	2.67–0.54 (d)	84.52 ± 3.68	10 (L/L)
	Slaughterhouse	3,500–32,000	1.75–16	24	68–95	0.04–0.28
	Dimethylformamide	3,000	1.63–4.22	18–48	92–97	0.557–1.270 (L/L)
	Cassava	18,000–20,000	15	N.A.	92.5	0.328
	Tapioca starch	16,700 ± 3,503	7.5–25	N.A.	70–82	0.23–0.34 (biogas)
EGSB	Cheese whey	46,348 ± 3,200	5.74–7.76	6–8 (d)	90–92	0.328–0.334
	Cephalosporin	5,981–91,870	3.60–9.96	24–25	72	0.24–0.33
	Corn alcohol	20,720	1.4	1.4–8 (d)	95.7	10.72 (L/L)
	Winery	1,250–1,327	15.32	26	70 ± 2	0.31 ± 0.07
	Leachate	4,909–7,200	14.8–21,8	8	70.8–91.6	0.07
	Piggery	780–2,970	0.32–6.13	12	86.2	0.215–0.299
	Tetracycline	6,000	5.06	24	67–79	0.099
IC	Pharmaceutical	16,500–18,950	4.3–6.3	3–4 (d)	82–92	0.3
	Dyeing	2,000–4,000	4.8–9.6	10	75–87	1.65 (L/L)
	Synthetic molasses	5,000–12,500	2.50–18.94	0.66–2	85–90	0.44–4.49 (L/L)

	Medicine	11,000	14.67	18	95	0.33
	Leachate	43,045–65,000	21.06–25.16	2.4 (d)	84.6–93.4	0.42–0.50 (biogas)
AnMBR	Municipal wastewater	755–1,403	0.6–1.28	25–41	86.7–93.1	0.07–0.169
		367–481	0.37–1.84	6–24	89.5–93.2	0.16–0.26 (biogas)
		576–6,510	0.25–1.96	2.4–3.6	90–96	0.214–0.322
		250.6–272.7	0.82–3.01	2–8	70.6–77.3	0.08–0.12
		300–600	0.41–0.53	16–24	90.5	0.21–0.23
		200–400	N.A.	6–12	85	0.31
		365 ± 101	0.72–1.52	6–12	88.7–89.3	0.21–0.24
		610 ± 260	1.3 ± 0.5	11 ± 3	88 ± 7	0.14 ± 0.06

nitrogen, so that during anaerobic degradation, they are normally transformed into ammonia. Exciting new developments in anaerobic biotechnology can now provide far more sustainable denitrification of ammonia, which includes SAD and the anammox processes, which also require either partial nitritation or nitrification-partial denitrification as pre-steps (Fig. 9.12). Nitrate or nitrite is removed from wastewater using sulfur compounds as electron donors and carbon dioxide as a carbon source. The SAD process can be quickly established, is easy to operate, and is a very stable and reliable system. It is ideal for wastewaters rich in NO_3 or NO_2, including leachate, and has also been successfully employed for treating groundwater (Liu *et al.*, 2021). The anammox process is a unique system that is highly efficient and cost-effective, which occurs under anoxic conditions where nitrite is converted to nitrogen gas by anammox bacteria using ammonium as an electron donor:

$$NH_4^+ + NO_2 \rightarrow N_2 + 2H_2O \tag{9.6}$$

Most commercial anammox processes utilize a separate nitritation process called partial nitritation (PN), where ammonium is oxidized to nitrite in order to ensure sufficient nitrogen dioxide for the process. So, while partial nitrification-anammox (PN/A) reduces some oxygen, which is reduced by over 60%, no extra carbon is required (Fig. 9.13).

The most widely used PN system is the SHARON® process, which is an acronym for **S**ingle reactor system for **H**igh activity **A**mmonium **R**emoval **O**ver **N**itrite, and is a process that produces the exact feed required for the Anammox® process, with the anammox bacteria housed in a separate reactor. The partial nitrification process provides the exact mixture of ammonium and nitrite (approximately 50:50 ammonium and nitrite), with pH control used to adjust the balance (van Donogen *et al.*, 2001).

$$NH_4^+ + HCO_3^- + 0.75\ O_2 \rightarrow 0.5\ NH_4^+ + 0.5\ NO_2^- + CO_2 + 1.5\ H_2O \tag{9.7}$$

The process is used to treat ammonia-rich effluents from anaerobic sludge digestion that is usually returned to the inlet works. The

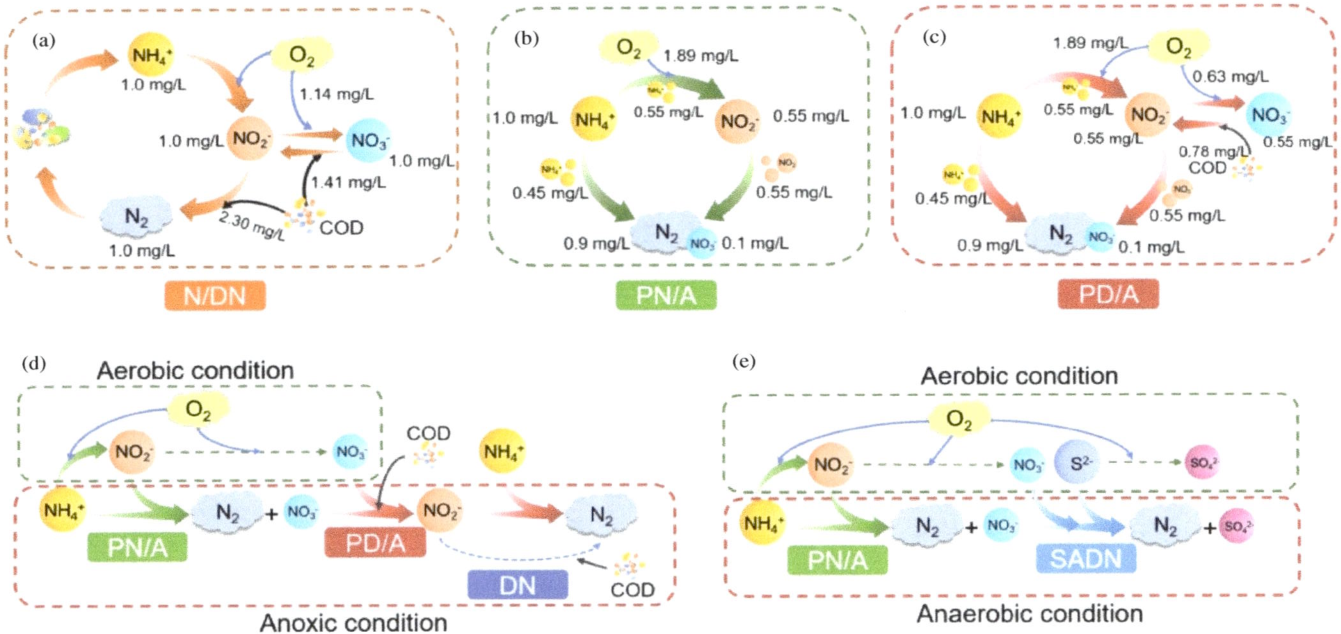

Figure 9.13. Conventional nitrification/denitrification (N/DN) and four anammox-based nitrogen removal pathways: (a) Conventional N/DN process, (b) partial nitrification/anammox (PN/A), (c) partial denitrification anammox (PD/A), (d) PN/A and PN/D combined with denitrification (DN), and (e) PN/A-sulfur autotrophic denitrification (SADN) (Deng *et al.*, 2025). Reproduced with permission of Elsevier Publishing.

stoichiometry implies that no extra addition of a base is required since sludge liquor from anaerobic digestion generally contains sufficient alkalinity in the form of bicarbonate to compensate for acid production, where only 50% of the ammonia is oxidized. The process operates without any biomass retention, so the SRT equals the HRT. Effluent concentration is only dependent on the growth rate (1/SRT) of the bacteria involved and is independent of the influent concentration. Fast-growing ammonium oxidizers are selected by the process when operated at temperatures >25°C. As these microorganisms have a low affinity for ammonia (affinity constant 20–40 mg NH_4-N L^{-1}), final effluents will always have a relatively high ammonia concentration (50–100 mg N L^{-1}). Thus, the SHARON® process is only applicable to wastewaters with high ammonia concentrations (>500 mg N L^{-1}) and where final effluent quality is non-critical. The microorganisms involved are diverse, with *Nitrosomonas* spp. identified using a 16S rRNA-targeted fluorescent oligonucleotide probe. The commonest species is very similar to *Nitrosomonas eutrophia*, which is known to be a fast-growing nitrifier able to grow at high ammonia and nitrite concentrations (Logemann *et al.*, 1998). When treating sludge digester effluents, the SHARON® process is operated at 30–40°C without any biomass retention. The dilution rate is set at such a rate that the ammonium oxidizers grow fast enough to stay within the reactor while the slower-growing nitrite oxidizers are washed out (Mulder *et al.*, 2001). The ratio of ammonium to nitrite can be altered by changing the reactor pH over the range of 6.5–7.5 (van Dongen *et al.*, 2001). Typical design data for full-scale reactors are given in Table 9.12.

The growth rate (doubling time: 11 days) and growth yield (0.11 g VSS g^{-1} NH_4-N) of anammox bacteria are very low, resulting in low sludge production. Therefore, efficient biomass retention is required to maintain suitable anammox biomass concentrations within reactors, therefore favoring AnSBR systems (Strous *et al.*, 1997). Anammox reactors are very compact due to a high maximum specific nitrogen consumption rate (0.82 g N g^{-1} VSS d^{-1}), a high affinity for ammonium and nitrite, and granular growth, allowing efficient biomass retention. The autotrophic bacteria responsible for the oxidation of ammonia and the simultaneous reduction of nitrite

Table 9.12. Dimensions of a full-scale SHARON®–Anammox® process for three different cases (van Dongen *et al.,* 2001). Reproduced with permission of IWA Publishing.

Reactor	Parameter	Unit	Case 1	Case 2	Case 3
General	N-load	kg N/day	1,200	1,200	1,200
	NH_4-N concentration	kg N/m^3	500	1,200	2,000
	Influent flow	m^3/day	2,400	1,000	600
SHARON-reactor	Volume	m^3	3,120	1,300	780
	Oxygen demand	kg O_2/day	2,181	2,181	2,181
	Air demand	Nm^3/day	56,000	56,000	56,000
Moving bed	Volume	m^3	450	450	450
Anammox-reactor	HRT	Hour	4.5	11	18
Granular sludge	Volume	m^3	75	75	75
Anammox-reactor	HRT	Hour	0.75	1.8	3

belong to the widely distributed phylum *Planctomycetes.* Not all the bacteria in this phylum are able to oxidize ammonium, with this being restricted to the Anammox-*Planctomycetes,* which have a characteristic membrane-bound cell compartment called the anammoxosome (Fuerst and Sagulenko, 2011). Ammonium is used as an energy source by the anammox bacteria and carbon dioxide as a carbon source, with NO_2^- , and NO_3^- , if present, used as electron acceptors. Kartel *et al.* (2011) have described the process in detail where the enzyme nitrite oxidoreductase reduces nitrite to nitrogen oxide, which then reacts with ammonium to form hydrazine (N_2H_2), enabled by the enzyme hydrazine synthase, which is oxidized by hydrazine dehydrogenase to nitrogen. The bacteria are extremely slow-growing and strict anaerobes, unable to function if the dissolved oxygen concentration rises above 0.5 mg L^{-1} (Sec. 4.4.7).

The Anammox® process has been used successfully at the Dokhaven Wastewater Treatment Plant in Rotterdam, where it is operated as a granular sludge AnSBR process. Ammonium conversion rates to nitrogen gas exceed 80% at loading rates of 1.2 kg N m^{-3} d^{-1} (van Dongen *et al.*, 2001). Typical design data for three different loading rates is given in Table 9.12.

Granulation creates inert centers on which bacteria can adhere. It is possible to operate the Anammox® process to simultaneously remove phosphorus, creating a complete nutrient removal process. Phosphorus is removed by biomineralization forming hydroxyapatite (HAP — $Ca_{10}(PO_4)_6(OH)_2$), which accumulates within the granules.

$$10Ca^{2+} + 6PO_4^{3-} + 2OH^- \rightarrow Ca_{10}(PO_4)_6(OH)_2 \qquad (9.8)$$

During the anammox reaction, H^+ ions are consumed, increasing the pH locally within the granule, so by adding calcium, calcium phosphate precipitation occurs, reaching supersaturation, causing HAP to become deposited within the granule. Over time, as the granule increases in size, the density of anammox bacteria within the granules declines due to the reduced substrate concentration and a high pH, causing an increase in calcium phosphate precipitation. This results in the creation of anammox-HAP granules, which have a unique structure, with an anammox biofilm surface layer for nitrogen removal and an inner inert HAP core for the biomineralization of phosphorus (Xue *et al.*, 2022; 2023). Apart from removing nitrogen from solution, the excess granule sludge provides a high-quality, easy-to-use phosphorus fertilizer.

There is no doubt that there will be a shift away from energy-hungry greenhouse gas-producing conventional BNR systems (AOA) to low-carbon footprint combined AnMBRs and anammox-HAP technology, enabling a whole range of byproduct recovery, especially biomethane and biohydrogen. The technology itself is resilient and reliable and is able to deliver the required treatment capacity currently unavailable at most centralized WWTPs, which are struggling to cope with ever-increasing flow rates, more complex wastewaters, increasing operational costs, and the inability to meet carbon-neutral or carbon-zero targets. Our increasing understanding of the role of QS signaling molecules in anaerobic treatment will be critical in enhancing overall performance and reactor stability, especially in relation to granulation and methanogenesis, whether this is by adding exogenous signaling molecules, adding QS microorganisms, or by the use of conductive materials. The key areas where QS will be able to make a significant difference are in promoting: (a) EPS development through regulation of amino acid synthesis and gluconeogenesis, (b) enzyme synthesis for all stages of anaerobic digestions (i.e., hydrolysis, acidogenesis, and methanogenesis), and (c) interspecies electron transfer by controlling electron carrier synthesis (Chen *et al.*, 2024). Performance data for a wide range of anaerobic wastewater treatment systems have been summarized by Stazi and Tomei (2018), who also discuss future prospects for anaerobic processes.

It is inevitable as we move toward using stricter sustainability indicators to measure WWTP performance, the role of anaerobic systems will begin to dominate future domestic as well as industrial WWTP designs, especially in relation to carbon footprint and nutrient and pathogen removal. This may also include eco-labeling specific processes using a tool such as ISO 14045 (ISO, 2024), as demonstrated by Lorenzo-Toja *et al.* (2016) and fully discussed in an earlier book in this series (Gray, 2022). However, Li *et al.* (2022) have shown that there is a significant difference in the way we assess WWTPs in terms of zero-carbon and the reality of wastewater treatment in terms of carbon neutrality. They propose that carbon footprint analysis should be carried out throughout the whole collection, treatment, and disposal cycle and proposed three boundaries for separate analysis: the actual wastewater treatment plant as defined by its boundary fence, the urban infrastructure as related to WWTP, and the human society and ecological system (Fig. 9.14). Within-the-fence boundary accounting for carbon emissions is restricted to when the wastewater starts its treatment and ends at the point it leaves the site as a discharge or sludge. This is the normal method of assessing the carbon footprint of wastewater treatment but fails to take into account the upstream and downstream impacts. So, using a secondary boundary, the urban infrastructure related to WWTPs, emissions could be included from the sewer network and also where other wastes are transported to the WWTP for treatment, including other materials such as food waste used in co-digestion. For example, anaerobic digestion is increasingly used to degrade food waste, which is diverting organic matter from landfills to produce useful biogas and digestion products. In the European Union alone, 88 million tonnes of food waste is produced, representing 55% of the total waste going

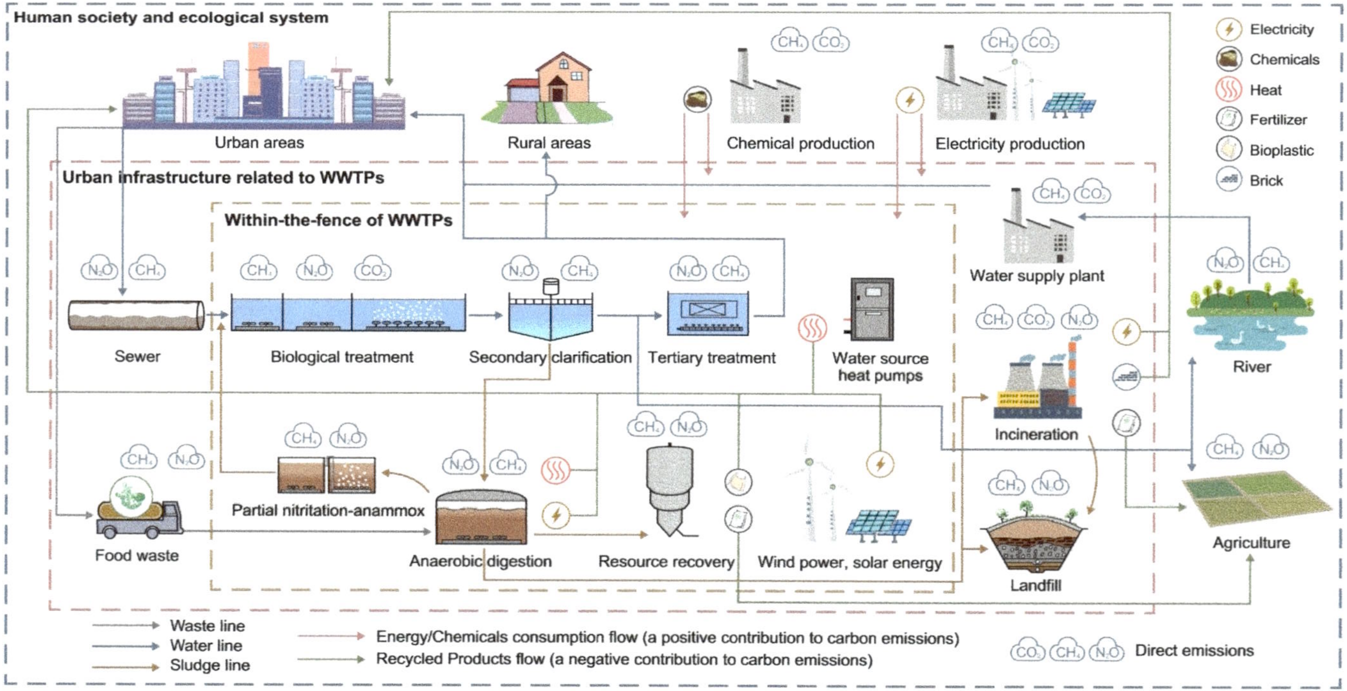

Figure 9.14. Proposed system for carbon accounting of wastewater treatment using three boundaries: within-the-fence of WWTPs (yellow-dashed line), urban infrastructure related to WWTPs (pink-dashed line), and human society and ecological system (blue-dashed line). Since the operation of most facilities in boundaries 1 and 2 requires energy and chemicals, their inputs are simplified as flows pointing into the boundary. Direct carbon emissions are presented by specific symbols. Indirect carbon emissions are implied in the flows of energy/chemicals consumption and recycled products, which represent positive and negative contributions to carbon emissions, respectively (Li *et al.*, 2022). Reproduced under the Creative Commons Attribution (CC BY) license (https://creativecommons.org/licenses/by/4.0/).

to landfill, while China, India, and the United States produce 108, 72, and 35 million tonnes of food waste, respectively, each year (Thi *et al.*, 2015). This represents an enormous opportunity to reduce greenhouse gas emissions through the generation of bioenergy. The third boundary, human society and ecological system, creates an inventory of how the treated wastewater can be used. For example, some wastewater effluents are reused for irrigation or for supply purposes, while the sludge can be used as a source of energy, nutrients, or soil conditioner. Recovered resources also benefit society by providing an alternative to high-energy production systems. When anaerobic systems are used instead of traditional aerobic-based treatment processes, the advantages can be directly related to significant reductions in carbon footprint and takes us ever closer to real carbon neutrality.

References

Agnihotri, S., Yin, D.M., Mahboubi, A., Sapmaz, T., Varjani, S., Qiao, W., Koseoglu-Imer, D.Y. and Taherzadeh, M.J. (2022) A glimpse of the world of volatile fatty acids production and application: A review. *Bioengineered*, **13**, 1249–1275. https://doi.org/10.1080/21655979.2021.1996044

Ahmad, A., K, R., Hasan, S.W., Show, P.L. and Banat, F. (2024) Biohydrogen production through dark fermentation: Recent trends and advances in transition to circular bioeconomy. *International Journal of Hydrogen Energy*, **52**, 335–357. https://doi.org/10.1016/j.ijhydene.2023.05.161

Aslam, A., Khan, S.J. and Shahzad, H.M.A. (2022) Anaerobic membrane bioreactors (AnMBRs) for municipal wastewater treatment- potential benefits, constraints, and future perspectives: An updated review. *Science of the Total Environment*, **802**, 149612. https://doi.org/10.1016/j.scitotenv.2021.149612

Atasoy, M., Eyice, O. and Cetecioglu, Z. (2020) A comprehensive study of volatile fatty acids production from batch reactor to anaerobic sequencing batch reactor using cheese processing wastewater. *Bioresource Technology*, **311**, 123529. https://doi.org/10.1016/j.biortech.2020.123529

Atasoy, M., Owusu-Agyeman, I., Plaza, E. and Cetecioglu, Z. (2018) Bio-based volatile fatty acid production and recovery from waste streams:

Current status and future challenges. *Bioresource Technology*, **268**, 773–786. https://doi.org/10.1016/j.biortech.2018.07.042.

Bhatia, S.K. and Yang, YH. (2017) Microbial production of volatile fatty acids: Current status and future perspectives. *Reviews in Environmental Science and Biotechnol*ogy, **16**, 327–345. https://doi.org/10.1007/s11157-017-9431-4

Bruus, J.H., Nielsen, P.H. and Keiding, K. (1992) On the stability of activated sludge flocs with implications to dewatering, *Water Research*, **12**, 1597–1604. https://doi.org/10.1016/0043-1354(92)90159-2

Carmona, M., Poblete-Castro, I., Rai, M. and Turner, R.J. (2023) Opportunities and obstacles in microbial synthesis of metal nanoparticles. *Microbial Biotechnology*, **16**, 871–876. https://doi.org/10.1111/1751-7915.14254

Chandler, J.R., Duerkop, B.A., Hinz, A., West, T.E., Herman, J.P., Churchill, M.E., Skerrett, S.J. and Greenberg, E.P. (2009) Mutational analysis of *Burkholderia thailandensis* quorum sensing and self-aggregation. *Journal of Bacteriol*ogy, **191**, 5901–5909. https://doi.org/10.1128/jb.00591-09

Chang, S. (2014) Anaerobic bioreactors (AnMBR) for wastewater treatment. *Advances in Chemical Engineering and Science*, **4**, 56–61. http://dx.doi.org/10.4236/aces.2014.41008

Chauhan, G., Stein, M., Seidel-Morgenstern, A., Pant, K.K. and Nigam, K.D. (2015) The thermodynamics and biodegradability of chelating agents upon metal extraction. *Chemical Engineering Science*, **137**, 768–785. https://doi.org/10.1016/j.ces.2015.07.028

Chen, L., Li, Y., Zhang, P., Li, Y., Zhang, G. and Lin, Y. (2024) Quorum sensing in anaerobic wastewater treatment: Potential microbiological mechanisms and enhancement strategies. *Chemical Engineering Journal*, **499**, 156150. https://doi.org/10.1016/j.cej.2024.156150

Cheryan, M. (2009) Acetic acid production. In: *Encyclopedia of Microbiology* (ed. M. Schaechter). Academic Press, Oxford, pp. 144–149. https://doi.org/10.1016/B978-012373944-5.00128-0

Chong, G., Kimyon, O., Rice, S.A., Kjelleberg, S. and Manefield, M. (2012) The presence and role of bacterial quorum sensing in activated sludge. *Microbial Biotechnology*, **5**, 621–633. https://doi.org/10.1111/j.1751-7915.2012.00348.x

Dahiya, S., Chatterjee, S., Sarkar, O. and Venkata Mohan, S. (2021) Renewable hydrogen production by dark-fermentation: Current status, challenges and perspectives. *Bioresource Technology*, **321**, 124354. https://doi.org/10.1016/j.biortech.2020.124354

Decho, A.W. (2000) Microbial biofilms in intertidal systems: An overview. *Continental Shelf Research*, **20**, 1257–1273. https://doi.org/10.1016/S0278-4343(00)00022-4

Deng, Z., Sun, C., Ma, G., Zhang, X., Guo, H., Zhang, T., Zhang, Y., Hu, Y., L. Y.-Y. and Kong, Z. (2025) Anaerobic treatment of nitrogenous industrial organic wastewater by carbon–neutral processes integrated with anaerobic digestion and partial nitritation/anammox: Critical review of current advances and future directions. *Bioresource Technology*, **415**, 131648. https://doi.org/10.1016/j.biortech.2024.131648

Elsayad, R.M., Sharshif, S.W., Khalil, A. and Basha, A.M. (2024) Recent advancements in wastewater treatment via anaerobic fermentation process: A systematic review. *Journal of Environmental Management*, **366**, 121724. https://doi.org/10.1016/j.jenvman.2024.121724

Ergal, İ., Gräf, O., Hasibar, B., Steiner, M., Vukotić, S., Bochmann, G., Fuchs, W. and Rittman, S. (2020) Biohydrogen production beyond the Thauer limit by precision design of artificial microbial consortia. *Communications Biology*, **3**, 443. https://doi.org/10.1038/s42003-020-01159-x

Feng, H., Ding, Y., Wang, M., Zhou, G., Zheng, X., He, H., Zhang, X., Shen, D. and Shentu, J. (2014) Where are signal molecules likely to be located in anaerobic granular sludge? *Water Research*, **50**, 1–9. https://doi.org/10.1016/j.watres.2013.11.021.

Fermoso, F.G., Bartacek, J., Jansen, S. and Lens, P. (2009) Metal supplementation to UASB bioreactors: From cell-metal interactions to full-scale application. *Science of the Total Environment*, **407**, 3652–3667. https://doi.org/10.1016/j.scitotenv.2008.10.043

Fermoso, F.G., Collins, G., Bartacek, J. and Lens, P. (2008b), Zinc deprivation of methanol fed anaerobic granular sludge bioreactors. *Journal of Industrial Microbiology and Biotechnology*, **35**, (6), 543–557. https://doi.org/10.1007/s10295-008-0315-z

Fermoso, F.G., Collins, G., Bartacek, J., O'Flaherty, V. and Lens, P. (2008a) Role of nickel in high rate methanol degradation in anaerobic granular sludge bioreactors. *Biodegradation*, **19**, (5), 725–737. https://link.springer.com/article/10.1007/s10532-008-9177-3

Flemming, H.C. and Wingender, J. (2010) The biofilm matrix. *Nature Reviews Microbiology*, **8**, 623–633. https://doi.org/10.1038/nrmicro2415

Fuerst, J.A. and Sagulenko, E. (2011) Beyond the bacterium: Planctomycetes challenge our concepts of microbial structure and function. *Nature Reviews Microbiology*, **9**, 403–413. https://doi.org/10.1038/nrmicro2578

Gray, N.F. (2022) *Natural Wastewater Treatment System and Sustainability*, World Scientific Publishing, Singapore.

Gray, N.F. (2023) *Activated Sludge: Developments and Sustainable Solutions*, World Scientific Publishing, Singapore.

Hamza, A., Rabii, A., Ezzahraoui, F.-Z., Morgan, G. and Iorhemen, O.T. (2022) A review of the state of development of aerobic granular sludge technology over the last 20 years: Full-scale applications and resource recovery. *Case Studies in Chemical and Environmental Engineering*, **5**, 100173. https://doi.org/10.1016/j.cscee.2021.100173

He, L. and Zhu, G. (2024) Regulation and application of quorum sensing on anaerobic digestion system. *Chemosphere*, **363**, 142983. https://doi.org/10.1016/j.chemosphere.2024.142983

Higgins, M.J. and Novak, J.T. (1997a) The effect of cations on the settling and dewatering of activated sludges: Laboratory results. *Water and Environmental Research*, **69**, 215–224. https://doi.org/10.2175/106143097X125371

Higgins, M.J., Tom, L.A. and Sobeck, D.C. (2004a) Case study I: Application of the divalent cation bridging theory to improve biofloc properties and industrial activated sludge system performance — direct addition of divalent cations. *Water and Environmental Research*, **76**, 344–352. https://doi.org/10.2175/106143004X141933

Higgins, M.J., Sobeck, D.C., Owens, S.J. and Szabo, L.M. (2004b) Case study II: Application of the divalent cation bridging theory to improve biofloc properties and industrial activated sludge system performance — using alternatives to sodium based chemicals. *Water and Environmental Research*, **76**, 353–359. https://doi.org/10.2175/106143004X141942

Hu, Y., Wang, J., Shi, J., Yang, Y., Ji, J. and Chen, R. (2024) A review of electro-conductive membrane enabled electrochemical anaerobic membrane bioreactor process for low-carbon wastewater treatment. *Journal of Environmental Chemical Engineering*, **12**, 113494. https://doi.org/10.1016/j.jece.2024.113494

Hulshoff-Pol, L.W. (1989) The phenomenon of granulation of anaerobic sludge. PhD thesis. Agricultural University of Wageningen, Wageningen, The Netherlands. https://www.proquest.com/openview/b22c194011

831d496584d2a091ec2109/1?pq-origsite=gscholar&cbl=2026366&diss=y

ISO (2024) Environmental management: Eco-efficiency assessment of product systems — Principles, requirements and guidelines. ISO 14045:2012 (reviewed and confirmed in 2024). International Standards Organization, Geneva. https://www.iso.org/standard/43262.html

Ji, J., Kakade, A., Yu, Z., Khan, A., Liu, P. and Li, X. (2020) Anaerobic Membrane bioreactors for treatment of emerging contaminants: A review. *Journal of Environmental Management*, **270**, 110913. https://doi.org/10.1016/j.jenvman.2020.110913

Kartal, B., Maalcke, W.J., de Almeida, N.M., Cirpus, I., Gloerich, J., Geerts, W., den Camp, H.J.M.O., Harhangi, H.R., Janssen-Megens, E.M., Francoijs, K.-J., Stunnenberg, H.G., Keltjens J.T., Jetten, M.S.M. and Strousm, M. (2011). Molecular mechanism of anaerobic ammonium oxidation. *Nature*, **479**, 127–130. https://doi.org/10.1038/nature10453

Kumar, M., Nandi, M. and Pakshirajan, K. (2021) Recent advances in heavy metal recovery from wastewater by biogenic sulfide precipitation. *Journal of Environmental Management*, **278**, 111555. https://doi.org/10.1016/j.jenvman.2020.111555

Lagoa-Costa, B., Kennes, C. and Veiga, M.C. (2020) Cheese whey fermentation into volatile fatty acids in an anaerobic sequencing batch reactor. *Bioresource Technology*, **308**, 123226. https://doi.org/10.1016/j.biortech.2020.123226

Lettinga, G., Van Velsen, A.F.M., Hobma, S.W., de Zeeuw, W. and Klapwijk, A. (1980) Use of the upflow sludge blanket (USB) reactor concept for biological waste water treatment especially for anaerobic treatment. *Biotechnology and Bioengineering*, **22**, 699–734. https://doi.org/10.1002/bit.260220402

Li, J., Chen, Q., Fan, Y., Wang, F. and Meng, J. (2023a) Improved methane production of two-phase anaerobic digester by cobalt: Efficiency and mechanism. *Bioresource Technology*, **381**, 129123. https://doi.org/10.1016/j.biortech.2023.129123

Li, J., Yan, H., Chen, Q., M, J., Li, J., Zhang, Y. and Jha, A.K. (2021) Performance of anaerobic sludge and the microbial social behaviors induced by quorum sensing in a UASB after a shock loading. *Bioresource Technology*, **330**, 124972 https://doi.org/10.1016/j.biortech.2021.124972

Li, L.Y., Zheng, M.Y., Ma, H.L., Gong, S.F., Ai, G.M., Liu, X.L., Li, J., Wang, K.J. and Dong, X.Z. (2015) Significant performance enhancement of a UASB reactor by using acyl homoserine lactones to facilitate the long filaments of *Methanosaeta harundinacea* 6Ac. *Applied Microbiology and Biotechnology*, **99**, (15), 6471–6480. https://doi.org/10.1007/s00253-015-6478-4

Li, Y., Ren, Y., Ji, J., Li, Y.-Y. and Kobayashi, T. (2023b) Anaerobic membrane bioreactors for municipal wastewater treatment, sewage sludge digestion and biogas upgrading: A review. *Sustainability* **2023**, (15), 15129. https://doi.org/10.3390/su152015129

Liao, B.Q.J., Kraemer, T. and Bagley, D.M. (2006) Anaerobic membrane bioreactors: applications and research directions. *Critical Review in Environmental Science and Technology*, **36**, (6), 489–530. http://dx.doi.org/10.1080/10643380600678146

Liu, H., Zeng, W., Zhan, M., Fan, Z., Li, J. and Peng, Y. (2021) Nitrogen removal performance of sulfur autotrophic denitrification under different $S_2O_3^{2-}$ additions using isotopic fractionation of nitrogen and oxygen. *Science of the Total Environment*, **794**, 148794. https://doi.org/10.1016/j.scitotenv.2021.148794

Liu, Y.-Q., Liu, Y. and Tay, J.-H. (2004) The effects of extracellular polymeric substances on the formation and stability of biogranules. *Applied Microbiology and Biotechnology*, **65**, (2), 143–148. https://doi.org/10.1007/s00253-004-1657-8

Logemann, S., Schantl, J., Bijvank, S., van Loosdrecht, M., Kuenen, J.G. and Jetten, M. (1988) Molecular microbial diversity in a nitrifying reactor system without sludge retention. *FEMS Micobiology Ecology*, **27**, (3), 239–249. https://doi.org/10.1111/j.1574-6941.1998.tb00540.x

Lorenzo-Toja, Y., Vázquez-Rowe, I., Amores, M.J., Termes-Rifé, M., Marin-Navarro, D., Moreira, M. and Feijoo, G. (2016) Benchmarking wastewater treatment plants under an eco-efficiency perspective. *Science of the Total Environment*, **566–567**, 468–479. https://doi.org/10.1016/j.scitotenv.2016.05.110

Ma, H.J., Wang, X.Z., Zhang, Y., Hu, H.D., Ren, H.Q., Geng, J.J. and Ding, L.L. (2018) The diversity, distribution and function of N-acyl-homoserine lactone (AHL) in industrial anaerobic granular sludge. *Bioresource Technology*, **247**, 116–124. https://doi.org/10.1016/j.biortech.2017.09.043

Maddela, N.R., Sheng, B., Yuan, S., Zhou, Z., Villamar-Torres, R. and Meng, F. (2019) Roles of quorum sensing in biological wastewater treatment: A critical review. *Chemosphere*, **221**, 616–629. https://doi.org/10.1016/j.chemosphere.2019.01.064

Mulder, J.W., van Loosdrecht, M., Hellinga, C. and van Kempen, R. (2001) Full-scale application of the SHARON process for treatment of rejection water of digested sludge dewatering. *Water Science and Technology,* **43**, 127–134. https://doi.org/10.2166/wst.2001.0675

Nakhate, S.P., Gulhane, M., Singh, A.K., Purohit, H.J., Shah, M.P. and Khardenavis, A.A. (2023) Trace metals as key controlling switches regulating the efficiencies of aerobic and anaerobic bioprocesses. *Biochemical Engineering Journal*, **198**, 108999. https://doi.org/10.1016/j.bej.2023.108999

Nancharaiah, Y.V. and Sarvajith, M. (2019) Aerobic granular sludge process: A fast growing biological treatment for sustainable wastewater treatment. *Current Opinion in Environmental Science and Health*, **12**, (December), 57–65. https://doi.org/10.1016/j.coesh.2019.09.011

Ozgun, H., Dereli, R.K., Ersahin, M.E., Kinaci, C., Spanjers, H. and Van Lier, J.B. (2013) A review of anaerobic membrane bioreactors for municipal wastewater treatment: Integration options, limitations and expectations. *Separation and Purification Technology*, **118**, 89–104. https://doi.org/10.1016/j.seppur.2013.06.036

Peeters, B., Dewil, R., Lechat, D. and Smets, I.Y. (2011) Quantification of the exchangeable calcium in activated sludge flocs and its implication to sludge settleability. *Separation and Purification Technology*, **83**, 1–8. https://doi.org/10.1016/j.seppur.2011.04.008

Qin, L., Liu, Y. and Tay, J.-H. (2004) Effect of settling time on aerobic granulation in sequencing batch reactor. *Biochemical Engineering Journal*, **21**, (1), 47–52. https://doi.org/10.1016/j.bej.2004.03.005

Rajbhandari, B.K. and Annachhatre, A.P. (2004) Anaerobic ponds treatment of starch wastewater: Case study in Thailand. *Bioresource Technology*, **95**, 135143. https://doi.org/10.1016/j.biortech.2004.01.017

Roussel, J. (2012) Metal behaviour in anaerobic sludge digesters supplemented with trace nutrients. PhD thesis, University of Birmingham, England. https://etheses.bham.ac.uk//id/eprint/4369/1/Roussel13Phd.pdf

Sakar, S., Yetilmezsoy, K. and Kocak, E. (2009) Anaerobic digestion technology in poultry and livestock waste treatment — A literature review. *Waste Management Research*, **27**, 3–18. https://doi.org/10.1177/0734242X07079060

Sarkar, O., Katakojwala, R. and Venkata Mohan, S. (2021.) Low carbon hydrogen production from waste based biorefinery system and environmental sustainability assessment. *Green Chemistry*, **23**, 561–574. https://doi.org/10.1039/D0GC03063E

Sarma, S.J., Tay, J.H. and Chu, A. (2017) Finding knowledge gaps in aerobic granulation technology. *Trends in Biotechnology*, **35**, (1), 66–78. https://doi.org/10.1016/j.tibtech.2016.07.003

Sato, T. and Ose, Y. (1980). Floc-forming substrates extracted from activated sludge by sodium hydroxide solution. *Water Research*, **14**, 333–338. https://doi.org/10.1016/0043-1354(80)90080-9

Schmidt, J.E. and Ahring, B.K. (1994) Extracellular polymers in granular sludge from different upflow anaerobic sludge blanket (UASB) reactors. *Applied Microbiology and Biotechnology*, **42**, 457–462. https://doi.org/10.1007/BF00902757

Schmidt, J.E. and Ahring, B.K. (1996) Granular sludge formation in upflow anaerobic sludge blanket (UASB) reactors. *Biotechnology and Bioengineering*, **49**, 229–246. https://doi.org/10.1002/(SICI)1097-0290(19960205)49:3<229::AID-BIT1>3.0.CO;2-M

Seviour, T., Pijuan, M., Nicholson, T., Keller, J. and Yuan, Z. (2009) Gel-forming exopolysaccharides explain basic differences between structures of aerobic sludge granules and floccular sludges. *Water Research*, **43**, 4469–4478. https://doi.org/10.1016/j.watres.2009.07.018

Shen, C.F., Kosaric, N. and Blaszczyk, R. (1993) The effect of selected heavy metals (Ni, Co and Fe) on anerobic granules and their extracellular polymeric substance (EPS). *Water Research*, **27**, 25–33. https://doi.org/10.1016/0043-1354(93)90191-J

Show, K.-Y., Zhang, Z.-P., Tay, J.-H., Liang, D.T., Lee, D.-J. and Jiang, W.-J. (2007) Production of hydrogen in a granular sludge-based anaerobic continuous stirred tank reactor. *International Journal of Hydrogen Energy*, **32**, 4744–4753. https://doi.org/10.1016/j.ijhydene.2007.07.005

Show, K.-Y., Yan, Y., Yao, H., Gu, H., Li, T., Show, De-Y., Chang, J.-S. and Lee, D.-J. (2020) Anaerobic granulation: A review of granulation hypotheses, bioreactor designs and emerging green applications. *Bioresource Technology*, **100**, 122751. https://doi.org/10.1016/j.biortech.2020.122751

Song, X.N., Cheng, Y.Y., Li, W.W., Li, B.B., Sheng, G.P., Fang, C.Y., Wang, Y.K., Li, X.Y., Yu, H.Q. (2014) Quorum quenching is responsible for the underestimated quorum sensing effects in biological wastewater treatment reactors. *Bioresource Technology,* **171**, 472–476. https://doi.org/10.1016/j.biortech.2014.08.027

Song, Y., Zhang, Z., Liu, Y., Peng, F. and Feng, Y. (2024) Enhancement of anaerobic treatment of antibiotic pharmaceutical wastewater through the development of iron-based and carbon-based materials: A critical review. *Journal of Hazardous Materials*, **479**, 135514. https://doi.org/10.1016/j.jhazmat.2024.135514

Stazi, V. and Tomei, M.C. (2018) Enhancing anaerobic treatment of domestic wastewater: State of the art, innovative technologies and future perspectives. *Science of the Total Environment*, **635**, 78–91. https://doi.org/10.1016/j.scitotenv.2018.04.071

Strous, M., van Gerven, E., Zheng, P., Kuenen, J.G. and Jetten, M.S.M. (1997) Ammonium removal from concentrated waste streams with the anaerobic ammonium oxidation (ANAMMOX) process in different reactor configurations. *Water Research,* **31**, (8), 1955–1962. https://doi.org/10.1016/S0043-1354(97)00055-9

Sun, Z.F., Zhao, L., Wu, K.K., Wang, Z.H., Wu, J.T., Chen, C., Yang, S.S., Wang, A.J. and Ren, N.Q. (2022) Overview of recent progress in exogenous hydrogen supply biogas upgrading and future perspective. *Science of the Total Environment,* **848**, 157824. https://doi.org/10.1016/j.scitotenv.2022.157824

Tan, C.H., Koh, K.S., Xie, C., Tay, M., Zhou, Y., Williams, R., Ng, W.J., Rice, S.A. and Kjelleberg, S. (2014) The role of quorum sensing signalling in EPS production and the assembly of a sludge community into aerobic granules. *International Society for Microbial Ecology Journal*, **8**, (6), 1186–1197.

Tansel, B. (2018) Morphology, composition and aggregation mechanisms of soft bioflocs in marine snow and activated sludge: A comparative review. *Journal of Environmental Management*, **205**, 231–243. https://doi.org/10.1016/j.jenvman.2017.09.082

Tay, J.H., Tay, T.L., Liu, Y., Show, K.Y. and Ivanov, V. (2006) *Biogranulation Technologies for Wastewater Treatment.* Elsevier Science, Oxford.

Thanwised, P., Wirojanagud, W and Reungsang, A. (2012) Effect of hydraulic retention time on hydrogen production and chemical oxygen demand removal from tapioca wastewater using anaerobic mixed cultures in anaerobic baffled reactor (ABR). *International Journal of Hydrogen Energy*, **37**, 15503–15510. https://doi.org/10.1016/j.ijhydene.2012.02.068

Thi, N.B.D., Kumar, G. and Lin, C.-Y. (2015) An overview of food waste management in developing countries: Current status and future perspective. *Journal of Environmental Management*, **157**, 220–229. https://doi.org/10.1016/j.jenvman.2015.04.022

van Dongen, U., Jetten, M.S.M. and van Loosdrecht, M.C.M. (2001) The SHARON®-Anammox® process for treatment of ammonium rich wastewater. *Water Science and Technology*, **44**, (1), 153–160. https://doi.org/10.2166/wst.2001.0037

Venkata Mohan, S., Amulya, K. and Modestra, J.A. (2020) Urban biocycles-closing metabolic loops for resilient and regenerative ecosystem: A perspective. *Bioresource Technology*, **306**, 123098. https://doi.org/10.1016/j.biortech.2020.123098

Venkata Mohan, S., Dahiya, S., Amulya, K., Katakojwala, R. and Vanitha, T.K. (2019) Can circular bioeconomy be fueled by waste biorefineries — A closer look. *Bioresource Technology Reports*, 7, 100277. https://doi.org/10.1016/j.biteb.2019.100277

Villa-Gomez, D., Ababneh, H., Papirio, S., Rousseau, D.P.L. and Lens, P.N.L. (2011) Effect of sulfide concentration on the location of the metal precipitates in inversed fluidized bed reactors. *Journal of Hazardous Materials*, **192**, 200–207. https://doi.org/10.1016/j.jhazmat.2011.05.002

Wang, G., Li, Q., Gao, X. and Wang, X.C. (2018) Synergetic promotion of syntrophic methane production from anaerobic digestion of complex organic wastes by biochar: Performance and associated mechanisms, *Bioresource Technology*, **250**, 812–820. https://doi.org/10.1016/j.biortech.2017.12.004

Wang, Y., Zhao, Q.B., Mu, Y., Yu, H.Q., Harada, H. and Li, Y.Y. (2008) Biohydrogen production with mixed anaerobic cultures in the presence of high-concentration acetate. *International Journal of hydrogen Energy*, **33**, 1164–1171. https://doi.org/10.1016/j.ijhydene.2007.12.018

Wilén, B.-M., Jin, B. and Lant, P. (2003) The influence of key chemical constituents in activated sludge on surface and flocculating properties, *Water Research*, **37**, 2127–2139. https://doi.org/10.1016/S0043-1354(02)00629-2

Wilschut, J. and Hoekstra, D. (1984) Membrane fusion: From liposome to biological membrane. *Trends in Biochemical Sciences*, **9**, 479–483. https://doi.org/10.1016/0968-0004(84)90316-5

Xiao, Y., Wang, X., Wang, P., Zhou, Z., Wang, H., Teng, T., Li, Y. and Yang, L. (2023) New insights into multi-strategies of sludge granulation in up-flow anaerobic sludge blanket reactors from community succession and interaction. *Bioresource Technology*, **377**, 128935. https://doi.org/10.1016/j.biortech.2023.128935

Xue, Y., Ma, H., Hu, Y. and Li, Y.-Y. (2023) Anammox-based granulation cycle for sustainable granular sludge biotechnology from mechanisms to strategies: A critical review. *Water Research*, **228**, 119353. https://doi.org/10.1016/j.watres.2022.119353

Xue, Y., Ma, H., Hu, Y., Kong, Z. and Li, Y.Y. (2022) Microstructure and granulation cycle mechanisms of anammox-HAP coupled granule in the anammox EGSB reactor. *Water Research*, **210**, 117968. https://doi.org/10.1016/j.watres.2021.117968

Yarimtepe, C.C., Oz, N.A. and Ince, O. (2017) Volatile fatty acid production dynamics during the acidification of pretreated olive mill wastewater. *Bioresource Technology*, **241**, 936–944. https://doi.org/10.1016/j.biortech.2017.05.173.

Yin, D.M., Uwineza, C., Sapmaz, T., Mahboubi, A., DeWever, H., Qiao, W. and Taherzadeh, M.J. (2022) Volatile fatty acids (VFA) production and recovery from chicken manure using a high-solid anaerobic membrane bioreactor (AnMBR). *Membranes*, **12**, 1133. https://doi.org/10.3390/membranes12111133

Yu, H.Q., Tay, J.H. and Fang, H.H.P. (2001) The role of calcium in sludge granulation during UASB reactor start-up. *Water Research*, **35**, 1052–1060. https://doi.org/10.1016/S0043-1354(00)00345-6

Zandvoort, M. H., van Hullebusch, E. D., Fermoso, F.G. and Lens, P. (2006) Trace metals in anaerobic granular sludge reactors: Bioavailability and dosing strategies. *Engineering in Life Sciences*, **6**, (3), 293–301. https://doi.org/10.1002/elsc.200620129

Zhang, D., Li, W., Hou, C., Shen, J., Jiang, X., Sun, X., Li, J., Han, W., Wang, L. and Liu, X. (2017) Aerobic granulation accelerated by biochar for the treatment of refractory wastewater. *Chemical Engineering Journal*, **314**, 88–97. https://doi.org/10.1016/j.cej.2016.12.128

Zhang, S., Tchalala, M.R., Cheng, H., Medina, J.S., Xiong, Y. and Hong, P.-Y. (2023) Anaerobic membrane bioreactor as the core technology for future decentralized wastewater treatment plants. *Current Opinion in Chemical Engineering*, **42**, 100975. https://doi.org/10.1016/j.coche.2023.100975

Zhang, W.L., Zhang, L. and Li, A.M. (2015) Enhanced anaerobic digestion of food waste by trace metal elements supplementation and reduced metals dosage by green chelating agent [S, S]-EDDS via improving metals bioavailability. *Water Research*, **84**, 266–277. https://doi.org/10.1016/j.watres.2015.07.010

Zhang, X., Hao, T., Zhang, T., Hu, Y., Lu, R., Li, D., Pan, Y., Li, Y.-Y. and Kong, Z. (2024) Towards energy conservation and carbon reduction for wastewater treatment processes: A review of carbon-neutral anaerobic biotechnologies. *Journal of Water Process Engineering*, **59**, 105026. https://doi.org/10.1016/j.jwpe.2024.105026

Zhang, Z., Yu, Z., Wang, Z., Ma, K., Xu, X., Alvarezc, P.J.J. and Zhu, L. (2019) Understanding of aerobic sludge granulation enhanced by sludge retention time in the aspect of quorum sensing. *Bioresource Technology*, **272**, 226–234. https://doi.org/10.1016/j.biortech.2018.10.027

Zhao, Y., He, J., Pang, H., Li, L., Cui, X., Liu, Y., Jiang, W. and Liu, X. (2024) Anaerobic digestion and biochar/hydrochar enhancement of antibiotic-containing wastewater: Current situation, mechanism and future prospects, *Environmental Research*, 120087 https://doi.org/10.1016/j.envres.2024.120087.

Zigová, J. and Šturdík, E. (2000) Advances in biotechnological production of butyric acid. *Journal of Industrial Microbiology and Biotechnology*, **24**, 152–160. https://doi.org/10.1038/sj.jim.2900795

Index

www.ingramcontent.com/pod-product-compliance
Ingram Content Group UK Ltd.
Pitfield, Milton Keynes, MK11 3LW, UK
UKHW021958190326
4861IPUK00001B/2